Technology Choice

Technology Choice

A Critique of the Appropriate Technology Movement

Kelvin W. Willoughby

Westview Press
BOULDER & SAN FRANCISCO

Intermediate Technology Publications
LONDON
1990

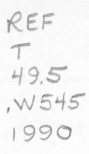

REF
T
49.5
,W545
1990

Published in 1990 in the United States of America by Westview Press, Inc., 5500 Central Avenue, Boulder, Colorado 80301

Published in 1990 in the United Kingdom by Intermediate Technology Publications, 103-105 Southampton Row, London WC1B 4HH, UK

Library of Congress Cataloging-in-Publication Data
Willoughby, Kelvin W.
  Technology choice.
  1. Appropriate technology. I. Title.
T49.5.W545 1990   338.9'27   89-24982
ISBN 0-8133-7806-0 (U.S.)
ISBN 1-85339-057-7 (U.K.)

A CIP catalog record for this book is available from the British Library.

Printed and bound in the United States of America

The paper used in this publication meets the requirements of the American National Standard for Permanence of Paper for Printed Library Materials Z39.48-1984.

10   9   8   7   6   5   4   3   2   1

This book is
dedicated to Anna, my daughter,
who has surprised me with
joy and hope.

Contents

Part Three
Prospects for Technology Choice

Preface

Technology is becoming the leitmotif of the modern world and a linchpin of the international economy. Businesses, governments, community organizations and individuals, seemingly everywhere, are looking to technology as a key to the attainment of their goals. In opposition, of course, there are those who are reluctant to join in with what they see as adulation of technology and who readily point to the technological causes of human and environmental problems. Even these "unbelievers" seem unable to avoid becoming embroiled in the new rhetoric, however, and unable to escape the technological milieu against which they protest.

This book breaks away from the impasse of the "pro-technology" versus "anti-technology" debate. Instead it concentrates on the idea that there are vital choices to be made in most fields of human endeavor - whether manufacturing, energy supply, transport, health care or food production - between different technological options with contrasting primary or secondary impacts; there is rarely, if ever, one "correct" technical solution to social, economic or environmental problems. It argues that a range of policy areas, perhaps not normally associated directly with technology, can be managed more effectively by paying special attention to the process of *technology choice*.

Technology choice emerged as a critical issue following the pathfinding work of economist E. F. Schumacher. He and his colleagues, with their concept of "appropriate technology", found themselves by the mid 1970s at the forefront of an international movement which offered a fresh approach to grappling with the technological dimensions of human and environmental problems. This book reviews the ideas and experiments of that movement. It examines the grounds for hope that policies based upon expanding the scope for enlightened technology choice might be feasible for both rich and poor communities, in both rural and urban contexts.

Much of the material in the chapters to follow consists of analysis of *ideas* on the nature of technology and its role within society. This book attempts to provide a theoretical framework for answering difficult questions evoked by the concept of technology choice. Its theory was not created from an academic vacuum, however, but was forged through personal experiences I gained while struggling directly with the practicalities of promoting environmentally sound local economic initiatives through community organizations, local businesses and public agencies in the region where I grew up - Western Australia. This study is international in scope and spirit, but it has been shaped by local experimentation. I would therefore like to thank my colleagues from the independent Australian organization *Apace* with whom I have been fortunate enough to share the task of testing these ideas.

Credit also belongs to Peter Newman and Brian Hill, my academic mentors at Murdoch University who have become my valued friends. Special acknowledgment is due to Suzanne, my wife, who made it possible for me to engage in voluntary community development work for two years as I grappled with applying theory in practice. Finally, I would also like to dedicate this book to the late Keith Roby, who was responsible for awakening my interest in the adventure of directing science and technology towards the needs of local communities, but whose tragic early death prevented him from seeing the results of his work.

Kelvin W. Willoughby
Berkeley, California

PART ONE

Conceptual Groundwork

1

Introduction:
The Concept of Technology Choice

Prolegomena: The Rise of Technology Studies

Every period is marked by its characteristic catchcries and commonplaces. "Technology" and "technological change" are marking our time as ubiquitous symbols and themes for debate.

Until recently the general public in the now industrialized countries tended to relegate technology to the professional domain of the engineer or technologist, or take it for granted as the benevolent provider of material wealth. In earlier historical periods common people were often more personally familiar than their modern counterparts with the workings of technology, albeit in a much simpler form; but while turbulent responses to the introduction of new technology did occur during earlier periods, as did large technology-related social changes, it would appear that popular attitudes towards technology *per se* are now more intense and deeply rooted.

The pervasiveness of technology appears to be increasing in urban-industrial societies and elsewhere - enough for the phrase "the technological society" to become common. Technology is frequently reified in popular rhetoric and has come to be viewed as a system or a form of rationality rather than just a collection of artefacts. Some view it as a kind of saviour, capable of solving most of the perennial problems of human existence; others view it as a kind of demon which threatens the health of human society and the environment. Some argue that a felicitous future will only be possible with continually increasing technological growth; others plead for the rejection of technology, arguing that it is intrinsically destructive. The ubiquity of technology in con-

3

temporary society means that most significant social, political, economic and environmental problems have a technological dimension to them. Technological issues tend to mirror the wider spectrum of issues around which various interest groups in society rally.[1]

Controversy over technological change has occurred amidst widespread controversy on a number of other fronts. Concern has been expressed about the basic prospects for human society and the natural environment. Many commentators point to the specters of catastrophic levels of natural resource depletion, irreparable damage to global and local ecosystems, structural economic malaise, mounting international military conflict, the explosion of human population levels, the spread of debilitating poverty, the degradation of human society and culture, and even the possibility of extinction through global nuclear confrontation. These problems are increasingly linked together by analysts on the understanding that they form a complex of mutually reinforcing factors, spawning such phrases as "the global problematique".

In contrast to a sanguine view of "progress" exhibited by previous generations, lack of confidence about the future is now endemic. Widely cited studies of the above global problems have been made by many writers and there is no need to duplicate their work here.[2] This book is grounded on the observation that technology, or technological change, is widely implicated in diagnoses of such problems by scholars, policymakers and others. By considering the nature of technology, the way in which it develops within society, and the manner in which it might be

---

[1] For a recent treatment of public attitudes to science and technology see R. Williams and S. Mills, eds., *Public Acceptance of New Technologies: An International Review* (London: Croom Helm, 1986). Examples of the many reviews of the contentious role of technology in society include: G. F. Jünger, *The Failure of Technology* (Chicago: Regnery, 1956); L. Mumford, *Technics and Civilization* (New York: Harcourt, Brace and Jovanovich, 1963); H. Marcuse, *One Dimensional Man* (Boston: Beacon Press, 1964); J. Ellul, *The Technological Society*, trans. J. Wilkinson (New York: Alfred A. Knopf, 1964); V. Ferkiss, *Technological Man: The Myth and the Reality* (London: Heinneman, 1969); E. Schuurman, *Technology and the Future: A Philosophical Challenge* (Toronto: Wedge,1977); E. Braun, *Wayward Technology* (London: Frances Pinter, 1984); B. Frankel, *The Post-Industrial Utopians* (Cambridge and Oxford: Polity Press in association with Basil Blackwell, 1987).

[2] E.g.: D. H. Meadows, et al., *The Limits to Growth* (New York: Universe Books, 1972); B. Ward, *Progress for a Small Planet* (Harmondsworth: Penguin, 1979); U. S. Council on Environmental Quality and the Department of State, *The Global 2000 Report to the President: Entering the Twenty-First Century* (Washington, D. C.: U. S. Government printing Office, 1980-1981); L. R. Brown, *et al., State of the World*, annual publication of the Worldwatch Institute, Washington, D. C. (New York: W. W. Norton, 1984, 1985, 1986, 1987, 1988). For the work of a more popular author see *Future Shock* (London: Pan Books, 1970) and *The Third Wave* (London: Pan Books, 1981) by A. Toffler.

managed by people, this book presents a fresh look at the prospects for addressing the constellation of problems mentioned above.

This book is an interdisciplinary study in the emerging field of technology studies. Technology studies is not yet well established as an academic discipline in its own right but a serious body of both scholarly and popular literature in the field is now emerging. The growth of academic activity centered on technology *sui generis* accompanies the increasing prominence of technological phenomena in society.[3] In recent decades this trend has been accompanied by increasing contention over the status and meaning of technology and its value to human beings. The contention over technology has been exacerbated by the increasing rate of technological change. Technological change, although probably always present throughout human history and not always greeted with equanimity, is now a controversial issue.

The Problem of Inappropriate Technology

It is in the above context that the concept of technology choice has emerged. It may be seen as an attempt to get beyond the simplistic options of either uncritical acceptance or uncritical rejection of technology. While some critics espouse a negative assessment of technology *per se*, a growing number have drawn attention to the notion that the adoption of *inappropriate* technology may undergird the contemporary constellation of critical issues. They claim that poor choices of technology are frequently made and that by placing more attention on the processes of technology choice the groundwork may be laid for better policy and practice. Adopting the *choice of technology* as a focus for analysis means recognizing that inappropriate technology choices can and do indeed occur.

The inappropriateness of technology may be reflected in the severity of unforeseen side effects, or by the gradual (undesirable) change in the dynamics and structures over time of a society through its dependence upon technology. The inappropriateness of technology might stem from its being deployed in a context quite different to that for which it was designed, or it might be manifested in the harmful effects of technology upon one class of people despite its appropriateness from

---

[3] The term "*sui generis*" is used here to emphasize the notion that technology ought to be viewed as something which is unique or peculiar, which possesses a status of its own and which is not merely a subcategory or aspect of something else, such as science, engineering or economics.

the point of view of another class of people. Technology might even be inappropriate for the specific purposes for which it was designed - due to technical incompetence of the designer or inability to effectively relate technical parameters to the real world of practice.

A school of thought has emerged which holds that the existence of inappropriate technology has become so ubiquitous as to constitute a major and urgent issue for policy and practice - at both the global level and the local level. The hallmark of this school of thought is its claim that concerted effort is required to ensure that the technology adopted in all fields of practice is as "appropriate" as possible. This claim is accompanied by the conviction that the appropriate choice of technology ought to be treated as a cardinal principle of human affairs and not just as a matter worthy of attention by specialists and enthusiasts.

During the previous two decades the number of publications, projects and organizations which have appeared as an outworking of this school of thought has burgeoned. This phenomenon has resulted in an international social movement operating under the rubric of "appropriate technology". The extent of the movement's efforts, as will be revealed later, is substantial.

The emphasis which the Appropriate Technology movement[4] places on "appropriateness" as a theme acknowledges the positive value of technology, as promoted by the proponents of technology, yet also responds to the shortcomings of technology, as stressed by its critics. In contrast to crude judgements as to whether or not technology should be developed, the concept of Appropriate Technology attempts to discriminate between different technologies according to their relative suitability for specific purposes or situations. As such it may be seen as a common sense approach to technology which may provide hope for the preeminence of human concerns in an increasingly technological world.

When approached in this common sense manner the importance of technology being appropriate seems rather obvious. It could therefore be argued that using the term "appropriate technology" achieves little more than cluttering our vocabulary with superfluous jargon. It may also be questioned whether there is much value in an extended discus-

---

[4] Hereafter the following convention will be adhered to: **Appropriate Technology** (initials capitalized) will denote the general concept - or the movement, innovation strategy or mode of technology-practice associated with the concept; **"appropriate technology"** (within inverted commas) will denote the term itself; and, **appropriate technology** (plain) will denote actual technology or technologies (i.e., artefacts). Italics will be used when special emphasis is warranted and inverted commas will sometimes be used when the meaning is ambiguous or contentious.

sion of something which appears so straightforward. In support of this view it might be suggested that no engineer would wish to design "inappropriate technology". It may be further suggested that the complexity and sophistication of modern economies, particularly of those with a competitive market, ensures that the technology employed is the most appropriate currently possible - otherwise it would not have been developed and put into use. The so-called "command" economies, it might also be claimed, have refined planning mechanisms to ensure that the best possible technology is employed. Furthermore, others may claim, "technology is technology!" and that a choice therefore exists only between "good" and "bad" technology, i.e., between technology which is well designed and constructed and that which is not: speaking of "appropriate technology" would thus be tautologous, like speaking of "true truth".

To foreshadow a thorough response to these objections in later chapters, the following rejoinder is offered here: there is now a solid record of evidence from a variety of fields that technologies deployed for specific purposes and circumstances are not automatically suitable for those same purposes and circumstances. In other words, in addition to the normal economic processes which occur in market oriented societies at present, or to the normal economic planning which currently occurs in socialist oriented societies, there is a need for *deliberate* and *specific* attention to be given to the peculiar difficulties of assessing and adopting technology. The Appropriate Technology notion points to the need for knowledge of a diversity of technical options for given purposes, careful analysis of the local human and natural environment, normative evaluation of alternative options, and the exercise of political and technological choice. Conscious human effort is required to ensure that technology is appropriate.

The factors which are implicated in technology choice extend beyond those which are normally taken into account by engineers and business managers under the rubrics of "efficiency" or "profit". In short, as this study will attempt to demonstrate, *the selection of technology within a given context is fraught with problems*. It is fraught with problems not only in the sense of many problems being presented which require solutions, but also in the sense that such solutions may not, in the main, be either straightforward or reducible to a technical format alone. Assurance of technology's appropriateness - from the point of view of people's interests or of certain environmental objectives - may only be gained if *special* effort is directed towards ensuring that it is so.

Unfortunately, despite its rather common sense nature, this important insight has been overlooked by the majority of commentators in the past because technology has generally been regarded as a neutral factor in social and economic policy. This has led to it either being ignored as a variable in its own right, or being viewed as strictly dependent upon fiscal or monetary factors, for example, in economics. Technologists and policy makers alike have tended to assume that, given a stated social or economic goal, there is always *one best way* to reach that goal and that technology choice is therefore not problematical.

The emerging field of social studies in science and technology has provided overwhelming evidence that technology is not neutral, that it does exert a significant determining influence in society (while at the same time reflecting social interests), and that choice from amongst technological alternatives is necessary.[5] It is not claimed here that technology is totally independent of other variables in society or the economy, but rather that it demands special attention in its own right, rather than as a merely dependent variable or as a given. The growing interest in technology studies as a discipline vindicates this perspective and points to the need for a more sophisticated approach to technological innovation and development than has previously been common.[6] The concept of Appropriate Technology may be viewed as an attempt to provide an adequate framework for dealing with the issues of technology choice just intimated.

The concept of technology choice may be defined as the concept that:

- there is frequently a range of alternative technological means available which are suitable for the attainment of primary objectives within a given field;
- the number of alternatives in the range may be increased over time by conscious human effort;
- alternative technological means of similar suitability, for the attainment of certain primary objectives, may vary widely in their suitability for the attainment of secondary objectives;
- the informed selection of technological means, taking into account secondary objectives as well as a primary objectives, com-

---

[5] The following are typical of the recent publications which have raised the profile of this matter within scholarly research: D. MacKenzie and J. Wajcman, eds., *The Social Shaping of Technology* (Milton Keynes and Philadelphia: Open University Press, 1985); N. Clark, *The Political Economy of Technology* (Oxford: Basil Blackwell, 1985); R. MacLeod, ed., *Technology and the Human Prospect* (London and Wolfeboro: Frances Pinter, 1986); W. E. Bijker, T. P. Hughes and T. Pinch, eds., *The Social Construction of Technological Systems* (Cambridge, Mass. and London: The MIT Press, 1987).

[6] A scholarly and reasonably comprehensive introduction to this field may be found in *A Guide to the Culture of Science, Technology and Medicine* (New York: The Free Press, 1984), edited by P. T. Durbin.

bined with long term efforts to expand the range of available alternatives, is an important element of social, economic and environmental policy.

This book seeks to demonstrate that the concept of "technology choice", and its conceptual partner, "Appropriate Technology", can be useful tools for elucidating the issues of central importance to technology studies and technology policy.

The Emergence of Technology Choice as a Policy Concept

While technology has always been viewed within economic theory as an important factor of production, it has until recently almost always been treated by economists as exogenous to the economic system itself.[7] The issue of technology choice was therefore not generally seen as very important. The state of technology in the economy was seen as something which was "given" at any particular period, changing from time to time as "breakthroughs" emerged from the supposedly independent activities of scientists and engineers. It was generally assumed that, under ideal conditions, normal economic forces would lead to the adoption of optimal production systems, given a particular stock of technology available at any particular time.

There have nevertheless been some non-dominant schools of thought within economics which have paid more attention to technology than normal. The most significant has probably been the "Schumpeterian" school which has placed great importance on the role of technological innovation in long term economic cycles.[8] There have been some economists in the neoclassical tradition who have gone to considerable length in seeking to explain the choice of technology in terms of the orthodox two-factor production function of neoclassical economics;[9] and there has been a long running debate about whether or not there are biases in the direction of technological change (e.g., towards

---

[7] Recently a number of respected economists have questioned this assumption and have begun to develop new economic theory in which technology is treated as an endogenous factor (see, e.g., N. Rosenberg, *Inside the Black Box: Technology and Economics* [Cambridge: Cambridge University Press, 1982]).

[8] This school followed on from the work of J. A. Schumpeter (*The Theory of Economic Development* [Cambridge, Mass.: Harvard University Press, 1934] and *Business Cycles* [New York: McGraw Hill, 1939]). An example of a contemporary economist working within this tradition is G. O. Mensch (*Stalemate in Technology: Innovations Overcome the Depression* [Cambridge, Mass.: Ballinger, 1979]).

[9] See A. K. Sen, *Choice of Techniques* (3rd. ed.; Oxford: Basil Blackwell, 1968).

capital intensive production methods).[10] This has been extended with considerable analytical sophistication.[11] On the whole, however, most of the scholarly publications of orthodox economists interested in technological change have not shown technology choice to be problematical. Those economists who have incorporated technology into their models as an endogenous rather than exogenous factor, while in one sense taking technology very seriously, have tended to reduce the process of technological change to something which is largely "determined" by economic processes.

There are some recent signs amongst development economists, however, that sophisticated economic models may be developed which both treat technology as endogenous to the economic system and also allow for the possibility that dynamic forces may be at work in local economies which may lead to sub-optimal choices of technology and constraints on economic development.[12] Within recent business-strategy studies there is a quite strong interest emerging in the importance of managerial decisions as determinants of technological innovation;[13] and even more recently some publications in economics have begun to directly address this perspective.[14]

While technology choice has not, in the main, been an important concept within economics, recent developments within the profession have begun to alter this situation. Nevertheless, the economics profession as such has not been the prime source from which the concept has emerged. It is the Appropriate Technology movement which has acted as the harbinger of technology choice as a policy concept and which has provided the main source of literature on the subject. In this study the concept of technology choice will therefore be explored primarily

---

10 Key contributions to this debate are: J. Hicks, *Theory of Wages* (London: Macmillan, 1932); W. Salter, *Productivity and Technological Change* (Cambridge: Cambridge University Press, 1960); H. Habakkuk, *American and British Technology in the 19th Century* (Cambridge: Cambridge University Press, 1962); S. Saul, ed., *Technological Change: The U.S. and Britain in the 19th Century* (London: Methuen, 1970).

11 P. A. David, *Technical Choice, Innovation and Economic Growth* (Cambridge: Cambridge University Press, 1975).

12 The most notable work along these lines is by agricultural economists Y. Hayami and V. Ruttan (*Agricultural Development: An International Perspective* [Revised and expanded edition; Baltimore and London: The Johns Hopkins University Press, 1985]).

13 E.g.: R. A. Burgelman and M. A. Maidique, *Strategic Management of Technology and Innovation* (Homewood, Ill.: Irwin, 1988); B. Twiss, *Managing Technological Innovation* (3rd. ed.; London and New York: Longman, 1986); O. Granstrand, *Technology, Management and Markets* (London: Frances Pinter, 1982).

14 See R. Coombs, P. Saviotti and V. Walsh (*Economics and Technological Change* [London: Macmillan, 1987]) for a review of this subject.

by conducting a review of the Appropriate Technology movement and its ideas and experiments. While many of the leaders in the movement have in fact been professional economists, their work has tended to have been promoted within the Appropriate Technology movement and its associated networks rather than through the mainstream fora of the economics profession.[15]

The Promises and Problems of Appropriate Technology

As technology has increasingly become implicated in diagnoses of contemporary global problems, technology policy has accordingly grown in importance as a potential source of solutions to these problems. Most national governments now have ministries of technology, or at least ministries in which technology has become a central focus of policies and programs, and the same trend is emerging amongst provincial and regional governments. Likewise, technology studies is beginning to emerge not only as a theme for research and teaching within the international academic community, but as a discipline in its own right.[16] Nevertheless, the embryonic discipline is still searching for unifying paradigms and theories. This book puts forward the concept of technology choice as a candidate for adoption as one of the ideas around which the new discipline may define itself. The study also suggests that giving more prominence within technology policy to technology choice will help to make that field of policy more potent as a device for managing the economy, environment and social complexity of communities.

Because the Appropriate Technology movement has championed the idea of technology choice, it is sensible to turn to that movement for insights as to what technology choice actually means and how it might be applied in practice.

---

[15] The most notable of these is the economist E. F. Schumacher (see Chapters Four and Five). One study (of the British economy) in which technology choice has been used as the central analytical concept more than in most other economic studies appeared in 1986 (J. Davis and A. Bollard, *As Though People Mattered* [London: Intermediate Technology Publications, 1986]). The two authors were both associates of one of the main international Appropriate Technology organizations, the Intermediate Technology Development Group Limited, in London.

[16] This is evidenced by the emergence of scholarly and professional journals such as the *International Journal of Technology Management*, the O.E.C.D.'s *Science Technology Industry Review*, and *Prometheus*, and the strengthening concentration over time on matters technological in established "science policy" journals such as *Research Policy* or *Social Studies of Science*.

As will be demonstrated in subsequent chapters, the Appropriate Technology movement is well established with an identifiable literature, and an extensive network of organizations, projects and field experiments. It has also achieved a modestly impressive track record of successful projects which lend weight to the movement's claims. Despite these facts, however, together with the appeal and common sense nature of the movement's core ideas, the movement has largely failed to evoke the transformation of industrial and technological practice in most countries in accordance with the principles of Appropriate Technology. In other words, while becoming a significant international movement Appropriate Technology has remained a minority theme within technology policy and practice.

The Appropriate Technology movement is thus an enigma. On one hand it may be seen as one of the most promising sources of hope that the constellation of contemporary global problems may be overcome. On the other hand the fact that after more than two decades it has failed to become the dominant mode of technological practice raises a shadow of pessimism over this hope. This enigma provides the underlying problem addressed in this book. The evidence and arguments in the following chapters will be woven together to grapple with the tension between the poles of hope and pessimism over the question of whether "appropriate" technology choice is feasible; that is, over the question of whether technology choice may be exercised on a large enough scale to influence the basic pattern of the economy internationally and locally.

A great deal of investigation has now been directed at the practicability of actual "appropriate" technologies. A solid core of evidence has been assembled for believing that, in the main, there are no insurmountable *technical* reasons why the Appropriate Technology notion might not be applicable. This study may not make any substantively new contributions in this area, beyond providing a new summary of evidence assembled in other studies. The technical feasibility of Appropriate Technology, however, does not appear to have made the movement's methods normal practice in the mainstream.

The limits in the uptake of the Appropriate Technology approach, with its emphasis on technology choice, arise partly from certain problems internal to the movement itself. Firstly, there is a great deal of confusion about the meaning of "Appropriate Technology". Secondly, it is not clear to what extent there are substantive common interests binding the various streams of the social movement together or to what extent the movement is merely a superficial collection of essentially unrelated interest groups and phenomena. Thirdly, it may be argued that many advocates and practitioners of Appropriate Technology have failed to pay sufficient attention to relevant institutional and political factors (although this charge has become less justified as the movement has matured). Fourthly, a significant reason for the limits in the influence of the movement would appear to lie with the lack of a clearly articulated formal theory, the salient features of which are both universally recognized by the movement and identifiable by those outside of the movement.

Systematic policies for the promotion of Appropriate Technology may not be readily developed without some clear conceptual framework to draw upon. As will be apparent in subsequent chapters, advocates of Appropriate Technology have produced and published theoretical statements relevant to the subject. Much of this theory, however, is not well developed and this is reflected in the multifarious nature of the Appropriate Technology movement and in the ease with which criticisms have been directed at it.

There is considerable inconsistency throughout the general technology studies literature as to the essential features of technology and no single theory of technology appears to have received general acceptance. The Appropriate Technology literature reflects this more general phenomenon. One obstacle to the development of an integrated theoretical framework for Appropriate Technology is, as this study will seek to demonstrate, the prevalence of confusion as to the nature and meaning of technology and as to the relationship between technology *per se* and other technology-related phenomena. There have been some attempts to rectify this situation, but the Appropriate Technology literature is still severely marred by weaknesses of this sort.

Two other important obstacles to the development of a comprehensive Appropriate Technology theory have been *inter alia*: the failure to clearly and systematically distinguish between what will be labelled in this study as "technology" and "technology-practice"; and, a failure to adequately incorporate all the relevant dimensions of Appropriate Technology into theorizing, planning and implementation. A significant gap in the field of technology studies is that of an integrated theoretical framework for technology choice and Appropriate Technology. The outlines of such a framework will be developed here in later chapters on the basis of material contained in the current literature. Such a framework is necessary, it will be demonstrated, for the effective analysis of criticisms of Appropriate Technology and for a realistic assessment of the prospects for the widespread adoption of technology choice as a guiding concept for technology policy.

A final problem with Appropriate Technology has been that because of the manner in which it has evolved many people consider Appropriate Technology to be an approach which is of relevance only to the so-called "less-developed" or poorer countries of the "South". This has been questioned by various proponents but the lack of a clear universally applicable theoretical framework has made the presentation of an acultural or country-neutral version of Appropriate Technology quite difficult. This study will help resolve this problem.

To conclude this chapter, there are four core hypotheses which this book proposes and investigates.

The first is that the capacity of technology policy to make useful contributions to solving critical economic, social and environmental problems will be enhanced greatly by treating *technology choice* as a cardinal policy concept.

The second is that the capacity of the technology choice concept to usefully enrich technology policy will be enhanced if the concept is defined in terms of certain principles embodied in the concept of *Appropriate Technology.*

The third is that the general applicability of Appropriate Technology will depend upon it being construed as a *mode of technology-practice* rather than as a particular collection of technologies; and upon this mode of technology-practice being characterized by the harmonious integration of technical-empirical, socio-political and ethical-personal factors.

The fourth hypothesis is that if Appropriate Technology is construed in the above manner then it has the potential to provide a framework for policy and implementation which is desirable and feasible for countries of both the *South* and the *North.*

2

Technological Semantics

Definition of Appropriate Technology

Comprehensive Definition

The concept of Appropriate Technology was first synthesized by the British economist E. F. Schumacher, drawing upon important foundations laid by Gandhi and others.[1] An appropriate technology is defined here as a *technology tailored to fit the psychosocial and biophysical context prevailing in a particular location and period.* This definition does not completely embrace the viewpoints which have emerged under the rubric of "appropriate technology" but is comprehensive enough to incorporate most of the definitions which have appeared in the literature, and it accords closely with the original ideas of Schumacher. Unless otherwise indicated the term will be used consistently with this meaning throughout the book. It will be explained shortly and amplified in subsequent chapters but certain difficulties with adopting a concise and accurate definition should first of all be discussed.

---

[1] "Help to Those Who Need it Most", paper presented to the International Seminar, "Paths to Economic Growth", 21st-28th January, 1961, Poona, India (published in *Roots of Economic Growth*, by E. F. Schumacher [Varanasi: Gandhian Institute of Studies, 1962], pp. 29-42); "How to Help Them Help Themselves", *Observer*, Weekend Review (London, 29th August, 1965). For discussion of Schumacher's debts to Gandhi and others see M. M. Hoda, "India's Experience and the Gandhian Tradition", in *Appropriate Technology: Problems and Promises*, ed. by N. Jéquier (Paris: Organization for Economic Cooperation and Development, 1976), pp. 144-155.

Definitional Problems

The term "appropriate technology" has been taken up by a plethora of organizations, interest groups, individuals and schools of thought, and its usage has consequently been loose and confusing. It is used variously to refer to particular philosophical approaches to technology,[2] to ideologies,[3] to a political-economic critique,[4] to social movements,[5] to economic development strategies,[6] to particular types of technical hardware,[7] or even to anti-technology activities.[8] The diffuse nature of the Appropriate Technology movement is illustrated by the comments of one of its veteran practitioners and critics who describes Appropriate Technology (dubbed "AT" by many of its proponents) as being part lay religion, part protest movement and part economic theory, and censures it for becoming what he terms a "bandwagon".[9] He, Rybczynski, writes as follows:[10]

[2] A. R. Drengson, "Four Philosophies of Technology", *Philosophy Today*, 26, 2/4 (1982), 103-117.

[3] D. E. Morrison, "Energy, Appropriate Technology and International Interdependence", paper presented to the Society for the Study of Social Problems, San Francisco (September, 1978).

[4] D. G. Lodwick and D. E. Morrison, "Research Issues in Appropriate Technology", paper presented to the Rural Sociological Society, Cornell University, Ithaca, New York, 20th-23rd August, 1980 (Michigan Agricultural Experiment Station Journal #9649); subsequently published in *Rural Society: Issues for the Nineteen Eighties*, ed. by D. A. Dillman and D. J. Hobbs (Boulder, Col.: Westview, 1982).

[5] L. Winner, "The Political Philosophy of Alternative Technology: Historical Roots and Present Prospects", *Technology in Society*, 1, 1 (1979), 75-86.

[6] A. Robinson, ed., *Appropriate Technologies for Third World Development* (London: MacMillan, 1979); R. K. Diwan and D. Livingstone, *Alternative Development Strategies and Appropriate Technology: Science Policy for an Equitable World Order* (New York: Pergamon, 1979).

[7] Canadian Hunger Foundation and Brace Research Institute, *A Handbook on Appropriate Technology* (Ottawa: Canadian Hunger Foundation, 1976); K. Darrow, K. Keller and R. Pam, *Appropriate Technology Sourcebook* (Compilation of 2 vols.; Stanford, Cal.: Volunteers in Asia, 1976); J. Magee, *Down to Business: An Analysis of Small Scale Enterprise and Appropriate Technology*, NCAT Brief #2 (Butte, Montana: National Center for Appropriate Technology, 1978).

[8] Cf., Office of Technology Assessment, *An Assessment of Technology for Local Development* (Washington, D. C.: U. S. Government Printing Office, 1981), p. 18.

[9] W. Rybczynski, *Paper Heroes: A Review of Appropriate Technology* (Dorchester: Prism, 1980), fwd.

[10] Ibid., pp. 28-29.

AT was a protest movement and for many it also became a True Belief. But it was more than that; AT developed into a bandwagon of Pullman-car proportions. And what a strange set of travelling companions one found: well dressed World Bank economists rubbing shoulders with Gandhians in metaphorical, if not actual, dhotis; environmentalists, Utopians, and bricoleurs; conventional politicians like President Jimmy Carter and less conventional politicians like Governor Jerry Brown of California, who had both met E. F. Schumacher (himself recently made a Companion of the British Empire by Queen Elizabeth II).

In actual practice the term "appropriate technology" is used to describe far more than just technology and, as Rybczynski's comments indicate, it is a symbol for a heterogeneous social movement which has not itself reached universal agreement on what the notion means. Any concise definition must therefore be stipulative and not merely descriptive.

Since Appropriate Technology was first publicly promoted by Schumacher and colleagues at a conference at Oxford University in 1968, and with its subsequent popularization through the publication of Schumacher's seminal book, *Small is Beautiful: A Study of Economics as if People Mattered,*[11] many related terms have come into widespread use. Examples include: (a) "alternative technology", (b) "appropriable technology", (c) "community technology", (d) "convivial tools", (e) "eco-technology", (f) "humanized technology", (g) "intermediate technology", (h) "liberatory technology", (i) "light-capital technology", (j) "modest technology", (k) "participatory technology", (l) "progressive technology", (m) "radical technology", (n) "soft technology", (o) "technology with a human face", (p) "utopian technology", (q) "vernacular technology", or (r) "village-level technology".[12] Each of these terms reflects their au-

---

[11] E. F. Schumacher (London: Blond and Briggs, 1973).

[12] (a): *Undercurrents* (London: 1972-1985); (b): P. de Pury, *People's Technologies and People's Participation* (Geneva: World Council of Churches, 1983); (c): K. Hess, *Community Technology* (New York: Harper and Row, 1979); N. Wade, "Karl Hess: Technology With a Human Face", *Science,* 187 (January 1975), 332-334; G. Boyle, *Community Technology* (Milton Keynes: Open University Press, 1978); (d): I. Illich, *Tools for Conviviality* (London: Calder and Boyars, 1973); (e): M. Bookchin, "The Concept of Ecotechnologies and Ecocommunities", *Habitat,* 2, 1/2 (1977), 73-85; G. Boyle, "A.T. is Dead - Long Live E.T.!", paper presented at the *A.T. in the Eighties* conference (London, 16th June, 1984); (f): E. Fromm, *The Revolution of Hope: Toward a Humanized Technology* (Perennial Library; New York: Harper and Row, 1968); (g): E. F.

thor's particular viewpoint, and there is considerable diversity amongst the meanings attached to them. This has led to some confusion and unproductive polemic. The definition adopted at the beginning of this chapter, however, is consistent with the original ideas of Schumacher, is comprehensive enough to embrace the majority of the ideas implied by the terms just listed and is specific enough to be of practical use.

In addition to the diversity of ideas associated with Appropriate Technology and the diversity of related terms, there has also been a variety of definitions of the term itself. There has been considerable discussion of the definitional problems involved, and two contrasting approaches have emerged: these may be labelled here as the *specific-characteristics* approach and the *general-principles* approach. The definition adopted here is of the latter type. Both approaches to defining Appropriate Technology have their own advantages and disadvantages as discussed below. Definitions of the former type tend to predominate within the so-called Appropriate Technology movement; and amongst those people outside the movement who are familiar with its main ideas, a concept of Appropriate Technology in keeping with the specific-characteristics approach also appears to predominate.[13]

Schumacher, *Roots of Economic Growth* (Varanasi: Gandhian Institute of Studies, 1962) and numerous other publications (to be surveyed in the following chapter); (h): M. Bookchin, "Toward a Liberatory Technology", in his *Post-Scarcity Anarchism* (San Francisco: Ramparts Press, 1971), pp. 83-139; (i): C. D. Long, *Congressional Record* (Washington, D.C.: U. S. Congress, 8th February, 1977); N. Jéquier and G. Blanc, *The World of Appropriate Technology* (Paris: Organization for Economic Cooperation and Development, 1983), p. 10; (j): R. Vacca, *Modest Technologies for a Complicated World* (Oxford: Pergamon, 1980); (k): J. D. Carrol, "Participatory Technology", *Science*, 171 (1971), 647-653; (l): K. Marsden, "Progressive Technologies for Developing Countries", *International Labour Review*, 101, 5 (1970), 475-502; (m): G. Boyle and P. Harper, eds., *Radical Technology* (Ringwood, Aust.: Penguin, 1976); (n): R. Clarke and D. Clarke, "Soft Technology: Blueprint for a Research Community", *Undercurrents*, #2 (1972); A. Lovins, "Soft Energy Technologies", *Annual Review of Energy*, 3 (1978), 477-517; Autrement, *Technologies Douces*, special edition of *Autrement* (Paris), #27 (October 1980); C. Norman, *Soft Technologies, Hard Choices*, Worldwatch Paper #21 (Washington, D. C.: Worldwatch Institute, 1978); (o): P. D. Dunn, *Appropriate Technology: Technology with a Human Face* (New York: Schocken, 1978); (p): D. Dickson, *Alternative Technology and the Politics of Technical Change* (London: Fontana/Collins, 1974); (q): I. Illich, "Vernacular Values", *The Schumacher Lectures*, ed. with an introduction by S. Kumar (London: Blond and Briggs, 1980), p. 77; (r): Lutheran World Service, *Village Technology Handbook* (Geneva: Lutheran World Service, 1977).

13 Examples of useful discussions of the definitional problems may be found in: C. Cooper, "A Summing Up of the Conference", in *Appropriate Technologies for Third World Development* (London: MacMillan, 1979), pp. 403-409; F. Stewart, "Macro-policies for Appropriate Technology: An Introductory Classification", *International Labour Review*, 122, 3 (1983), 279-293; A. K. N. Reddy, *Technology, Development and the Environment: A Re-appraisal* (Nairobi: United Nations Environment Programme, 1979), pp. 15-23; A.

General-Principles Approach

The general-principles approach has the advantage that it abides by the normal conventions of language by keeping to the commonly accepted meaning of words. The word "appropriate", when used as an adjective, conventionally means that something (technology, in this case) is specially fitting, suitable, proper or applicable for or to some special purpose or use.[14] Used in this way, the adjective places emphasis on technology as a means to certain ends and on the importance of articulating the ends in each case. It raises the question, "Appropriate to what?" The general-principles approach to defining Appropriate Technology contains no specific and tangible content. Rather, it emphasizes the universal importance of examining the appropriateness of technology in each set of circumstances.

A difficulty with this type of definition is that it is very formal. While this fact gives it validity, its all-embracing nature makes it somewhat vague and amorphous. Consequently, it is possible for opposing interest groups in a community to adopt the rhetoric of Appropriate Technology while in practice promoting radically different types of technology.[15] An example is Robertson's use of the term "appropriate technology" as part of his advocacy of the CANDU nuclear reactor system, in stark contrast to the pronounced anti-nuclear stance of the Appropriate Technology movement in general.[16]

Thomas and M. Lockett, *Choosing Appropriate Technology* (Milton Keynes: Open University Press, 1979), pp. 10-48; N. Jéquier, "The Major Policy Issues", in *Appropriate Technology: Problems and Promises* (Paris: Organization for Economic Cooperation and Development, 1976), pp. 16-42; N. Jéquier, "Appropriate Technology: Some Criteria", in *Towards Global Action for Appropriate Technology*, ed. by A. S. Bhalla (Oxford: Pergamon, 1979), pp. 1-22; D. W. J. Miles, *Appropriate Technology for Rural Development: The I.T.D.G. Experience*, I.T.D.G. Occasional Paper #2 (London: Intermediate Technology Publications Ltd., 1982).

[14] J. A. H. Murray, et al., *The Oxford English Dictionary* (Oxford: Oxford University Press, 1933); W. W. Skeat, *An Etymological Dictionary of the English Language* (4th ed. [originally published in 1879-1882]; Oxford: Oxford University Press, 1910); J. B. Sykes, ed., *The Concise Oxford Dictionary of Current English* (6th ed. [originally published in 1964]; Oxford: Oxford University Press, 1976).

[15] This is demonstrated by the difference of approach between Roby and Swinkels in the seminar, "The Appropriate Levels of Technology for Western Australia", at the ANZAAS conference, "Prospect 2000", Perth, May 1979 (*Prospect 2000* [Perth: ANZAAS, 1979]); Roby argues a case for a range of low-cost, small-scale technology, while Swinkels advocates costly, large-scale technology - with both people employing the same rubric.

[16] J. A. L. Robertson, "The CANDU Reactor System: An Appropriate Technology", *Science*, 199, 4329 (1978), 657-664; cf., Institute for Local Self Reliance, "Appropriate

The lack of specific criteria and parameters leaves the abstract type of definition open to diverging interpretations at the practical level. This is because all technology must be "appropriate" for something - irrespective of the possible absurdity of that "something". A further problem with the general-principles type of definition is that the widespread usage of the term "appropriate technology" in fact invokes connotations of specific characteristics, even when it is defined in general terms - this leads towards trivial and sometimes fatuous usage of the term.[17]

Specific-Characteristics Approach

The specific-characteristics approach avoids the above difficulties by assigning specific and tangible operational criteria to the definition. In this way the specific-characteristics definition is more than a concept about the nature of technology and the way it relates to ends. It is simultaneously a normative statement (because it assumes priority for certain ends rather than others) and an empirical statement (because the practical criteria of appropriateness must be based upon some assessment of which technical means generally best serve the ends in question). Whereas the general-principles approach tends to leave the evaluation of ends and means relatively open, the specific-characteristics approach embodies the results of previous efforts to evaluate both of these factors. The specific-characteristics type of definition is therefore of more immediate practical use because it contains various "signposts" for planning and decision making.[18] It contains substantive judgements about the real nature of certain technologies in certain contexts.

Despite these advantages, the rhetoric of the specific-characteristics approach has been employed for trivial and misleading purposes as

Nukes?", *Self Reliance*, 25 (1981), 2. The anti-nuclear stance stance of the Appropriate Technology movement is represented in: T. Bender, *Sharing Smaller Pies* (Portland: RAIN, 1975); G. Coe, *Present Value: Constructing a Sustainable Future* (San Francisco: Friends of the Earth, 1979); A. Lovins and J. H. Price, *Non-Nuclear Futures: The Case for an Ethical Energy Strategy* (San Francisco and Cambridge, Mass.: Friends of the Earth International and Ballinger, 1975); R. Merrill and T. Gage, *Energy Primer* (updated and revised from 1974 ed.; Sydney: Second Back Row Press, 1977); Schumacher, *Small is Beautiful*, esp. pp. 124-135.

[17] The fatuous usage of "appropriate technology" rhetoric is illustrated by the article which appeared in the journal of the American Association for the Advancement of Science entitled "Pen Registers: The 'Appropriate Technology' Approach to Bugging" (D. Shapley, *Science*, 199 [1978], 749-751).

[18] Cf., Stewart, "Macro-policies", p. 281.

much as that of the general-principles approach. For example, in deference to the reputed advantages of small-scale and low-cost technology, Schumacher's phrase "small is beautiful" has become something of a symbol of the Appropriate Technology movement; the phrase has now been used with reference to the possible emergence of smaller and cheaper fusion reactor technology - despite its unproven nature and its high cost relative to other forms of energy technology and its inability to measure up easily against most of the commonly used criteria of Appropriate Technology.[19]

Comparison of Definitions

Some examples from the literature will illustrate these two approaches. The following three quotes from an analyst of Third World affairs, a report from a United Nations Industrial Development Organization conference and a philosopher of technology, respectively, are examples of the general-principles approach to defining appropriate technology:[20]

'Appropriate technology' means simply any technology that makes the most economical use of a country's natural resources and its relative proportions of capital, labour and skills, and that furthers national and social goals. Fostering AT means consciously encouraging the right choice of technology, not simply letting business men make the decision for you.

The concept of appropriate technology was viewed as being the technology mix contributing most to economic, social and environmental objectives, in relation to resource endowments and conditions of application in each country. Appropriate technology was stressed as being a dynamic and flexible concept which must be responsive to varying conditions and changing situations in different countries.

'Appropriate [technology]' here refers to the right and artful fit between technique, tool and human and environmental limits.

---

[19] M. M. Waldrop, "Compact Fusion: Small is Beautiful", *Science*, 219 (1983), 154.

[20] (1) P. Harrison, *The Third World Tomorrow* (Harmondsworth: Penguin, 1980), p. 140; (2) United Nations Industrial Development Organization, *Conceptual and Policy Framework for Appropriate Industrial Technology*, Monographs on Appropriate Industrial Technology, #1 (New York: United Nations, 1979), p. 4; (3) Drengson, "Four Philosophies", p. 103.

The next three quotes, from an O.E.C.D. economist, a brochure from an American Appropriate Technology organization and from a science journalist, respectively, illustrate the specific-characteristics approach.[21]

Appropriate technology (AT) is now recognized as the generic term for a wide range of technologies characterized by any one or several of the following characteristics: low investment cost per workplace, low capital investment per unit of output, organizational simplicity, high adaptability to a peculiar social or cultural environment, sparing use of natural resources, low cost of final product or high potential for employment.

An appropriate technology is relatively inexpensive and simple to build, maintain and operate; uses renewable resources rather than fossil fuels, and does not require high energy concentrations; relies primarily on people's skills, not on automated machinery; encourages human scale operations, small businesses and community cohesion; is protective of human health, and is ecologically sound.

Appropriate technology differs from the other kind in being labour-intensive, accessible to its users, frugal of scarce resources, unintrusive on the natural ambience, and manageable by the individual or small groups.

The first three definitions are generalized formulations emphasizing the achievement of a good fit between technology and its context, and they avoid explicit normative assertions. The latter three clearly exhibit elements of advocacy and stipulation and they also give the reader a better comprehension of the actual type of technologies which are typically associated with the Appropriate Technology movement.

In some cases, as illustrated by the following quote from a study of the commercial prospects of "AT" in the United States, the specific-characteristics approach is taken to extremes by defining appropriate technology in terms of the hardware involved.[22]

---

[21]   (1) N. Jéquier and G. Blanc, *The World of Appropriate Technology* (Paris: Organization for Economic Cooperation and Development, 1983), p. 10; (2) National Center for Appropriate Technology, Information Brochure (Butte, Montana: National Center for Appropriate Technology, 1981), p. 1; (3) N. Wade, "Appropriate Technology and the Too High Outhouse", *Science*, 207 (1980), p. 40.

[22] Magee, *Down to Business*, pp. 2-3.

In order to concentrate on specific aspects of small AT businesses, the author is defining appropriate technology in terms of products and technical systems - solar collectors, composting toilets, recycling, organic agriculture, wood stove manufacturing, small-scale hydropower, energy conservation, methane, greenhouses, adobe, and so on.

While the author of the study just mentioned provides reasons for taking such an approach in that particular case, the definition exemplifies a fundamental problem with the specific-characteristics definitions. They break the conventions of normal language by giving the word "appropriate" a substantive meaning it does not have in general usage. The phrase "appropriate technology", a noun preceded by a qualifying adjective, becomes a compound word "appropriate-technology", which amounts to being just a noun but with the adjectival connotations associated with the word "appropriate". The resulting semantic confusion is significant for at least two reasons. Firstly, it results in a lack of rigor in the analyses and action programmes of people within the Appropriate Technology movement, which can lead to a considerable waste of scarce resources. Secondly, it can do damage to the way in which the general-principles approach to Appropriate Technology is received by policy makers and the broader public - who may ignore its value as a policy making tool because of its association with some community activities and technical experiments with very limited applicability or of isolated relevance.

In addition to the above problems, the adoption by a country or political group of policies for the promotion of Appropriate Technology, where such policies were based upon a specific-characteristics definition, could lead to the adoption of technology which was in fact inferior or did not best serve stated social, economic or environmental objectives. The more specific the characteristics in the definition become, the more static the definition becomes, with the result that the responsiveness of the associated policies to the dynamic environment (both natural and human) may also be reduced. When discussing this problem a scholar of technology and development economics demonstrates the semantic dilemma which has been created by the wide usage of the specific-characteristics approach within the Appropriate Technology movement. By adopting the terminology of this approach in her analysis she is forced to state that, in effect, "inappropriate" technology might sometimes be the most "appropriate"! Viz.:[23]

---

[23] Stewart, "Macro-policies", p. 280 [Emphasis added].

This [i.e., the adoption of so-called "appropriate technology" which does not serve a region's actual needs] might arise because the technology with appropriate characteristics was very inefficient in a technical sense (of low productivity) compared with one with inappropriate characteristics. In some cases, the technology with appropriate characteristics might still be the best choice if its effects on some objectives...outweighed its low productivity. But in others this might not be so, and therefore *the technology with inappropriate characteristics should be preferred.*

Despite its attractiveness to engineers and practitioners who desire straightforward technical design criteria, the specific-characteristics approach to defining Appropriate Technology exhibits severe shortcomings. The semantic difficulties, which themselves create problems, illustrate that the specific-characteristics definition may only be valid when applied to specific circumstances. The general-principles approach, in contrast, is suitable for universal application; it provides a basis for the development of a specific-characteristics definition within specified circumstances.

Preferred Type of Definition

As the balance of this study will attempt to demonstrate, the substantive ideas of the Appropriate Technology movement, of both the normative and empirical kind, may make a valuable contribution to policy formulation and action. As the foregoing discussion reveals, however, the lack of consistency in the literature of the movement creates obstacles to the achievement of this goal. Consequently, the general-principles approach to defining Appropriate Technology is preferred in this book over the specific-characteristics approach. Nevertheless, the insights and practical ideas which emanate from the specific-characteristics approach must still be incorporated into any study which purports to be objective and comprehensive.

The theory, policy and implementation of Appropriate Technology would be enhanced by employing only the general-principles approach for general definitions and by restricting the use of a specific-characteristics definition to specific contexts for which the circumstances have been clearly defined.

Definition of Technology

Dictionary Definitions

Throughout the foregoing discussion the term "technology" was used without being defined. No discourse on this subject would be adequate without such a definition because the idea of Appropriate Technology is fundamentally *inter alia* an idea about technology. Providing a consistent and workable definition of technology is, however, fraught with difficulties.

The *Concise Oxford Dictionary*, in a similar manner to other dictionaries, defines technology as the "(science of) practical or industrial art(s); [the] ethnological study of the development of such arts; [the] application of science."[24] Definitions of this type are of limited value, however, because the meaning and use of the word "technology" has changed over time, it is used differently by different schools of thought and between different languages, its common use is haphazard, and the definition does not convey much of the complexity of meaning attributed to the term in the literature. A number of different approaches to defining "technology" should therefore be examined.

Historical Approach

Technology evolves historically and may therefore only be fully understood from a historical viewpoint. Technology is not static and many different technologies have been developed and superseded in a variety of times and places. This has led many scholars to adopt a historical approach to the study of technology[25] and has encouraged others to speak only of technologies and groups of technologies, in the conviction that it is not possible to discern such a thing as technology itself.[26] This latter view receives considerable criticism in the current

---

[24] J. B. Sykes, ed., *The Concise Oxford Dictionary of Current English* (6th ed. [originally published in 1964]; Oxford: Oxford University Press, 1976), p. 1188.

[25] E.g., D. Landes, *The Unbound Prometheus* (Cambridge: Cambridge University Press, 1969).

[26] Cf., F. Rapp, *Analytical Philosophy of Technology* (Dordrecht: D. Reidel, 1981), pp. 24-25.

literature.[27] It reflects the eighteenth and nineteenth century perspective, as embodied in the dictionary definitions. Referring to this period Winner has written:[28]

> Technology, in fact, was not an important term in descriptions of that part of the world we now call technological. Most people spoke directly of machines, tools, factories, industry, crafts, and engineering and did not worry about 'technology' as a distinctive phenomenon.

Such a perspective lends itself to an approach which gives preference to the historiography of tangible technological change, and which avoids the awkward task of defining technology in systematic or philosophically precise terms.

While it is true that a historical study of technology would be required to do justice to its diversity, it does not follow that some common elements may not be observed throughout that diversity, thereby justifying the place of a systematic analysis of technology in addition to a historical analysis.[29] Evidence for the continuity of technological phenomena throughout history has been surveyed elsewhere and need not be duplicated here.[30] It should be stressed, however, that the validity of *Appropriate Technology* as a theoretical category depends itself upon the validity of *technology* as a theoretical category.

Ambiguity of English Terminology

"Technology" is employed in the English language to denote and connote a mixture of phenomena and concepts; it is therefore impossible to provide a precise and universal definition of the term without it becoming specialized jargon. Some other languages, in contrast, are less ambiguous. French, for example, distinguishes between "*technologie*"

---

27 Cf., J. Ellul, *The Technological System* (New York: Continuum, 1980), pp. 23-33.

28 L. Winner, *Autonomous Technology: Technics-out-of-Control as a Theme in Political Thought* (Cambridge, Mass.: Massachusetts Institute of Technology Press, 1977), p. 8.

29 Rapp, *Analytical Philosophy*, pp. 24-25.

30 E.g.: L. Mumford, *Technics and Civilization* (New York: Harcourt Brace Jovanovich, 1963; first published in 1934); A. Pacey, *The Maze of Ingenuity* (New York: Holmes and Meier, 1975); C. Singer, et al., eds., *History of Technology*, 5 vols. (Oxford: Oxford University Press, 1954-1958); E. Shuurman, *Technology and the Future: A Philosophical Challenge*, trans. by H. D. Morton (Toronto: Wedge, 1980). Note: while there may be much novelty in modern technological phenomena compared with earlier types, it does not necessarily follow that there is no continuity between the two.

and "*technique*";[31] *technologie* is the science of analyzing and describing *technique* or *techniques*, whereas the latter are the object of study of the former.[32] Thus, "*technologie*" (French) conforms with the dictionary definitions of "technology" (English) but not with common English language usage of the term. *Techniques* (French) are the individual technical means, either processes or objects, and *technique* (French) is the general phenomenon of which *techniques* (French) are particular examples. English, in contrast, uses the word "technology" to denote both the science and its object of study. This lack of discrimination in the English language usage of "technology" may partly explain why it is so frequently defined as applied science - rather than as technology *sui generis*. It may also explain why "technics" has become popular amongst American commentators as a general term for all things technological.[33]

The rapid expansion of the role of technology in modern society, both in the level and scope of its deployment, has led to a closer integration of technological phenomena with other factors and to the spread of new phrases such as "the technological order", "the technological society", "the technocratic society", "technological man", "the technostructure" or "the technetronic age".[34] In the rhetoric of the counterculture of the 1960s and early 1970s the ubiquitous influence of technology was symbolized in an extreme form simply as "the system".[35] One schclar has observed that there is a tendency among those who write or talk about technology in our time to conclude that "technology is everything and everything is technology."[36] It appears

---

[31] Ellul, *Technological System*, pp. 23-33.

[32] In German a similar distinction is made between "*technologie*" (technological science) and "*Technik*" (technology, or technical things); cf., E. Ströker's lucid paper, "Philosophy of Technology: Problems of a Philosophical Discipline", in *Philosophy and Technology*, ed. by P. T. Durbin and F. Rapp (Dordrecht: D. Reidel, 1983), pp. 323-336.

[33] E.g., Winner, *Autonomous Technology*.

[34] See the following sources respectively: J. Ellul, "The Technological Order", paper presented to the Encyclopaedia Britannica conference on the Technological Order, March, 1962, Santa Barbara, California; published in *The Technological Order*, ed. by C. F. Stover (Detroit: Wayne State University Press, 1963), pp. 10-37; Ellul, *Technological Society*; T. Roszak, *The Making of a Counter Culture: Reflections on the Technocratic Society and its Youthful Opposition* (London: Faber and Faber, 1969); V. Ferkiss, *Technological Man: The Myth and the Reality* (London: Heinemann, 1969); J. K. Galbraith, *The New Industrial State* (2nd ed., first published in 1967; Harmondsworth: Penguin, 1972); Z. Brzezinski, *Between Two Ages: America's Role in the Technetronic Era* (New York: Viking Press, 1970).

[35] O. , *The Dust of Death: A Critique of the Counterculture* (Downer's Grove, Ill.: Inter Varsity Press, 1973).

[36] Winner, *Autonomous Technology*, pp. 9-10.

that during the last century the term "technology" has grown from something with a quite limited meaning to become an all-embracing symbol. For the purpose of analysis a more discrete definition of technology is required; but such a definition must of necessity be stipulative rather than merely descriptive, due to the lack of uniformity in general usage.

Broad Scope of "Technology"

A selection of definitions will now be considered so as to establish consistent terminology for this study. One important observation of the definitions which are propounded in most studies is that they tend to portray technology as something much more than the hardware, machines or individual apparatus normally associated with "technology" in popular thinking.

The dictionary definitions refer to technology as a group of arts or as the science of such arts - in other words, as a form of human skill or activity. This concurs with the main etymological root of the term, the Greek "techné", which denotes art, craft, skill, or practical knowledge. "Techné" was used by the Pre-Socratics to denote a process, rather than a set of objects, and in such a way as to emphasize the unity of action and knowledge ("epistémé", Greek). The other main etymological root of "technology" is the Greek word "logos" which denotes the ideas of word, reason or principle.[37] The modern usage of the term "technology" normally embraces some reference to social or cultural institutions; this is illustrated by the term "technological society", a phrase first coined in France in 1938 by Georges Friedman ("societé technicienne").[38] There is a growing trend in the serious literature for technology and society

---

[37] J. Burnet, *Early Greek Philosophy* (4th ed., first published in 1892; London: Adam and Charles Black, 1930); M. Heidegger, "The Question Concerning Technology", trans. by W. Lovitt from the text of a lecture delivered before the Bavarian Academy of Fine Arts, June 6th , 1950, entitled "*Die Frage nach der Technik* ", published in *The Question Concerning Technology - and Other Essays*, ed. by W. Lovitt (New York: Harper Colophon, 1977), pp. 1-35; E. B. Koenker, "The Being of the Material and the Immaterial in Heidegger's Thought", *Philosophy Today* (Spring 1980), 54-61; D. D. Runes, et al., *Dictionary of Philosophy* (Totowa, N. J.: Littlefield Adams, 1962), pp. 93, 183-184, 314. Note: "*Technologia*" came to be used in Greek to denote systematic treatment.

[38] Friedman's seminal ideas are published in his book, *Probléms Humains du Machinisme Industriel* (22nd ed.; Paris: Gallimard, 1956); a subsequent publication, *The Anatomy of Work: The Implications of Specialization* (translated by W. Rawson from the French , "Le Travail en Miettes, Spécialisation et Loisers") (London: Heinemann, 1961), has made Friedman's ideas more known in the English speaking world; Ellul (*Technological System*, p. 12) acknowledges Friedman as the progenitor of the term "*societé technicienne*".

not to be viewed as discrete phenomena which interact, but rather as overlapping and mutually determining phenomena: society becomes "technologized" and technology reflects the structures and interests of society. The debate over this issue is complex and a thorough discussion may not be properly conducted here but we may concur with Johnston and Gummett who have concluded:[39] "Technology is not merely a matter of physical and social hardware, but a force which permeates our political, economic and social systems." Recognizing the broad scope and influence of technology does not, however, guarantee us a clear grasp of its distinguishing characteristics. Further discussion is required.

Technique and Structure

Galtung has produced a formula which echoes some of the above observations and which draws upon the distinction, as stressed in French, between *technologie* and *technique*: technology = technique + structure.[40] He describes *technique* as the "visible tip of the iceberg": the tools and the know-how (or, skills and knowledge). The structure is defined as the social relations or "mode of production" within which the tools become operational, and the cognitive structure within which the knowledge becomes meaningful. Galtung's notion of *structure*, while narrower and differing from it in a number of ways, corresponds to the notion of *psychosocial and biophysical context* in our definition of appropriate technology proposed earlier. While it usefully emphasizes the social context of *techniques*, the formula uses the English "technology" with a broader and more equivocal meaning than the French "*technologie*". The term "*technique*" is given a broader meaning than is normal in English by including technical apparatus as part of the concept; it thus accords closely with the French usage of the term.

Galtung's approach stresses how the social relations surrounding the application of technical knowledge substantially determine the nature of the resulting technology. Nevertheless, the formula does not provide us with an actual definition of technology.

---

[39] R. Johnston and P. Gummett, eds., *Directing Technology* (London: Croom Helm, 1979), p. 9.

[40] J. Galtung, *Development, Environment and Technology: Towards a Technology for Self-Reliance*, report to the United Nations Conference on Trade and Development, with the support of the United Nations Environment Programme, TD/B/C.6/23/Rev.1 (New York: United Nations, 1979), p. 15.

Humanity, Nature and Economics

In addition to the foregoing formula Galtung provides a functional definition of technology as the modification of natural or ecological cycles into economic cycles.[41] In this way he appears to define technology as a *process* (for modifying cycles) rather than as *apparatus*; he also portrays technology as a form of mediation between people and non-human reality, drawing upon a conception of people as primarily economic beings. This theme of the modification of nature by people is also emphasized in our paraphrase of Rapp's definition:[42]

> Technology is the refined totality of procedures and instruments which aims at the domination of nature through transformation of the outside material world, and which is based on action according to the engineering sciences and on scientific knowledge.

While Rapp chooses to limit the definition of technology to those things which affect "the outside material world", he does this simply out of a desire to limit the scope of his analysis and to approximate certain popular viewpoints, rather than on etymological or logical grounds.[43] There appears to be no reason for not applying his notion of "procedures and instruments for the domination of nature" to the transformation of the "inner world" of human experience.

For Karl Marx also, technology concerns the interaction of people with nature. In *Capital* he writes:[44]

> Technology discloses man's mode of dealing with nature, the processes of production by which he sustains his life, and thereby also lays bare the mode of formation of his social relations, and of the mental conceptions that flow from them.

---

[41] Galtung, *Development, Environment and Technology*, pp. 5 and 29.

[42] Rapp, *Analytical Philosophy*, pp. 33-36. Note: this paraphrase is the present author's and draws upon several partial definitions provided by Rapp. Cf. L. Tondl, "On the Concept of 'Technology' and 'Technological Sciences' ", in *Contributions to a Philosophy of Technology*, ed. by F. Rapp (Dordrecht: D. Reidel, 1974).

[43] Rapp, *Analytical Philosophy*, pp. 35-36.

[44] K. Marx, *Capital*, Vol. I (London: Lawrence and Wishart, 1954; first published in English in 1887), p. 352.

Marx relates technology here to what he viewed as a definitive characteristic of human existence, *production*; this perspective appears to undergird Galtung's emphasis on economic cycles vis-a-vis technology. A striking feature of Marx's conception is his view that technology *reveals* things.

The German metaphysician Martin Heidegger, along with Marx, has also emphasized technology's capacity for revealing things - nature in particular. In fact, he describes technology as *a mode of revealing*. By taking this approach Heidegger contrasts certain modern technology which, in revealing nature, works *against* it by dominating it and transforming it (to echo Rapp's terminology) and certain pre-modern technology which tends to reveal nature by working *with* nature's inner principles and by cultivating it. An example of such pre-modern technology would be a windmill which causes very little apparent disruption to nature; while an example of the other type would be a nuclear fission reactor which, according to Heidegger, works by "extracting from" nature. Heidegger's perspective is important to note here because it illustrates how domination is not the only mode of relationship between people and nature which may be invoked by the term "technology".[45]

Despite their differences, the thinkers just considered each emphasize the role of technology in mediating human relationships and activities, both in relation to nature and as part of economic life. It is still necessary to inquire, however, what special characteristics technology gives to this mediating role.

Efficient Means

The reference to "the engineering sciences and scientific knowledge" in Rapp's definition is meant by him to indicate that technology always involves efficient goal-oriented activity.[46] Accordingly, Skolimowski has noted how an increase in the efficiency of technological procedures has been a defining characteristic of technological progress since the industrial revolution.[47] The French sociologist, Jacques Ellul, has been amongst the strongest proponents of the view

---

[45] See: Heidegger, "Question Concerning Technology", p. 12, pp. 15-16.

[46] Rapp, *Analytical Philosophy*, p. 32.

[47] H. Skolimowski, "The Structure of Thinking in Technology", *Technology and Culture*, 7, 3 (1966), 371-383.

that efficiency has been the overriding feature of technology ("*La Technique*", French) since its origins, viz.:[48]

[Technology is] the ensemble of the absolutely most efficient means at a given moment...Wherever there is research and application of new means as a criterion of efficiency, one can say that there is a technology.

This emphasis on efficient means as the distinctive feature of technology accords with Habermas' view of technology as purposive-rational action[49], accords with conceptions which emphasize technology's mediating role and either amplifies or is consistent with the other definitions and formulae outlined above. It therefore appears that such an emphasis should be used to provide substantive content to the definition of technology adopted earlier.

Knowledge

Other scholars choose to emphasize knowledge as the defining characteristic of technology - an emphasis which, as previously indicated, is consistent with its etymology. MacDonald, for example, writes:[50]

Technology is really the sum of knowledge - of received information - which allows things to be done, a role which frequently requires the use of machines, and the information they incorporate, but conceivably may not.

He argues against those who treat technology's characteristics as no more than those of the machine. He employs the term "techniques" to denote what he refers to as the tools of technology (e.g., machines). This emphasis on technology as knowledge has grown with the widespread deployment of technologies for processing and transmitting information;[51] the term "technology" has even come to be used as a

---

[48] Ellul, *Technological System*, p. 26.

[49] J. Habermas, "Technology and Science as 'Ideology'", in his *Toward a Rational Society*, trans. by J. Shapiro (London: Heinemann, 1971), pp. 81-122.

[50] S. MacDonald, "Technology Beyond Machines", in *The Trouble with Technology*, ed. by S. MacDonald, et al. (London: Frances Pinter, 1983), p. 27.

[51] Cf., T. Stonier, *The Wealth of Information: A Profile of the Post-Industrial Economy* (London: Thames Methuen, 1983).

synonym for "information technology".[52] The literature on the "sociology of knowledge" also tends to view technology as a special form of knowledge.[53]

While knowledge is, with some justification, recognized by this school of thought as an essential aspect of technology, the tendency by MacDonald and others to define technology as a particular category of knowledge *per se* may be questioned. Firstly, popular English language usage of "technology" normally refers to tangible manifestations of technology - machines, apparatus, tools and technical artefacts etc. Secondly, MacDonald's definition does not fit easily with the dictionary portrayals of technology as a category of human activity; the "sum of knowledge" is surely a product of human activity rather than human activity as such.[54] Thirdly, by limiting "technology" to mean knowledge, we are forced to use the English term "technique" to denote artefacts as well as human skill and method - which sits rather awkwardly with common English parlance. Fourthly, any approach which separates the knowledge aspects of technology too rigidly from the operational and tangible aspects, runs the risk of overemphasizing the role of explicit knowledge as opposed to implicit knowledge in the historical development of technology. The development of technology from predominantly explicit, scientific knowledge (e.g., the nuclear fission bomb) is historically far less common than the development of technologies from the gradual accretion of practical experience (e.g., indigenously developed efficient agricultural implements such as the Mexican *sembradora*)[55] and from the practical innovativeness of technology-users and engineers.[56] Fifthly, as we will subsequently attempt

---

[52] This is apparent, for example, in I. Reinecke's, *Micro Invaders: How the New World of Technology Works* (Ringwood, Aust.: Penguin, 1982).

[53] The following are representative of publications of the "sociology of knowledge": K. D. Knorr-Cetina, *The Manufacture of Knowledge* (Oxford: Pergamon, 1981); T. Jagtenberg, *The Social Construction of Science* (Dordrecht: D. Reidel, 1983).

[54] Some commentators resolve this particular difficulty by referring to technology as the *application* of knowledge rather than just as knowledge; Schnaiberg, for example, drawing upon the work of Schooler (D. Schooler, *Science, Scientists and Public Policy* [New York: Free Press, 1971]), defines technology as the "application of knowledge in the processes of social production" (A. Schnaiberg, "Responses to Devall", *Humboldt Journal of Social Relations*, 7, 2 [1980], 278), and in doing so he links the knowledge aspect of technology with the economics and political-relations aspects emphasized by Galtung, drawing on Marx (see earlier comments).

[55] B. De Walt, "Appropriate Technology in Rural Mexico: Antecedents and Consequences of an Indigenous Peasant Innovation", *Technology and Culture*, 19 (1978), 32-52.

[56] Evidence for the historical role of technology prior to and independent of scientific or explicit knowledge may be found in Lynn White Jr.'s book, *Medieval*

to demonstrate, it becomes difficult to discuss the problems of "appropriate" versus "inappropriate" technology in a given context if technology is viewed simply as knowledge to the exclusion of artefacts or the tangible embodiment of knowledge.

In conclusion, there are some etymological reasons and some indications from other languages which may provide a *prima facie* case for defining technology as a particular form of knowledge; and, it appears possible - in principle - to construct an internally consistent taxonomy based upon such an approach. This approach will not be adopted here, however, because to do so would *inter alia* break with certain entrenched conventions of the English language - and also depart from the common use of "technology" in English language literature on technology. In raising a definition of technology it is nevertheless still important to incorporate the useful insights of the "technology-as-knowledge" school of thought, but with a different semantic convention.

Artefacts

Some commentators avoid the difficulties of the "technology as knowledge" precept by emphasizing *artefacts* as characteristic of technology. Scriven, for example, defines technology as "the systematic process, and the product, of designing, developing, maintaining and producing artefacts."[57] Artefacts, according to the *Concise Oxford Dictionary*, are the products of human art and workmanship.[58]

This approach avoids the semantic problems cited above; it includes much of the meaning attributed to "*technique*" (French), "*technik*" (German) or "technique" (English[59], as per Galtung, MacDonald etc.) under the rubric of "technology". Scriven qualifies his definition with the comment that "technology is artefacts *and* knowl-

---

Technology and Social Change (Oxford: Oxford University Press, 1962); these themes have been developed by D. Ihde (e.g., *Technics and Praxis* [Dordrecht: D. Reidel, 1979], and "The Historical-Ontological Priority of Technology Over Science", in *Philosophy and Technology*, ed. by P. T. Durban and F. Rapp [Dordrecht: D. Reidel, 1983], pp. 235-252); Ihde argues that the "historical-ontological priority" of technology over science is as true for modern technology as for traditional technology.

57 M. Scriven, et al., *Education and Technology in Western Australia*, report of the Working Party on Education and Technology, Western Australian Science, Industry and Technology Council (Perth: W. A. Science, Industry and Technology Council, 1985), p. 25.

58 Sykes, ed., *Oxford Dictionary*, p. 52.

59 When used in this sense "technique" ought perhaps to be labelled as pseudo-English rather than as English in the strict sense.

edge (implicit or explicit) in the service of artefact production,"[60] and makes the distinction between the *application skills of technology* (meaning roughly equivalent to "technique" [English, as per *Oxford Dictionary*] , the *embodiment of technology* (artefacts) and *technological theory* (either explicit or implicit).[61] He avers that techniques (English) may be thought of as technology when they are employed in using or creating artefacts.[62]

Scriven's definition is a useful advance on some of the artefact-free definitions of technology within the English language, but it exhibits a weakness. The part of the definition which refers to technology as the *product* of producing artefacts in effect makes "technology" equivalent to "artefacts"; this gives technology the ubiquitous status of being all things produced by human art and workmanship. Given the meaning of the term "artefacts", the other part of his definition (i.e., the *process* of "designing, developing, maintaining and producing artefacts") may be logically reduced to mean, roughly, "systematic human art and workmanship". Such a formula would include crop rotation and modern managerial practice,[63] but Scriven explicitly rules these out as examples of technology.[64] Scriven's definition does not provide a sufficiently clear indication of the distinguishing features of technology - in addition to those of artefacts. His emphasis on the *process of producing* artefacts does however imply concurrence with the Marxist economic depiction of technology as means of production.

Technology-practice

Most discussions of technology are consistent with Galtung's formula of technology as technique plus structure, in that they consider other factors in addition to specific technical means themselves. Some scholars have attempted to maintain this syncretic approach and strengthen it with more precise definitions - thereby incorporating both the knowledge emphasis and the artefacts emphasis while hopefully avoiding semantic ambiguity. For example, Winner includes the following under the rubric of "technology": *apparatus* (the physical de-

---

[60] Scriven, et al., *Education and Technology*, p. 40.

[61] *Ibid.*, p. 27.

[62] *Ibid.*, p. 39.

[63] MacDonald ("Technology Beyond Machines", pp. 26-27) includes these two activities as part of technology.

[64] Scriven, et al., *Education and Technology*, p. 26.

vices of technical performance); *technique* (human activities characterized by their purposive, rational, step-by-step manner); and, *organization* (social arrangements of a technical form).[65] The Dutch engineer and philosopher, Schuurman, makes a broad distinction between
practical activity ("*techniek*", translated from Dutch as "*technology*")
and *theoretical activity* ("*technische wetenschap*", translated from
Dutch as "*technological science*" or "*technicology*").[66] He then divides
technology (i.e., *techniek*) up into three categories: *technological objects* (things or processes which are put to use in technology); *technological form-giving* (the execution and operation of technology); and
technological designing (the preparation for technological objects and
form-giving).[67] Schuurman's taxonomy brings helpful precision and
semantic consistency to the task of explicating the scope of technology.
It does so, however, at the price of introducing new jargon which may be
cumbersome and have difficulty becoming adopted beyond the confines
of specialized scholarship.

In an attempt to synthesize the various perspectives on technology
illustrated by the above examples, in such a way that semantic consistency is achieved without giving the word "technology" a meaning it
does not possess in popular usage, Pacey has coined the compound word
"*technology-practice*". He defines it as "the application of scientific
and other knowledge to practical tasks by ordered systems that involve
people and organizations, living things and machines."[68] Pacey outlines three aspects of technology-practice: the technical aspect
(equivalent to technique in Galtung's formula), the cultural aspect and
the organizational aspect (which together correspond to structure in
Galtung's formula).[69] The concept of technology-practice covers all the
content normally indicated by "technological" in both popular and specialist discourse; while it corresponds closely to some usages of
"technology" it should nevertheless be viewed as possessing a broader
meaning than is normally given to "technology".

Pacey's schema is not very precise for analytical purposes and it
contains a number of problematical elements but it illustrates the diver-

---

[65] Winner, *Autonomous Technology*, p. 12.

[66] E. Schuurman, *Technology and the Future: A Philosophical Challenge*, trans. by
H. D. Morton (Toronto: Wedge, 1980), p. 377. Note: these notions correspond to the notions
of *technique* and *technologie*, respectively, in French.

[67] *Ibid.*, pp. 8-50.

[68] A. Pacey, *The Culture of Technology* (Oxford: Basil Blackwell, 1983), p. 6
(emphasis added).

[69] *Ibid.*, pp. 4-7.

sity of factors associated with technology-practice and provides a convenient starting point for the development of a more effective nomenclature. It may act as a kind of map for locating technology-related concepts. For example, *Appropriate Technology* (as generally understood throughout the literature) may be viewed as a particular type of technology-practice, while a particular *appropriate technology* would be included as part of the technical aspect; Schuurman's *technological objects* would fall within the technical aspect of technology-practice, while his *technological form-giving, technological designing* and *technicology* (or technological science) would embrace the organizational and cultural aspects as well; Winner's notions of *apparatus* and *technique* would fall within the technical aspect, while his notion of *organization* corresponds to the organizational aspect; and, the definitions of both Heidegger and MacDonald, different though they are, embrace all three aspects of technology-practice.

Pacey's notion points clearly to the cultural and organizational context which always surrounds the technical aspects of technology. By including the cultural aspect as a fundamental component of technology-practice Pacey focuses attention on the normative dimension of technology. While normative factors may have been implicit in the other definitions considered here, his concept encourages explicit examination of such factors. This perspective will be seen as important for elucidating the concept of Appropriate Technology.

Pacey's "technology-practice" accords with the increasing reference by European critics to the notion of *technical praxis*[70] and to the use by American writers of the term "*technics*"[71] in discussing the general subject of technology. His emphasis on "the application of scientific and other knowledge", while not strictly at odds with it however, does not appear to adequately address the growing emphasis, amongst philosopher's and critics of technology, on the distinction between science and technology.[72] Despite the usefulness of "technology-practice" for discussing technological matters in general Pacey's notion does not provide substantive content to the term "technology" itself. An informative and reliable *definition* of technology (in addition to *statements* about technology) is still needed here.

---

[70] E.g., Ströker, *Philosophy of Technology*, pp. 323-331.

[71] E.g., Winner, *Autonomous Technology*; Ihde, *Technics and Praxis*.

[72] See, e.g., J. Agassi, "The Confusion Between Science and Technology in the Standard Philosophies of Science", *Technology and Culture*, 7, 3 (1966), 348-367.

Nomenclature

Need for Stipulative Definition

There is no universally agreed upon definition of technology in the technology studies literature; although, as the foregoing discussion indicates, there is much overlap in the definitions raised and there is a general consensus that a field of study may be identified under the rubric of "technology". It is therefore necessary in this study to stipulate a definition. The semantic rules outlined below could be debated at some length, but an exhaustive analysis of all relevant issues is beyond the scope of the present study. The conventions adopted herein, although based upon a extensive survey of the literature, ought to therefore be viewed as stipulative.

Technology has relevance to most fields of human endeavor and appears to be deeply enmeshed in most aspects of modern society. It is important that this be recognized; but, at the same time time, it is important to avoid adopting a definition which is so general as to render the concept indistinguishable from other concepts. A spectrum exists in schools of thought vis-a-vis technology between those who adopt a broad, all-inclusive definition and those who adopt a narrower, more discriminating definition. The narrower approach is adopted here. The main reasons are that this sits more easily with popular English usage, it assists in resolving some of the confusion in the technology studies literature, it allows a more lucid discussion of the relationship between technology and other factors, and it is to some extent a necessary condition for a fruitful analysis of the concepts of technology choice and Appropriate Technology. This last point will become more apparent later on when the meaning of technology "fitting" its context is considered in detail.

"Technology"

Technology is defined here as the *ensemble of artefacts intended to function as relatively efficient means.*

This definition has the advantage of resolving the difficulties which result from reducing technology to a category of knowledge alone and it also goes beyond the limitations of the artefact-based definition discussed earlier. By including "the function of being relatively efficient means" it is possible to distinguish between artefacts-qua-technology and artefacts-qua-artefacts. "Artefacts" denotes the products of human art and workmanship, and hence need not refer only to physical apparatus and machines; our definition is therefore capable of embracing such less tangible technologies as computer software or cybernetic control systems. Our emphasis on artefacts incorporates the knowledge aspects of technology but stresses that such knowledge needs to be realized, incarnated, embodied or objectified if it is to be deemed "technology". The inclusion of "artefacts" in the definition also indicates the human or social element of technology, but without reducing technology to being simply an aspect of society.[73] In this way it is possible to avoid treating technology as either completely discrete and autonomous, at one extreme, or as lacking endogenous characteristics and a dynamism of its own, at the other extreme.

The use of the term "means" as part of the definition emphasizes technology's instrumental function, and the inclusion of efficiency as part of the definition makes possible a distinction between means-qua-technology and means-qua-means. The term *"relatively* efficient means" is used in preference to just "efficient means" in acknowledgement that: the efficiency of technology may vary geographically and over time, due to innovation and other factors; perfect efficiency may in fact never be attainable[74]; and, the degree of efficiency also relates to such things as the degree of specialization of the technology's function, the nature of the task for which it is in fact used, and the manner in which it is used. It also reflects the recognition that although technical efficiency may be an important factor for technologists and others it is not the only factor influencing the adoption of technology.

In accordance with the above definition, individual technologies, in contrast to technology in general, may be defined as artefacts intended to function as relatively efficient means.

---

[73] Many political economy writers exhibit a trenchant resistance to discussing technology *sui generis*, because of a fear that this would amount to "reifying" or "fetishising" particular technologies (e.g.: H. Thompson, "The Social Significance of Technological Change", *The Journal of Australian Political Economy*, 8 [July 1980], 57-58; L. Levidow, "We Won't Be Fooled Again? Economic Planning and Left Strategies", *Radical Science Journal*, 13 [1983], 28-38).

[74] An amusing survey, by Papanek and Hennessey, of technologies which are poorly designed and ineffective, provides palpable evidence that in the context of the "real world" as opposed to the "drawing board", efficiency in technology-practice is often elusive (V. Papanek and J. Hennessey, *How Things Don't Work* [New York: Pantheon, 1977]).

There are other characteristics of technology which deserve attention, such as its systemic tendencies, but for reasons of space these will be addressed later. For the duration of this study the following terms will be used as outlined below.

"Technological", "Technology-practice"

"Technological" will be used to qualify all operations, activities, situations or phenomena which involve technology. "Technological" shall to be distinguished from "technical", which has a more specific and concise content. "Technology-practice" will be used as the noun equivalent of "technological", and will cover the scope of meanings raised by Pacey; i.e., "technology-practice" will denote all things technological, and will possess a similar scope of meaning associated with Galtung's use of "technology". It will be used on occasions when, as in many (if not most) popular discussions of technology, an all-inclusive meaning is intended and semantic precision is not considered to be important.

"Technical", "Technicity"

"Technical" will be used as the adjective or adverb to qualify phenomena (either human or non-human) dedicated to efficient, rational, instrumental, specific, precise and goal-oriented operations. Unfortunately there appears to be no commonly used English word suitable to be employed unequivocally as the noun corresponding to "technical" - i.e., to denote the factor or quality itself which makes something technical. We shall therefore use the term "technicity" for this purpose.[75]

Technicity is thus a defining feature of technology; but technology, as we have defined it, and as generally understood, may involve other features besides technicity (in the same way, for example, that "humanity" is not an exact substitute for the noun "human"). Accordingly, some technologies may be more technical than others - i.e., the degree of technicity may vary between different technologies.

---

[75] For the purposes of rigorous scholarship there appears no alternative to the introduction of a rarely used word here, "technicity", to enable us to achieve adequate differentiation of concepts. "Technicity" has been used elsewhere in an attempt to translate some of Heidegger's ideas into the English idiom (see, Schuurman, *Technology and the Future*, pp. 87-95).

The crucial importance of this observation will become apparent later, but an example here may help. Horticulture may be considered to be a technological activity because it involves the use of technology (e.g., drip-trickle irrigation systems or rotary hoes) but it might not always generally be conducted in a highly technical manner. Some aspects of modern horticulture (e.g., cloning of hybrid plants, or automated computer control of temperature and fertilizer levels in hydroponic systems) may, however, be highly technical. Nevertheless, just because some horticulturalists do not employ highly technical procedures or highly technical apparatus - preferring instead to work with so-called "low" or less complicated technologies - it does not follow that their operation is therefore not technological.

"Technique"

"Technique" will denote *human skill* which involves a significant technical element. Used in this way the term therefore corresponds closely to the meaning of Scriven's term "application skills of technology". It should be realized, however, that not all skills in *using* technology involve a significant degree of technicity (e.g., storing food in a refrigerator). The normal English meaning of the word "technique" will therefore be adopted here, and other terms such as "technology", "technicity" or other phrases, will be used when appropriate to denote the meaning given to *"technique"* in French (or to equivalents from other languages). In contrast to the tendency of some thinkers to employ "techniques" to denote particular examples of technological artefacts we will use "technologies" for this purpose. As implied above, and in parallel to the situation of "technology", some techniques may be more technical than others.

"Technological Science"

As discussed earlier, *"technologie"* is used in continental European languages (e.g., in French, German and Dutch) to denote an essentially scientific or theoretical activity, but in such a manner as to avoid confusion between the science itself and its object (i.e., *technique* [French] , *technik* [German] or *techniek* [Dutch]). We will therefore avoid using the term "technology" in this sense, and will instead consistently employ "technological science" as the English equivalent of *"technologie"*. The term may denote either the scientific study of technological matters or scientific practice which involves a significant amount of technology.

"Technology" and "Appropriate Technology"

By defining technology as the ensemble of artefacts intended to function as relatively efficient means, a narrower and more precise formula has been adopted than is often found in the literature on Appropriate Technology. For example, Hibbard and Hosticka write:[76]

> Technologies are thus not only objects and material processes, but also organizational forms, methods of knowledge production, decision-making techniques and so on. For example, the common organizational form of the pyramidal bureaucracy is as much a technology as the word-processors, microcomputers, filing systems and calculators which facilitate the work of the organization. One only has to consider the current search for more effective and efficient organizational forms to appreciate this.

To help them address the broad concerns of the Appropriate Technology movement, the authors of this statement have defined technology to incorporate factors normally considered to be part of the environment of the technology.[77]

The issues embraced by Appropriate Technology, both as a concept and as a movement, are quite eclectic and questions of organizational form, social relations, political bias and human experience, amongst others, are integral to an explanation of the concept. It does not follow that in order for these factors to be adequately addressed they must be included as part of technology. It also does not follow that because they may not be included as part of technology they do not exhibit some intrinsic relationship to technology. Technology will therefore be construed throughout this book as technology *sui generis*, and related factors or phenomena will be referred to in their own right and not just as

---

[76] M. Hibbard and C. J. Hosticka, "Socially Appropriate Technology: Philosophy in Action", *Humboldt Journal of Social Relations*, 9, 2 (1982), 1-10.

[77] Jacques Ellul has coined the term "*La Technique*" to denote "the totality of methods rationally arrived at and having absolute efficiency (for a given stage of development) in every field of human activity" (*Technological Society*, p. xxv). This is a much broader and more contentious concept than technology, as we have defined it, and does incorporate such human products as organizational forms; however, "*La Technique*" (as per Ellul), although related to it, is different in meaning to "*technique*" as normally used in French.

part of technology: e.g., organizational structure will be denoted by the term "organizational structure" (or something similar) rather than by "technology".

A definition of appropriate technology was adopted earlier as technology tailored to fit the psychosocial and biophysical context prevailing in a particular location and period. The foregoing discussion of the nature of technology may now be drawn upon to amplify this definition of appropriate technology to: *artefacts which have been tailored to function as relatively efficient means and to fit the psychosocial and biophysical context prevailing in a particular location and period.* The following formula could be used as a "shorthand definition": an appropriate technology is an artefact which functions as a relatively efficient means and which is compatible with its context.

A number of corollaries may be derived from the above definition (in its various forms) and a number of ambiguities might be identified which warrant clarification. These matters will be addressed in later chapters. Having developed a working definition of appropriate technology it is now possible to survey the Appropriate Technology movement. The definitions raised so far are gathered together in Table 2.1 for simple reference.

Table 2.1   Technology-related Nomenclature

Term	Definition / Explanation
"technology"	The ensemble of artefacts intended to function as relatively efficient means.
"technology-practice"	The ensemble of operations, activities, situations or phenomena which involve technology to a significant extent.
"technological"	A term used to qualify operations, activities, situations or phenomena which involve technology to a significant extent (i.e., the adjectival form of "technology-practice").
"technical"	The adjective or adverb used to qualify phenomena (either human or non-human) dedicated to efficient, rational, instrumental, specific, precise and goal-oriented operations.
"technicity"	The distinguishing factor or quality which makes a phenomenon technical (i.e., the noun equivalent of "technical").
"technique"	Human skill which involves a significant technical element.
"technological science"	The scientific study of technological matters, or scientific practice which involves a significant amount of technology.
"appropriate technology"	Artefacts which have been tailored to function as relatively efficient means and to fit the psychosocial and biophysical context prevailing in a particular location and period (i.e., technology which is compatible with its context).
"Appropriate Technology"	A mode of technology-practice aimed at ensuring that technology is compatible with its psychosocial and biophysical context. The term may also be used to denote the general concept, social movement or innovation strategy associated with this mode of technology-practice.

3

Appropriate Technology
as a Social Movement

Existing Literature on the Movement

Here and in the following chapters "Appropriate Technology" will be considered in its capacity as the rubric of an emergent social movement. As indicated in the previous chapter, there is a diversity of definitions in the literature and a diversity of ways in which the definitions may be applied. For example, "Appropriate Technology" has been used as a slogan, a cultural symbol, or as a label for a particular set of technologies; and some writers have used it to denote a philosophical framework, an economic theory, an ideology, a form of dogma, or an approach to innovation. While each of these may be valid approaches to the subject, however, it appears that the relevant literature may be summarized most effectively by viewing it from the perspective of the social movement with which it is associated.

While the literature on Appropriate Technology has burgeoned since the mid 1970s, the movement is relatively young and has not been subject to extensive systematic analysis as a phenomenon in its own right. Some studies of this type have, however, begun to emerge.[1]

---

[1] F. A. Long and A. Oleson, eds., *Appropriate Technology and Social Values: A Critical Appraisal* (Cambridge, Mass.: Ballinger, 1980); M. Carr, ed., *The AT Reader: Theory and Practice in Appropriate Technology* (London: Intermediate Technology Publications Ltd., 1985); M. Hollick, "The Appropriate Technology Movement and its Literature", *Technology in Society*, 4, 3 (1982), 213-229; M. Howes, "Appropriate Technology: A Critical Evaluation of the Concept and the Movement", *Development and Change*, 10 (1979), 115-124; M. Hibbard and C. J. Hosticka, eds., "Socially Appropriate Technology", special issue of *Humboldt Journal of Social Relations*, 9, 2 (1982); G. McRobie, *Small is Possible* (London: Jonathan Cape, 1981); D. E. Morrison, "Energy,

Common to most of them is a view which has been stated succinctly by the sociologist Denton Morrison:[2] "It is fundamentally important to keep in mind that AT, whatever else it is or isn't, is a social movement."

In its capacity as a social movement Appropriate Technology is very difficult to describe concisely. It is emergent and evolving rather than static; it is multifarious and open ended rather than discrete; it is both subjective and objective in nature; and it operates through both informal and formal institutions.

Jéquier and Blanc have attempted an empirical study and statistical analysis of the Appropriate Technology movement but their work has serious limitations.[3] There are difficulties and ambiguities in their selection of suitable organizations for inclusion in their data base, and the connection between the measurable statistics of those organizations and the parameters of the broader social movement are complex and often tenuous. Furthermore, their quantitative work is based upon a once-off static survey which does not embrace the dynamic nature of the

Appropriate Technology and International Interdependence", Paper presented to the Society for the Study of Social Problems, San Francisco, September 1978; D. E. Morrison, "The Soft, Cutting Edge of Environmentalism: Why and How the Appropriate Technology Notion is Changing the Movement", *Natural Resources Journal*, 20, 2 (1980), 275-298; D. E. Morrison and D. G. Lodwick, "The Social Impacts of Soft and Hard Energy Systems: The Lovins' Claims as a Social Science Challenge", *Annual Review of Energy*, 6 (1981), 357-378; N. Jéquier, *Appropriate Technology: Problems and Promises* (Paris: Organization for Economic Cooperation and Development, 1976); "Innovations in Appropriate Technology: The Systems Dimension", in *Systems Models for Decision Making*, ed. by N. Sharif and P. Adulbhan (Bangkok: Asian Institute of Technology, 1978), pp. 287-323; Statement before the Subcommittee on Domestic and International Scientific Planning, Analysis and Cooperation of the Committee on Science and Technology of the U.S. House of Representatives, Ninety Fifth Congress, Second Session, July 25-27, 1978, #110 (Washington, D.C.: U.S. Government Printing Office, 1978), pp. 81-99; "Appropriate Technology: Some Criteria", in *Towards Global Action for Appropriate Technology*, ed. by A. S. Bhalla (Oxford: Pergamon, 1979), pp. 1-22; "Appropriate Technology: The Challenge of the Second Generation", *Proc. R. Soc. Lond. B*, 209 (1980), 7-14; W. Rybczynski, *Paper Heroes: A Review of Appropriate Technology* (Dorchester: Prism, 1980); K. W. Willoughby, "Appropriate Technology for Australian Industry: A Challenge to Philosophical Analysis and Policy Formulation", a paper presented to Section 37, History, Philosophy and Sociology of Science, ANZAAS Congress, Macquarie University, May, 1982 (Sydney: ANZAAS, 1982); L. Winner, "The Political Philosophy of Alternative Technology: Historical Roots and Present Prospects", *Technology in Society*, 1 (1979), 75-86; "Building a Better Mousetrap: Appropriate Technology as a Social Movement", in *Appropriate Technology and Social Values: A Critical Appraisal*, ed. by F. A. Long and A. Oleson (Cambridge, Mass.: Ballinger, 1980), pp. 27-51.

[2] Morrison, "Energy, Appropriate Technology", p. 5.

[3] This research was published in two books by N. Jéquier and G. Blanc (*Appropriate Technology Directory* [Paris: Organization for Economic Cooperation and Development, 1979] and *The World of Appropriate Technology* [Paris: Organization for Economic Cooperation and Development, 1983]).

movement. Consequently, while their data do provide a useful picture of the range of organizations in some way connected with Appropriate Technology work during the late 1970s, many of the statistical correlations which might be drawn from the data are of limited use.[4]

Elsewhere in his work, and presumably in recognition of the inadequacy of a static and quantitative analysis, Jéquier has stressed the emergent and evolving nature of the Appropriate Technology movement together with its non-technical aspects.[5] He has described the movement as a *cultural revolution* in the process of transition from the "first generation" to a "second generation". The former is characterized by minority philosophical, polemical and experimental activities at the "grass roots" level, while the latter is characterized by large scale diffusion and acceptance as part of the status quo. He uses the term "cultural revolution" to denote a broad shift in sociocultural paradigms[6] in the same way that Kuhn[7] has used the term "scientific revolution" to denote broad paradigm shifts within the scientific community. This concurs with the conception of Appropriate Technology as a social movement provided by Lodwick and Morrison:[8]

'Appropriate technology' is not a narrow notion about technology *per se*. It is a broad political-economic critique and proposal for a fundamental revision of the total *sociotechnical system* as it now exists in industrialized nations.

In view of the foregoing it may be concluded that the studies which produce the most useful knowledge about the Appropriate Technology movement are those which are interdisciplinary in nature, which are prepared to look at a variety of data (such as pamphlets, polemical material, philosophical documents, verbal communications and other non-formal records), and which draw upon direct involvement in and personal knowledge of the movement. The studies of McRobie,

---

[4] This point is admitted by the authors: *World of Appropriate Technology.*, p. 22.

[5] Jéquier: "Policy Issues"; "Innovations in A.T."; U.S. Congress Statement; "Some Criteria"; "Second Generation".

[6] E.g., Jéquier, "Second Generation", p. 7.

[7] T. S. Kuhn, *The Structure of Scientific Revolutions* (2nd ed.; Chicago: University of Chicago Press, 1970).

[8] D. G. Lodwick and D. E. Morrison, "Research Issues in Appropriate Technology", a paper presented to the Rural Sociological Society, Cornell University, Ithaca, New York, August 20-23, 1980 (Michigan Agricultural Experiment Station Journal Article #9649), p. 3.

Rybczynski and Whitcomb and Carr embody these elements.[9] The most sophisticated sociological descriptions to date of the movement are those of Morrison and Lodwick mentioned above; the cogency of their work was achieved by their inclusion of a much broader range of evidence than quantitative data alone.

Distinguishing Between the Movement and Its Artefacts

A number of commentators, as indicated above, use the term "appropriate technology" to denote a social and cultural phenomenon. In this study, in contrast, "appropriate technology" (lower case initials) has been defined as a particular category of artefacts, rather than as a social phenomenon *per se*. Technology involves cultural and organizational aspects (as argued earlier) and these two usages of "appropriate technology" are therefore not discrete; nevertheless, this does not deny that they are in fact different notions. This semantic ambiguity is unhelpful. Consequently, the distinction between the *concept* of Appropriate Technology, the *social movement* associated with the concept, and the *views of society* implied by the concept and promoted by the movement, should be stressed. The concept of Appropriate Technology will be developed *analytically* in subsequent chapters. This task may nevertheless be distinguished from the present task of *describing* the origins and manifestations of the social movement and its ideas.

The characteristics imputed to the Appropriate Technology movement by Lodwick and Morrison in the passage quoted above are also prominent in the studies of other authors.[10] There would appear to be a number of advocates of Appropriate Technology, however, who would fall short of advocating a "fundamental revision of the total sociotechnical system as it now exists in industrialized nations".[11] By reserving

---

[9] McRobie, *Small is Possible*; Rybczynski, *Paper Heroes*; Whitcomb and Carr, *A.T. Institutions*.

[10] E.g.: G. Boyle, *Community Technology* (Milton Keynes: Open University Press, 1978); P. H. De Forest, "Technology Choice in the Context of Social Values: A Problem of Definition", in *Appropriate Technology and Social Values: A Critical Appraisal*, ed. by F. A. Long and A. Oleson (Cambridge, Mass.: Ballinger, 1980), pp. 11-25; R. Dorf and Y. Hunter, eds., *Appropriate Visions* (San Francisco: Boyd and Fraser, 1978); A. Thomas and M. Lockett, *Choosing Appropriate Technology* (Milton Keynes: Open University Press, 1979).

[11] E.g.: H. Brookes, "A Critique of the Concept of Appropriate Technology", in *Appropriate Technology and Social Values: A Critical Appraisal*, ed. by F. A. Long and

the term "appropriate technology" for usage according to the definition stipulated in Chapter Two, and by referring to the movement as a movement rather than as part of technology itself, it is possible to study the distinctive "political-economic critique" of the Appropriate Technology movement without excluding various viewpoints which do not fit comfortably within Lodwick and Morrison's framework.[12]

One of the perplexing characteristics of the Appropriate Technology movement is that it embodies a large range of differing viewpoints and differing interest groups. It is therefore not readily apparent to the casual observer just what it is that makes the movement an identifiable social phenomenon. While the concept of Appropriate Technology has flourished in some sections of the academic world, it has developed primarily in an action-oriented context. The successful *development of appropriate technology* has been of greater concern to the movement than the *articulation of systematic theory*. Consequently, the emerging conceptual framework has not been developed extensively and cogently in any one particular place.[13] The analytical potential of the concept has not yet been widely realized.

Despite these difficulties, practitioners and writers within the movement appear to have few problems in identifying the Appropriate Technology approach in broad terms. Furthermore, the movement appears to be evolving a distinctive ideology which draws upon a variety of traditional ideologies, but in such a way that it may not be effectively described in orthodox theoretical categories. The peculiar characteristics of this inchoate synthesis have been described by Morrison in the following way:[14]

> Indeed, one of the more interesting and important aspects of AT is the way that it selectively combines elements of various challenge [ideologies] and established ideologies into a package that is coherent and unique without being fully compatible or fully incompatible with any of its "sources".

A. Oleson (Cambridge, Mass.: Ballinger, 1980), pp. 53-78.; P. D. Dunn, *Appropriate Technology*.

[12] As per Lodwick and Morrison, "Research Issues", p. 3.

[13] One helpful recent contribution to beginning to rectify this situation - with reference to the context of the South - may be found in the work of W. Riedijk ("Appropriate Technology for Developing Countries: Toward a General Theory of Appropriate Technology", in *Appropriate Technology for Developing Countries*, ed. by W. Riedijk (Delft: Delft University Press, 1984), pp. 3-20.

[14] Morrison, "Energy, Appropriate Technology", p. 15.

Various lists and charts illustrating the salient features of this emerging "ideological package" have been produced and reproduced in the literature.[15] Later in this study we will propound a theoretical framework as a synthesis of these component ideologies and notions.

Despite its youth and diffuse nature, the Appropriate Technology movement is a significant international phenomenon. For example, according to a survey conducted by the London based Intermediate Technology Development Group Limited, the number of institutions in the world operating under the rubric of "appropriate technology" grew from a handful in the early 1970s, to over five hundred (of one form or the other) by 1977, to an estimated one thousand by 1980.[16] It is beyond the scope of this study to survey these institutions in detail. Institutional descriptions of the Appropriate Technology movement have also been conducted by Jéquier and Blanc, Herrera and by Reddy.[17] Further evidence for the spread of the movement is provided by the range of directories of Appropriate Technology institutions which have been published by authorities such as the United Nations Environment Programme (UNEP)[18], the Canadian Freedom from Hunger Campaign[19], the Commonwealth Secretariat[20], the International

---

[15] E.g.: R. Clarke, "Some Utopian Characteristics of Soft Technology", notes for Biotechnic Research and Development, United Kingdom, published in *Alternative Technology and the Politics of Technical Change*, by D. Dickson (London: Fontana/Collins, 1974), pp. 103-104, and frequently throughout the literature; cf., Morrison, "Soft, Cutting Edge", pp. 290-291.

[16] R. Whitcombe and M. Carr, *Appropriate Technology Institutions: A Review*, I.T.D.G. Occasional Paper #7 (London: Intermediate Technology Publications Ltd., 1982), p. 2.

[17] Jéquier, "Some Criteria"; Jéquier and Blanc, *World of A.T.*; A. O. Herrera, *The Generation and Dissemination of Appropriate Technologies in Developing Countries*, World Employment Programme, Working Paper #51 (Geneva: International Labour Office, 1979); A. K. N. Reddy, "National and Regional Technology Groups and Institutions", in *Towards Global Action for Appropriate Technology*, ed. by A. S. Bhalla (Geneva: International Labour Office, 1979); W. Rohwedder, *Appropriate Technology in Transition: An Organizational Analysis*, doctoral dissertation, University of California at Berkeley, 1987.

[18] United Nations Environment Programme, *Institutions and Individuals Active in Environmentally Sound and Appropriate Technologies* (Nairobi: United Nations Environment Programme, 1978).

[19] Canadian Hunger Foundation and Brace Research Institute, *A Handbook on Appropriate Technology* (Ottawa: Canadian Hunger Foundation, 1976), Section C4.

[20] Commonwealth Secretariat, *Rural Technology in the Commonwealth: A Directory of Organizations* (London: Commonwealth Secretariat Food Production and Rural Development Division, 1980).

Labour Organization (ILO)[21], the Organization for Economic Cooperation and Development (OECD)[22] and Volunteers in Technical Assistance (VITA)[23]. International periodicals[24], such as *La Lettre du GRET, TRANET Newsletter* and *Appropriate Technology*, also provide regular updates on Appropriate Technology organizations. Part Two of this book will provide a thematic overview of the movement, referring to particular representative institutions as warranted.

The Appropriate Technology movement has evolved in two distinct but interrelated streams. One is concerned predominantly with the so called "less developed" countries (L.D.C.'s), and with their problems of economic development. These countries are sometimes referred to as "underdeveloped", "poorer", the "periphery", "developing" or as the "Third World". In this study they will normally be referred to collectively as the "South", in keeping with recent internationally accepted terminology.[25] The other stream of the movement is based predominantly in the wealthier urban-industrialized countries, to be known herein as the "North". Proponents of Appropriate Technology in the North tend to be disenchanted with certain aspects of mainstream industrial culture and advocate either a transformation or a renewal of technological growth and technological systems. The former stream is mainly concerned with the *attainment* of increased economic development, while the latter tends to be concerned with the *social value and environmental impact* of economic development. The distinction between the two streams is not rigid and there are indications that they are converging in substantive policies, general perspectives and in their geopolitical foci.

---

[21] H. Singer,*Technologies for Basic Needs* (Geneva: International Labour Office, 1978).

[22] Jéquier and Blanc, *A.T. Directory.*

[23] B. Mathur, *International Directory of Appropriate Technology Resources* (Washington, D.C.: Volunteers in Technical Assistance, 1978).

[24] These periodicals are published by the following three organizations respectively: Groupe de Recherche et d'Echanges Technologique (GRET), Paris; Transnational Network for Appropriate Technology (TRANET), Maine, U.S.A.; Intermediate Technology Development Group Ltd. (ITDG), London.

[25] Cf.: W. Brandt, et. al., *North-South: A Programme for Survival*, report of the Independent Commission on International Development Issues under the Chairmanship of Willy Brandt (London: Pan Books, 1980); GJW Government Relations with P. Stephenson, *Handbook of World Development* (Harlow: Longman, 1981).

PART TWO

The Appropriate
Technology Movement

4

Schumacher and Intermediate Technology: Foundations

Schumacher: Progenitor of Intermediate Technology

The most important starting point for a description of the Appropriate Technology movement is the work of Ernst Friedrich Schumacher and the Intermediate Technology Development Group Limited (I.T.D.G.), the organization he founded in 1965 in London, along with George McRobie and Julia Porter. Schumacher and his colleagues were a seminal influence on the whole movement, both through spreading the concept of Intermediate Technology and by demonstrating its practicability through tangible applications and field trials. The concept of Intermediate Technology[1] is only one manifestation of the broader Appropriate Technology concept, but it is an exemplary one. Schumacher, who was normally referred to as Fritz Schumacher, was the progenitor of "intermediate technology" and is generally regarded as the grandfather of the Appropriate Technology movement as a whole.

The broader scope of Schumacher's thought will be considered further on. The present discussion will focus on the narrower concept of Intermediate Technology. This concept is important for at least three

---

[1] Hereafter the following convention will be adhered to: Intermediate Technology (initials capitalized) will denote the general concept or movement associated with the concept, "intermediate technology" (in inverted commas) will denote the term itself, and intermediate technology (plain) will denote actual technology or technologies. Italics will be used when special emphasis is warranted and inverted commas will sometimes be used when the meaning is ambiguous or contentious.

reasons. Firstly, it was the main stimulus for the growth of an international interest in Appropriate Technology. Secondly, much of the semantic confusion and theoretical disagreement in the literature has stemmed from the misapplication and misunderstanding of "intermediate technology". Thirdly, as will be demonstrated shortly, Schumacher's use of "intermediate technology" is an excellent example of a theoretically consistent approach within the terms laid down in Chapter Two: it is an example of the specific-characteristics approach to Appropriate Technology (for specific contexts in which the circumstances have been clearly defined) and it also accords with the general-principles approach.

Schumacher was a high level professional economist and the idea of Intermediate Technology was essentially an economic concept.

Born in Bonn in 1911, Schumacher was son of the acclaimed German economist Professor Hermann Schumacher. During the early 1930s, and following in the footsteps of his father, Fritz Schumacher studied economics - at Oxford University in England and at Colombia University in New York where, because of his quickly recognized brilliance, he was appointed to the academic staff at the age of only twenty two. During this time he was also employed in a number of banking and financial positions as an economist. During the latter half of the 1930s he worked in private commerce, wishing to firmly ground his academic theory in practical experience. His most successful work during that period was with a German based syndicate, known as *Syndikat zur Schaffung Zusätzlicher Ausfuhr*, which organized international trading arrangements. It was extremely profitable financially for Schumacher but, in the growing shadow of Nazism, he was forced to leave Germany for England where he became involved in commerce in London.[2]

Being German born and living in England during the Second World War, he was interned as an alien in 1940 and sent to work as a farm laborer at Eydon in Northamptonshire for two years.[3] He maintained his pre-War interests in economics, however, and despite his very difficult circumstances, continued to study and write. Using his evenings to the fullest, he developed an impressive range of ideas on international economic relations and on the rejuvenation of depressed and underemployed economies. Through a fortunate series of acquaintances Schumacher managed to make friends with such prominent economists

---

[2] B. Wood, *Alias Papa: A Life of Fritz Schumacher* (London: Jonathan Cape, 1984), pp. 1-104.

[3] *Ibid.*, pp. 105-131.

in England at that time as John Maynard Keynes, who was something of a mentor to Schumacher, and his ideas found an audience with such influential people as Sir Stafford Cripps, the Chancellor of the Exchequer. This led to appointments at Oxford University (in 1942) and the British Treasury (in 1944).[4]

Schumacher became one of the major intellectual influences on national and international economic planning during and after the War. His influence on Keynes was significant, so much so that, shortly before his death, Keynes pronounced to Sir Wilfred Eady of the Treasury: "If my mantle is to fall on anyone, it could only be Otto Clarke or Fritz Schumacher. Otto Clarke can do anything with figures, but Schumacher can make them sing."[5] In fact, as Schumacher himself confirmed candidly after the death of his mentor-cum-friend, Keynes had plagiarized much of his own material.[6] The so called "Keynes Plan", presented to the historic Bretton Woods conference (which led to the establishment of the International Monetary Fund), consisted largely of Schumacher's work.[7] Schumacher published his own piece on multilateral financial clearing mechanisms for the post-War period at virtually the same time as Keynes, in a 1943 issue of *Economica*, subsequent to Keynes' repeated advice that he delay publication.[8]

---

[4] *Ibid.*, pp. 132-167.

[5] Quoted by Schumacher in a letter to his parents, 16th April 1949; cf.,Wood, *Alias Papa*, p. 135.

[6] Personal communication with Schumacher's colleague, George McRobie, November 1982; this description is corroborated by evidence recounted by Wood in *Alias Papa*, pp. 132-134.

[7] The "Keynes Plan" was first publicized through an article in the 13th March 1943 issue of *The Economist* ("Post-war Currency Plans", 144, 5194 [1943], 330-331); Keynes subsequently published it as a White Paper of the British Government (London: H.M.S.O., 1943; Cmd. 6437) under the title *Proposals for an International Clearing Union* during the week leading up to 10th April 1943 [cf.,*The Economist*, 144, 5198 (1943), 452]. Note: the pages from *The Economist* on which the article "Post-war Currency Plans" were printed were missing from copies of the journal in six major libraries in Australia consulted during research for this book.

[8] E. F. Schumacher, "Multilateral Clearing ", *Economica*, 10, 38 (1943), 150-165 (issued in May 1943); although formally published two months after Keynes had taken credit for "his" plan, Schumacher's article had been in private circulation since November 1942 (cf., *Economica*, 10, 38 [1943], 150, footnote) and an earlier version, entitled *Free Access to Trade* (London: Royal Institute of International Affairs, 1942) had been in circulation since March 1942; thus, Schumacher's work had been released one year prior to the so-called "Keynes Plan". Keynes, having read Schumacher's work, advised him not to publish his ideas because the time was supposedly not "opportune" (see Wood, *Alias Papa*, pp. 132-134; cf., G. McRobie, "The Philosophy and Work of E. F. Schumacher", Keynote Address at the *E. F. Schumacher Memorial Conference on Appropriate Technology*, 1st December 1982, Macquarie University, Sydney). Note: while the "Keynes Plan" was substantially plagiarized from Schumacher's work, it

Schumacher's contributions in this field achieved further influence through two additional publications during 1943, one entitled *Export Policy and Full Employment* [9] and another, written with the distinguished Polish economist Michal Kalecki, who was by then at Oxford, entitled *New Plans for International Trade*.[10]

In 1944 Schumacher became the chief economic leader writer for *The Times* in London.[11] He also worked vigorously, both at Oxford and with the Treasury, on programs for attaining full employment through public financial mechanisms; he published a treatise on this topic in 1945.[12] Furthermore, he was the main author of the influential report published the previous year, *Full Employment in a Free Society*, officially accredited to Sir William Beveridge.[13] Once again, his German origins combined with his desire for the successful application of his ideas more than for his own self aggrandizement, led to public credit not going where it was very much due. During the same period Schumacher also published material on the role of planning in private enterprise economies, focussing on the economic dimensions of land use and housing policies.[14]

In 1945 Schumacher published, along with Walter Fliess, a treatise on the revitalization and reform of the German financial system.[15] This was followed in 1946, after being granted British citizenship, with an appointment as Economic Advisor to the Economic Sub-Commission of the British Control Commission in Germany, where he remained for several years.[16]

---

differed from Schumacher's work on a number of points, and Schumacher came into conflict with Keynes over the final outcome of the Bretton Woods agreement; furthermore, Schumacher always remained highly critical of Keynes' (perhaps inconsistently held) view that capital accumulation would eventually lead to wealth and leisure for all (see I. Low, "Humane Economist", an interview with E. F. Schumacher, *New Scientist*, 63, 914 [1974], 656-657).

[9] London: Victor Gollancz and Fabian Publications, 1943.

[10] M. Kalecki and E. F. Schumacher (Oxford: Basil Blackwell, 1943).

[11] Low, "Humane economist", p. 656.

[12] E. F. Schumacher, "Public Finance - Its Relation to Full Employment", in *The Economics of Full Employment*, a publication of the Oxford University Institute of Statistics (Oxford: Basil Blackwell, 1945).

[13] W. H. Beveridge (London: George Allen and Unwin Ltd., 1944; originally published in London, earlier in 1944, by the Liberal Publication Department). Cf. Wood, *Alias Papa*, pp. 161-167.

[14] E. F. Schumacher, *What Will Planning Mean in Terms of Money?* (Cheam, U. K.: The Architectural Press, 1944).

[15] E. F. Schumacher, *Betrachtungen zur Deutschen Finanzreform* (London: St. Clement's Press, 1945).

[16] Wood, *Alias Papa*, pp. 187-204.

In 1950 Schumacher was appointed as Economic Advisor to the National Coal Board in Britain, where he remained for over two decades. Before his retirement from that post in the early 1970s he added the posts of Director of Planning and Director of Statistics to his responsibilities.[17] He thus spent over twenty years in a demanding and influential position where he applied his considerable expertise in economic planning from the center of one of Britain's massive nationalized industries. He also published a series of papers on energy and resource economics.[18]

When, in the 1960s and 1970s, Schumacher spent an increasing amount of his energy developing, promoting and applying his concept of Intermediate Technology, he did so as an internationally reputed economist with an impressive track record. The profession of economics was his forte. The apparent lack of orthodoxy of Intermediate Technology could not be put down to ignorance or naivety on Schumacher's behalf. The concept had its roots, rather, in his considerable proven capacity for incisiveness, rigor and imagination as an economist - both in theory and in practice.

Intermediate Technology: Origins of the Concept

Before explaining Intermediate Technology the historical background to the concept should be identified. It is useful to distinguish between its immediate origins and the deeper structural or cultural influences which prepared the ground for the concept's eventual popularization. These latter influences will be adduced elsewhere throughout the book.

---

[17] *Ibid.*, pp. 240-252, 268-346.

[18] E.g., E. F. Schumacher: "Coal - the Next Fifty Years", in *Britain's Coal*, report of the study conference of the National Union of Mineworkers, 25th - 26th March 1960 (London: National Union of Mineworkers, 1960); *Prospect for Coal* (London: National Coal Board, 1961); *Clean Air and Future Energy - Economics and Conservation* , the 1967 Des Voeux Memorial Lecture (London: National Society for Clean Air, 1967); "The Use of the Land", *Resurgence*, 2, 7 (1969), 6-9; "Energy and Man", in *Energy, Man and the Environment*, proceedings an international symposium, 3rd - 5th February, 1972, Rüschlikon, Switzerland (Zurich: Gottleib Duttweiler Institute for Economic and Social Studies, 1972); "Western Europe's Energy Crisis - A Problem of Lifestyles", *Ambio*, 2, 6 (1973), 228-232; "No Future for Megalopolis", *Resurgence*, 5, 6 (1974), 12-16; *Think About Land* (London: Catholic Housing Aid Society, 1974); "Western Europe's Energy Crisis: Where are we heading? What can we do?" *Resurgence*, 5, 3 (1974), 11-12; a large collection of Schumacher's published and unpublished writings on energy, in addition to the above, has been assembled and edited by G. Kirk under the title *Schumacher On Energy* (London: Abacus, 1983).

The most conspicuous of the immediate origins of Intermediate Technology occurred in the context of the economic development difficulties of the South. It was the realization by people in both aid-giving and aid-receiving countries that the development aid of recent decades, and the associated attempts at accelerated industrialization through capital-intensive technology imported from the North, had largely failed as means of solving the basic problems of economic development in the South. Such methods had not been able to fulfill the hopes which had been placed in them. The O.E.C.D. economist, Nicolas Jéquier, comments as follows:[19]

> This problem has been vividly expressed by Dr E. F. Schumacher in his influential book, *Small is Beautiful*, which perhaps more than any other, has contributed to popularize the concept of intermediate technology, both in the developing countries and in the industrialized nations.

Intermediate Technology was the product of Schumacher's genius. It is to his credit that, despite its initial rejection by economists, the concept's poignant yet common sense nature, gave it widespread appeal as a fresh approach to solving global problems of poverty - which had remained enigmatic despite the proliferation of professional experts in the field of economic development.

Many of the ideas behind Intermediate Technology were not unique to Schumacher. His unique contribution was to synthesize a wide array of material into a simple package with a broad scope for application. One tradition which made a direct contribution to Schumacher's thinking was that of Gandhi and the Indian community development, or "Sarvodaya", movement. This tradition, which dates back to the late nineteenth century, had developed reasonable sophistication in its theories of development by the mid 1950s, under the leadership of people such as J. C. Kumarappa, D. R. Gadgil, A. V. Bhave and J. P. Narayan.[20] A scholarly study of the economics of this tradition, pub-

[19] N. Jéquier, "The Major Policy Issues", in *Appropriate Technology: Problems and Promises* , ed. by N. Jéquier (Paris: Organization for Economic Cooperation and Development, 1976), p. 25.

[20] M. M. Hoda, "India's Experience and the Gandhian Tradition", in *Appropriate Technology: Problems and Promises* , ed. by N. Jéquier (Paris: Organization for Economic Cooperation and Development, 1976).

lished in 1958 by Richard Gregg, had certainly been read at some stage by Schumacher, as had Kumarappa's *Economy of Permanence.*[21]

Gandhi's influence is quite evident in much of Schmacher's writing.[22] Schumacher had studied the writings and speeches of Gandhi, whom he considered one of the greatest men to have lived during this century;[23] Gandhi, along with R. H. Tawney, had exerted more influence on the longer term development of his economic thinking than any other writer.[24] M. M. Hoda, head of the Appropriate Technology Development Unit, and an acquaintance of Schumacher, makes the following comment:[25]

> In 1963,[26] Dr E.F. Schumacher...visited India at the invitation of the Planning Commission and Jayaprakash Narayan. He was influenced by the Gandhian ideas of industrialization and technology, adapted them to modern needs and turned intermediate technology into a worldwide movement. ...Schumacher's movement of intermediate technology gave a new lease of life to the concept of village development and the Gandhian movement, reinforced as expected by Schumacher's ideas, took a lead in giving a new meaning and a scientific backing to the rural development program.

A major event in Schumacher's life, which led to his influence during the early 1960s on Indian thinking about rural economic develop-

---

[21] R. B. Gregg, *A Philosophy of Indian Economic Development* (Ahmedabad: Navajivan Publishing House, 1958); J. C. Kumarappa (4th ed.; Rajghat, Kashi: Sarva-Seva-Sangh-Publication, 1958). See E. F. Schumacher, "Buddhist Economics", *Resurgence,* 1, 11 (1968), 37-39.

[22] E.g.: E. F. Schumacher: *Economic Development and Poverty* (London: Africa Bureau, 1966), p.11; *Small is Beautiful* , pp. 31,34,143; "Economics Should Begin With People, Not Goods", *The Futurist* (December 1974), 275; "Conscious Culture of Poverty", *Resurgence,* 6, 1 (1975), 5.

[23] See E. F. Schumacher, "Economics in a Buddhist Country", paper written for the Government of the Union of Burma, Rangoon, 1955, published in *Roots of Economic Growth,* by E. F. Schumacher (Varanasi: Gandhian Institute of Studies, 1962), p. 3; cf. Wood, *Alias Papa,* p. 243.

[24] Personal communication with George McRobie, November, 1982; Wood, *Alias Papa,* pp. 292-294; P. Gillingham, "The Making of Good Work", epilogue in *Good Work,* by E. F. Schumacher (New York: Harper and Row, 1979), p. 203.

[25] Hoda, "India's Experience", p. 147.

[26] Other evidence points to the year as 1962, e.g.: G. McRobie, *Small is Possible* (London: Jonathan Cape, 1981), p. 19; Wood, *Alias Papa,* pp. 320-321; E. F. Schumacher, *Reflections on the Problem of Bringing Industry to Rural Areas* (New Delhi: Indian Planning Commission, 1962). It appears that Schumacher remained in India until early 1963.

ment, was a period he spent in 1955 as an economic adviser to the Government of the Union of Burma. He was funded by the United Nations, as a high level economist with considerable expertise in the production and execution of economic development plans and a specialized knowledge of modern fiscal theory and practice, to advise U Nu, the Burmese Prime Minister. Schumacher was confronted directly by conditions quite different to those under which he was used to applying his economic knowledge. This provoked him to question how useful the orthodox forms of economic planning practiced in the industrialized countries would be to a country such as Burma.[27] He conveyed something of his concern in a letter to his wife, Muschi, back in England:[28]

> There is an innocence here which I have never seen before, - the exact contrary of what disquieted me in New York. In their gay dances with their dignified and composed manners, they are lovable; and one really wants to help them, if one but knew how. Even some of the Americans here say: "How can we help them, when they are much happier and much nicer than we are ourselves?"

Schumacher's formal training in economics told him that the very low per capita income figures exhibited by Burma reflected a desperate level of poverty; however, his real life observations pointed to a very different picture.

This experience of Burma crystallized thoughts in Schumacher's mind which he had been evolving for some time. He understood more clearly that economics was not an independent body of knowledge, but was rather derived from presuppositions which normally remained beyond the thinking of most economists in their capacity as economists.[29] Different presuppositions inherent in different philosophies of life or different cultures would, as Schumacher now argued, lead logically to different types of economic systems. In the first of six reports to the Economic and Social Council of Burma, Schumacher introduced this insight in the following way:[30]

---

[27] Wood, *Alias Papa*, p. 243.

[28] *Ibid.*, p. 244.

[29] This insight was subsequently developed more extensively throughout Schumacher's writings, where he made the distinction between "economics" and "metaeconomics"; eg., E. F. Schumacher, "Does Economics Help? An Exploration of Metaeconomics", paper presented to Section F (Economics) at the 1972 Annual Meeting of the British Association for the Advancement of Science, published in *After Keynes*, ed. by J. Robinson (Oxford: Basil Blackwell, 1973), pp. 26-36.

[30] Schumacher, "Economics in Buddhist Country", p. 1.

All actions of government have an economic aspect. In view of the universality of the "economic aspect" it is not surprising, neither is it abnormal, that a "science", a systematic "body of thought", should have grown up, commonly called Economics. But one thing is surprising, and is indeed abnormal, namely, that there should only be one "science", only one body of thought called Economics. Because people's ideas of the purpose and meaning of life vary very much; and when different people attach different meanings to life, this must inevitably also affect their ideas about any particular aspect of life. The whole is greater than any of its parts or aspects. Well, let us say it straight out: What today is looked upon as the science of economics is based upon one particular outlook on life, on one only, the outlook of the Materialist.

The centrality of Buddhism in Burmese life had a profound influence on Schumacher and led him to attempt an outline of a framework for an economic system which accorded with the basic principles of Buddhism. He went on to write:[31]

Because Economics, up to a point, can rightly claim universal validity, it has been accepted as possessing universal validity throughout. ...The essence of Materialism is not its concern with material wants, but the total absence of any idea of Limit or Measure. The materialist's idea of progress is an idea of progress without limit. ...Economics, as taught today throughout the world - before the iron curtain and behind, - recognizes no limit of any kind. It is, therefore, the Economics of Materialism and nothing else.

In contrast to the "Economics of Materialism" Schumacher then advocated a form of economics which aimed for *progress up to a point*, where the parameters of such progress were defined by the central precepts of Buddhism:[32]

When, then, shall we get a system of thought that could be called Buddhist Economics? When will people at last realize and understand that the Economics of Materialism is not of universal validity, that any ordering of life in accordance with its precepts will be utterly incompatible with, and inimical to, the Buddhist way of life? ...When

---

[31] *Ibid.*, pp. 2-3.
[32] *Ibid.*, p. 3.

will they take cognizance and admit that other systems of Economics are possible and necessary and are even already available in rudimentary form?

Schumacher drew upon foundations laid by Gandhi to spell out what he saw as the three main characteristics of a system of economics in keeping with Burma's cultural heritage. These were: the acceptance of certain limits to material economic growth; a distinction between renewable resources and non-renewable resources; and, the recognition of differences in the value of materials other than differences in price.[33]

In the remaining five reports to the Economic and Social Council, before leaving Burma, Schumacher expanded these principles in some detail and in jargon which was generally more acceptable to economists. He emphasized: the role of developing indigenous economic expertise amongst the Burmese, as opposed to over-reliance upon Western advisers; the importance of rural development programs; the need for balance between the production of goods and the development of infrastructure (such as modern transport systems); and, he stressed the need for a strategy of economic self-reliance rather than one of dependency on foreign economies. He did not receive a positive response from the Burmese officials.[34]

Despite the cool response his advice received in Burma, the experience confirmed in Schumacher's mind that different countries or regions, different types of economic problems and different value systems, require the implementation of different economic strategies. He expanded this line of argument in a 1960 article concerned with two problems: viz., "how to conduct international affairs in such a manner that there is never again a resort to large scale violence", and "how to conduct economic affairs in a manner that is compatible with both peace and permanence". He concluded:[35]

> The West can indeed help the others [i.e.,"underdeveloped nations"], as the rich can always help the poor. But it is not an easy matter, expressible in terms of money alone. It demands a deep respect for the indigenous culture of those that are to be helped - maybe even a deeper respect than is possessed by many of them themselves. Above

---

[33] *Ibid.*, pp. 5-8.

[34] Private communication with George McRobie, November 1982; cf., Wood, *Alias Papa*, pp. 240-252.

[35] E. F. Schumacher, "Non-violent Economics", *The Observer* (London), Weekend Review (21 August 1960).

all, it would seem, it must be based on a clear understanding that the present situation of mankind demands the evolution of a non-violent way of political and economic life.

The need for non-violence in political and economic life became a cornerstone of Schumacher's future concerns with the role of technology in society and the natural environment.

In the meantime Schumacher had made friends with Jayaprakash Narayan, the leading Gandhian from India, who had visited London in 1958 and 1959. Narayan, who was very impressed with Schumacher's approach, circulated copies of Schumacher's papers amongst leaders in the Indian Government, including Prime Minister Nehru.[36] This led to Schumacher's attendance at an international conference on "Paths to Economic Growth" at Poona in India in January 1961[37] and to the preparation of two papers on Indian economic problems for the Gandhian Institute of Studies in Varanasi in April and July the same year.[38] Drawing upon his Burmese experience, combined with his long interest in employment-related economics and more recent experience in resource economics in Britain, Schumacher expounded the choices confronting India as to pathways for economic development; he focussed strongly on questions of *social dynamism* rather than on those of *finance*. He argued that successful economic development for India would depend primarily on whether ways could be found to mobilize the creative power of people. For example:[39]

> While it is formally true to say that economic development depends on "the accumulation of capital", on "savings" and "investment", what really matters is their material source, which is labour, and which, realistically speaking, must be primarily indigenous labour. The labour power of the indigenous populations is the great potential source of "capital", "savings", "investment" and so forth, as it is the source of income. How can it be mobilized and usefully applied?

---

[36] McRobie, *Small is Possible*, p. 21.

[37] See E. F. Schumacher, "Paths to Economic Growth", and "Help to Those Who Need it Most", papers presented to international seminar, "Paths to Economic Growth", 21st - 28th January, 1961, Poona, India, published in *Roots of Economic Growth*, by E. F. Schumacher (Varanasi: Gandhian Institute of Studies, 1961), pp. 14-28 and pp. 29-42.

[38] E. F. Schumacher, "Notes on Indian Development Problems", report to the Gandhian Institute of Studies in April 1961, and "Levels of Technology", report to the Gandhian Institute of Studies in July 1961, published in *Roots of Economic Growth*, by E. F. Schumacher (Varanasi: Gandhian Institute of Studies, 1961), pp. 43-48, and pp. 49-56.

This "people-centered" approach to economics led Schumacher to conclude that forms of industry or technology which discouraged the spontaneous mobilization of labour power, no matter how acceptable they appeared from the perspective of mainstream economic theory, would ultimately hinder economic development where it was most needed.[40]

On the basis of the above perspective Schumacher developed his first formal statement of Intermediate Technology, in a July 1961 paper for the Gandhian Institute of Studies, where he concluded:[41]

> Economic development is obviously impossible without the introduc-
> tion of "better methods", "higher technology", "improved equipment" -
> call it what you like. But ... all development, like all learning, is like a
> process of stretching. If you attempt to stretch too much, you get a
> rupture instead of a stretch, or you lose contact and nothing happens
> at all. ... The only hope, I should hold, lies in a broadly based, decen-
> tralized crusade to support and improve the productive efforts of the
> people as they are struggling for their livelihoods now. "Find out what
> they are doing and help them to do it better. Study their needs and
> help them to help themselves."

The concept did not receive widespread international acclaim until after the publication in 1973 of Schumacher's best selling book, *Small is Beautiful: A Study of Economics as if People Mattered*, with its chapter entitled "Social and Economic Problems calling for the Development of Intermediate Technology".[42] It is interesting to note, however, that the notion had been formally articulated twelve years earlier. Other publications on the topic had begun to appear by the end of the 1960s.[43] The work of Schumacher and colleagues will be the focus here, how- ever, because of its more seminal historical influence.

---

39 Schumacher, "Help to Those", pp. 32-33.
40 *Ibid.*, passim.; Schumacher, "Indian Development".
41 Schumacher, *Levels of Technology*, pp. 55-56.
42 (London: Blond and Briggs, 1973), pp. 159-177.
43 The most notable examples are those of: K. Marsden, *Appropriate Technologies For Developing Countries* (Geneva: International Labour Office, 1966), "Towards a Synthesis of Economic Growth and Social Justice", *International Labour Review*, 100, 5 (1969), "Progressive Technologies for Developing Countries", *International Labour Review*, 101, 5 (1970), 475-502; R. K. Vepa, *Appropriate Technology for a Decentralized Economy* (Bombay: Vora and Co., 1969); Volunteers in Technical Assistance, *Village Technology Handbook* (Revised from 1970 edition; Mt. Ranier, U.S.A.: V.I.T.A., 1977).

His work during 1961 led to Schumacher's invitation back to India the following year by the Prime Minister, Pandit Nehru, to advise the Indian Government on economic development policies.[44] Under the auspices of the Indian Planning Commission Schumacher travelled the country extensively, making detailed practical observations as a basis for improved policies on rural industrialization. He used this period, along with the research he conducted, to refine the ideas in his earlier paper, *Levels of Technology*. This crystallized in a report to the Planning Commission entitled, *Reflections on the Problem of Bringing Industry to Rural Areas.*[45] Unfortunately for Schumacher the Planning Commission did not embrace his plans; although, unlike the Burmese seven years earlier, a minority of Indian planners were enthused by his approach.[46] The Report to the Indian Planning Commission was eventually published in London in 1964, under the title, *Rural Industries*, by the Overseas Development Institute.[47] Schumacher further developed his arguments in an attempt to win over the British and international economics elite, in a paper entitled *Industrialization through Intermediate Technology*; this was presented in 1964 to the Cambridge University Conference on Rural Development, a gathering attended by many of the world's leading development economists, and it was subsequently published.[48]

The concept of Intermediate Technology proved to require too much of a volte-face by most professional development planners, both in Britain and abroad, and was rejected by the majority of them.[49] Most critics feared that "intermediate" implied "regressive" or "inferior". This was not, of course, the interpretation placed on the concept by Schumacher, who viewed Intermediate Technology as the *best* means for the attainment of economic development amongst the majority of the poor in the South.

Despite widespread antipathy to his ideas on technology and development, Schumacher's Intermediate Technology was taken seriously by a small network of friends and colleagues, including David Astor

---

[44] McRobie, *Small is Possible*, p. 19.

[45] E. F. Schumacher (New Delhi: Indian Planning Commission, 1962).

[46] McRobie, *Small is Possible*, pp. 22-23.

[47] Published in *India at Midpassage*, ed. by W. Clark (London: Overseas Development Institute, 1964).

[48] Published in: *Resurgence*, 1, 2 (1966), 6-11; *Minerals and Industries* (Calcutta), 1, 4 (1964); and, *Developing the Third World: The Experience of the 1960s*, ed. by R. Robinson (Cambridge: Cambridge University Press, 1971).

[49] McRobie, *Small is Possible*, pp. 22-24; Wood, *Alias Papa*, pp. 321-324.

(Editor of London's *Observer*), Lord Robens (Chairman of the National Coal Board of Britain) and Julia Porter (Secretary of the African Development Trust).[50] One of the closest of these colleagues was the Scottish economist, originally from the London School of Economics, George McRobie. McRobie had worked as a partner with Schumacher since he joined him at the Coal Board in 1956, and remained one of his most dependable and hardest working allies until Schumacher's death in 1977. McRobie had, in fact, by then (1964) spent some time in India assisting the development of Intermediate Technology activities - for which some interest had begun to grow since the initial rebuttal by the Planning Commission in 1962.[51]

Following the rejection of Intermediate Technology by the academic economists at the Cambridge conference, Schumacher and McRobie attempted to gain support for the idea from the Ministry of Overseas Development. While the Minister, Barbara Castle, apparently had little trouble accepting the validity of their arguments, the Ministry was not prepared to back the concept until a practical and effective demand for such technology from the poor in the South could be demonstrated. In effect, this meant that the professional politicians were no more helpful for the promotion of Schumacher's approach than were the professional-cum-academic economists.[52]

The reticence of the development professionals towards Intermediate Technology led Schumacher and McRobie towards a realization which profoundly affected their work from then on. The concept was based upon the supposition that "all peoples - with exceptions that merely prove the rule - have always known how to help themselves..."[53] Accordingly, they realized that they would have to do something themselves to implement Intermediate Technology, together with whichever sympathetic supporters they could find, without waiting for the backing of the "authorities". Barbara Wood, Schumacher's daughter and biographer, describes her father's situation in the following terms:[54]

> He had spent many lonely years in which the solutions he had proposed for major problems in the world had remained unrecognized, or

---

[50] Julia Porter was known as Julia Canning Cook at that time.

[51] Wood, *Alias Papa*, pp. 276-279, 312-326; personal communication with George McRobie, November 1982.

[52] Personal communication with George McRobie, November 1982.

[53] Schumacher, "Help to Those", p. 37.

[54] Wood *Alias Papa*, pp. 328-329.

too controversial to be acceptable to those who had the power to implement them. Keynes, Stafford Cripps, Cecil Weir, the Coal Board, the Burmese Government, the Indian Government, the economists of the 1960s and the Ministry of Overseas Development, all of these had failed him and rendered him powerless. The concept of intermediate technology was another such world improvement plan. Again he tried to go to the top to get it implemented, again those who had the power failed him. And then the ideas which [he had] developed... had their own liberating effect on him. He perceived that this plan was different. In the past his plans had depended on government action, on the changing of 'the system', on structural alterations. The concept of intermediate technology was free from this necessity. The earliest slogan he had coined held the answer: 'Find out what the people are doing and help them to do it better.' Action would result not from government intervention but from the people themselves. Here too lay the great power and appeal of intermediate technology. In it most people could find hope that they could raise themselves above grinding poverty.

In this spirit Schumacher and McRobie went about the absorbing task of trying to generate public support for their work. Progress was difficult, but the publication during August 1965 in London's *Observer*, of an article entitled *How to Help Them Help Themselves* was a real fillip for their cause.[55] A tremendous expression of support was received from a large range of people. Encouraged by the growing interest, Fritz Schumacher, George McRobie and Julia Porter established an organization - the Intermediate Technology Development Group - which eventually led to the worldwide network referred to earlier in Chapter Three. In 1966, with a solid base of core supporters, the Group was formally incorporated and commenced work - its first task being the production of catalogues of currently existing equipment which could be well suited to small scale rural development projects but which was not widely known in the South.[56]

The work of the Group will be explained in more detail further on. It should be stressed here, however, that the rapid international growth in experiments with intermediate technologies, and the eventual admission of the validity of the Intermediate Technology ap-

---

[55] E. F. Schumacher, *Observer*, Weekend Review (29th August 1965).

[56] Cf.: E. F. Schumacher, *Good Work*, ed. by P. N. Gillingham (New York: Harper Colophon, 1979), pp. 83-84; McRobie, *Small is Possible*, pp. 24-38; Wood, *Alias Papa*, pp. 324-329.

proach by economic development professionals who had previously scorned the idea, stemmed from the independent initiative of those people who were convinced enough of the truth of Schumacher's arguments to act without the backing of powerful authorities and world opinion.[57]   In a 1972 publication of the International Labour Organization Schumacher averred:[58]

> The creation of the Intermediate Technology Development Group in 1966 was the result of an initiative by people from the professions and industry in the United Kingdom, all with extensive overseas experience, who found a common basis for action in the approach of "intermediate technology". This group, which is keenly aware of the worldwide dangers inherent in the build-up of unemployment taking place in virtually every poor country, is a company limited by guarantee and a registered charity endeavoring to furnish the poor and the unemployed in developing countries with the means to work themselves out of poverty.

It was very much a "people's" or "grass roots" movement - although it certainly attracted the participation of professionals and technical specialists.

A large space has been allocated here to the work of Schumacher and colleagues in describing the origins of Intermediate Technology. This may be considered apt in view of the inordinate influence Schumacher actually exerted on its development and on the broader Appropriate Technology movement.

This section on the origins of Intermediate Technology may be concluded by summarizing the five major reasons why Schumacher generated the concept and succeeded in promoting its application.

- The first was his solid background as a high level professional economist with extensive experience at both a theoretical and a practical level.
- Secondly, he possessed a practical familiarity "on the ground" with the actual circumstances in the rural areas of the South; and his conception of poverty was based upon direct observation.

---

[57] Cf.: Robinson, ed., *Third World*, p. 4; A. Robinson, ed., *Appropriate Technologies for Third World Development* (London: MacMillan, 1979).

[58] E. F. Schumacher, "The Work of the Intermediate Technology Development Group in Africa", *International Labour Review*, 106, 1 (1972), 75.

- Thirdly, he maintained an ethical commitment to justice and the priority of satisfying human needs, as guiding motivations for most of his technical work in economics.

- Fourthly, he cultivated an active appreciation of the "human" dimension of both poverty and development; and he resisted subjugating this dimension (which is by nature very difficult to quantify) under the abstract theoretical frameworks of formal economics, with their emphasis on finance and quantification.

- Finally, Intermediate Technology was the counterpart of a popularist or "grass roots" development strategy which did not rely for its implementation upon the full backing of powerful elites (despite the helpful role that, in principle, they might play).[59]

In the following three sections the actual *meaning* of Intermediate Technology will be examined. The proliferation of literature on the subject, of both a serious academic and popular kind, has led to much duplication and inconsistency. It is therefore didactically helpful to draw primarily upon the work of Schumacher and the I.T.D.G.

While a large number of Schumacher's reports, articles, books and transcripts of lectures and interviews have been published, Schumacher wrote about Intermediate Technology primarily as a practitioner rather than as an academic - he was, after all, a full time civil servant for the British Government and even his Intermediate Technology work was a largely "part time" activity. Consequently, his writings are not systematic; even his well known *Small is Beautiful: A Study of Economics as if People Mattered* is a collection of different essays rather than a treatise. There is considerable overlap between his publications but it is necessary to survey a great deal of his material from disparate sources (much of which is not well known - particularly the material prior to the 1970s) to obtain a comprehensive view. The following review of Intermediate Technology will take the form of a paraphrase of Schumacher's thought; an attempt will thus be made to demonstrate that his analysis is systematic, even though it may not be published in a systematic form. Therefore, while the ideas following may essentially be derived from Schumacher's work, the way they are articulated is the work of the *present author*.

---

[59] R. H. Brown, "Appropriate Technology and the Grass Roots: Toward a Development Strategy from the Bottom Up", *The Developing Economies*, 15, 3 (1977), 253-279.

Intermediate Technology: Statement of Problems

The Development Problematique

By examining Schumacher's writings six major mutually interdependent problems may be adduced which call for the development of Intermediate Technology: the prevalence of extreme poverty on a massive scale; the widespread threat or actuality of malnutrition and starvation; long-term mass unemployment; mass urban migration; intranational political conflict; and, international political conflict. The relative emphasis given by Schumacher to each problem varies in different parts of his writings, but each problem may be viewed as sufficient in itself to evoke the concept of Intermediate Technology. The problems may be loosely grouped into three orders: poverty (1st order), maldevelopment (2nd order), and political conflict (3rd order). The relationships between the problems may be illustrated as in Figure 4.1. Schumacher himself does not portray his ideas in the concise form of Figure 4.1 in any one place; the Figure, and this discussion, is a synthesis by the present author of the material in Schumacher's writings.

Extreme mass poverty and starvation are grouped together here as "1st order" problems because of their very basic nature and human immediacy, and also because they were the most fundamental issues of concern to Schumacher when he developed the Intermediate Technology concept. Mass unemployment and mass urban migration were of equal interest to Schumacher as the first two problems, but are labelled here as "2nd order" problems because they are, in some ways, less acute. The "3rd order" problems concerned with political conflict were major influences on Intermediate Technology, but in an *indirect* manner. The four problems of extreme poverty, starvation, unemployment and urban migration, all at a mass level, were the *direct* foci of the concept.

The six problems just cited are grouped as in Figure 4.1, rather than merely listed, to reflect the insight that they form a self-reinforcing syndrome rather than a collection of discrete and autonomous issues. As a syndrome these problems may be viewed as the *development problematique* which confronts most countries of the South. The problems within each order are portrayed as mutually reinforcing, as are the different orders themselves.

Figure 4.1 *The Development Problematique*

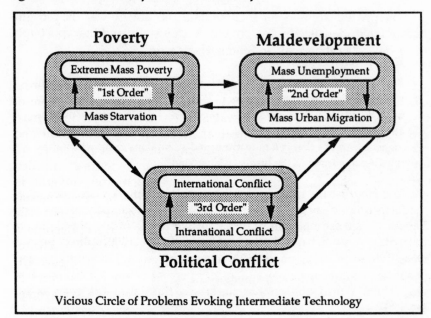

Vicious Circle of Problems Evoking Intermediate Technology

An understanding of the dynamic interactions between the "3rd order" political problems of political economy and others had been present in Schumacher's thinking since he first began discussing the problems of Germany, while studying at Oxford during the early nineteen thirties.[60] It was brought out more fully in his latter writings published during the nineteen seventies[61] under the broader rubric of "Appropriate Technology". The more specific concept of Intermediate Technology was initially directed at the first two orders of problems. Schumacher emphasized the political economy conflicts addressed by Intermediate Technology more explicitly, however, once the concept was established.

---

[60] Wood, *Alias Papa*, pp. 14-40.

[61] E.g., Schumacher, *Good Work*, esp. Chapters 1 and 3.

Extreme Mass Poverty

The pivotal stimulus for Intermediate Technology was the recognition of institutionalized mass poverty in the South, and that this ought to be viewed as abnormal. In Schumacher's words:[62]

> Poor peasants and artisans have existed from time immemorial; but miserable and destitute villagers in their thousands and urban pavement dwellers in their hundreds of thousands - not in wartime or as an aftermath of war, but in the midst of peace and as a seemingly permanent feature - that is a monstrous and scandalous thing which is altogether abnormal in the history of mankind.

The fact that the majority of the world's population are poor made the focus on poverty even more apt. By 1976, for example, International Labour Office estimates indicated that two thirds of the population of the non-communist developing countries (some 1200 million people) were living in serious poverty;[63] seven hundred million of these were destitute, with incomes more than fifty percent below the poverty line. One Third World analyst has described this latter category as follows:[64]

> These are the absolutely poor, and their low income is simply the numerical expression of a series of multiple deprivations that, taken together, add up to lives that are hardly worth living. These are people with poor land, little land or no land, people with not enough work, people on inadequate diets, people with perpetual illnesses, people without the most basic knowledge and skills to improve their lives. They are people who have been denied the right to develop their full human potential.

Schumacher states explicitly that the starting point for all his considerations in development economics was the poverty of this extreme degree and that his first task was to recognize and understand the boundaries and limitations which such a degree of poverty imposed.

---

[62] Schumacher, "Help to Those", p. 37.
[63] Cited in P. Harrison, *The Third World Tomorrow* (Harmondsworth: Penguin, 1980), p. 287.
[64] *Ibid.*

Another aspect of Schumacher's approach, as is apparent from the above remarks, is his distinction between two types of poverty. He observes that there is a kind of poverty when people have enough to "keep body and soul together but little to spare" and another extreme kind for which the word "poverty" is not adequate. He used the term "misery" to describe such poverty where people "cannot keep body and soul together, and even the soul suffers deprivation." It was this latter kind, misery, which Schumacher had in mind when he wrote his seminal article, "How to Help Them Help Themselves," in 1965.[65]

Thirdly, Schumacher holds that the *main* purpose for economic development programs ought to be the eradication of the extreme poverty of destitution and misery.[66] He considers that this is more important than the pursuit of economic growth *per se*.[67] Schumacher believes such a purpose to be a moral imperative; during a lecture tour of the United States he said:[68]

> We must do what we conceive to be the right thing ... My only business, the only real job we have, is to look after, to the best of our ability, the little people who can't help themselves. If education and the advantages that we have had from society are only so that we might form a sort of trade union of the privileged, then our soul is so burdened with darkness that life is not worth living.

The majority of Schumacher's ideas in the area of technology and development may be seen as stemming from his struggle with the enigma of poverty.

Mass Starvation

To the general population in the relatively wealthy North, mass starvation is the most visible expression of poverty in the South. The contributions of Schumacher and the I.T.D.G. lie not in the conduct of research on the levels and patterns of malnutrition and starvation, a task which is actively pursued by other authorities, but in highlighting the ways such problems are interconnected with other aspects of the total

---

[65] E. F. Schumacher, "Conscious Culture of Poverty", *Resurgence*, 6, 1 (1975), 10. The preceding quotes in this paragraph also come from this source.

[66] E. F. Schumacher, *Economic Development and Poverty* (London: Africa Bureau, 1966).

[67] See, e.g., Schumacher, *Small is Beautiful*, pp. 10-47.

[68] Schumacher, *Good Work*, p. 100.

development problematique. From the mid 1960s Schumacher and colleagues insisted that unless self-sustaining efforts to eradicate extreme poverty in the rural areas of the South were evolved, the poor in the South would be faced with the persistent threat (or actuality) of mass starvation.[69] Perceiving that Africa was probably faced with the severest of threats in this regard,[70] soon after the Group's establishment the I.T.D.G. mounted a series of programs within, or in cooperation with, various African countries.[71] The massive levels of famine and starvation which have recently surfaced in Africa vindicate Schumacher's foresight and analysis.[72]

Mass Unemployment

While mass starvation and extreme poverty were the "1st order" stimuli for Intermediate Technology, the main *functional focus* for Schumacher's analysis is the growth and intensification of mass unemployment in the South. In his paper to a 1965 UNESCO Conference, Schumacher describes the typical condition of the poor in most of the "so-called developing countries" [sic] in the following words:[73]

> Their work opportunities are so restricted that they cannot work their way out of misery. They are under-employed or totally unemployed, and when they do find occasional work their productivity is exceedingly low.

---

[69] See E. F. Schumacher, *Intermediate Technology - Its Meaning and Purpose*, mimeo (London: Intermediate Technology Development Group, 1973); McRobie, *Small is Possible*, p. 33.

[70] A survey of global agricultural trends in the 1960s and 1970s is presented by the United Nations' Food and Agriculture Organization (*Agriculture: Toward 2000* [Rome: Food and Agriculture Organization, 1981], esp. pp. 17-28).

[71] E. F. Schumacher, "The Work of the Intermediate Technology Development Group in Africa", *International Labour Review*, 106, 1 (1972), 75-92.

[72] *Africa in Crisis*, ed. by J. Tinker (London: Earthscan, 1985), both surveys the nature and extent of famine in Africa and shows it to be firmly linked to the issue of development strategies identified by Schumacher, rather than just to natural phenomena such as drought; analyses of this type have featured prominently in I.T.D.G. publications (e.g., R. Baker, "Crisis in the Sahel: Causes and Solutions", *Appropriate Technology*, 1, 1 [1974], 19-21).

[73] "Social and Economic Problems Calling for the Development of Intermediate Technology", paper presented to the Conference on the Application of Science and Technology to the Development of Latin America (Santiago, Chile, September, 1965), organized by the United Nations Educational, Scientific and Cultural Organization with the cooperation of the Economic Commission for Latin America; edited version published in Schumacher, *Small is Beautiful*, pp. 159-177.

As this quote indicates, Schumacher sees a close connection between the continuation of mass unemployment and the continuation of mass poverty. He defines unemployment as the non-utilization or gross under-utilization of available labour.[74]

Schumacher himself never published detailed empirical material justifying his views concerning mass unemployment. He considered that the severity of the problem would be readily apparent to even the casual observer of Third World affairs.[75]

It is very difficult to obtain reliable statistics about employment levels in the South for a number of reasons such as: the lack of adequate raw data; the degree of hidden or disguised unemployment; the high proportion of work which takes place in the informal economy; and, the sheer enormity of the problem. Nevertheless, there is sufficient published research on the subject to make the following general claims:

- many people are frustrated by lack of employment opportunities; they include both those without work and those who have jobs but want to work longer hours or more intensively;
- a large fraction of the labour force, both urban and rural, lack a source of income both reliable and adequate for the basic needs of themselves and their dependants;
- a considerable volume of unutilized or under-utilized labour forms a potential productive resource, which ought to be brought into use.[76]

The available empirical evidence, in short, supports Schumacher's assertion that in the South "great numbers of people do not work or work only intermittently, and that they are therefore poor and helpless and often desperate..."[77]

---

[74] Schumacher, *Small is Beautiful*, p. 192.

[75] Cf.: R. Williams, interview with E. F. Schumacher, *The Science Show*, Radio 2 (Sydney: Australian Broadcasting Commission, 1977); Schumacher, *Small is Beautiful*, esp. pp. 155-204, passim.

[76] This is the conclusion reached by D. Seers, et.al., in *Towards Full Employment*, a programme for Colombia prepared by an inter-agency team organized by the International Labour Office (Geneva: International Labour Office, 1970). Cf.: A. S. Bhalla, ed., *Technology and Employment in Industry* (Geneva: International Labour Office, 1975), p. 309; F. Stewart, *Technology and Underdevelopment* (London: MacMillan, 1977), pp. 32-57; J. Mouly and E. Costa, *Employment Policies in Developing Countries* (London: Allen and Unwin, 1974); R. Jolly, et al., *Third World Employment* (Harmondsworth: Penguin, 1973); A. Sen, *Employment, Technology and Development* (Oxford: Clarendon Press, 1975); International Labour Office, *Employment, Incomes and Equality* (Geneva: International Labour Office, 1972).

[77] Schumacher, *Small is Beautiful*, p. 160.

Unemployment is viewed by Schumacher as a key problem in the South not only because of its pervasiveness, but also because it leads to the growth of extreme poverty and has damaging effects on the psyche of the unemployed and on their cultural environment. He also believes productive work - of the appropriate kind - to be of value in itself and not just in its capacity as a means of producing an income for those who work.[78]

Mass Urban Migration

The second of the two factors which make up the "2nd order" problems - maldevelopment - is the massive drift of the South's rural population to the rapidly growing cities, and, in particular, to the even more rapidly growing slums on the fringes of those cities. Schumacher observes that, in the main, the development planning in the South and aid programs from the North have been biased towards urban based industry, but that the vast majority of the population in the South dwells in rural areas.

Schumacher's work in Burma and India led him to observe first hand what he later called the decay of rural life.[79] For a number of reasons, millions of rural dwellers were finding it necessary to leave the countryside and migrate to the urban areas in search of better opportunities. Mass poverty and mass unemployment in rural areas, combined with the cultural stagnation which tends to accompany economic stagnation, were forcing peasants and others to abandon their traditional homelands in attempts to find improved employment opportunities in the urban areas, thereby swelling the urban population. As later studies have shown, while part of the urban population growth is due to the overall growth of population in the South (consequent upon reduced infant mortality rates and other factors), it is mostly due to rural-urban migration; and, while the rapid population growth in Africa during the 1960s, when Schumacher was formulating the Intermediate Technology concept, was typically in the vicinity of three per cent per

---

[78] These themes are apparent in various places throughout the following works by Schumacher: *Good Work*, esp. Prologue; *Small is Beautiful*; "Insane Work Cannot Produce a Sane Society"; *Resurgence*, 5, 2 (1974), 9 - 10; "Economics Should Begin with People, Not Goods", *The Futurist* (December 1974), 274-275; "Education for Leisure and Wholesome Work", *Resurgence*, 5, 1 (1974), 10, 19; "Buddhist Economics", *Frontier*, 8 (Autumn 1965), 201-205 (Further versions of "Buddhist Economics" appeared subsequently in: *Asia: A Handbook*, ed. by Guy Wint [London: Anthony Blond, 1966]; *Resurgence*, 1, 11 [January 1968]; and in *Small is Beautiful*, Part I, Ch. 4).

[79] E. F. Schumacher, "Patterns of Human Settlement", *Ambio*, 5, 3 (1976), 93.

annum, urban growth rates of seven to ten percent were a common phenomenon.[80]

Much orthodox Western economic theory emphasizes the gradual transformation of an economy from one with a rural, agrarian base to one with an urban, industrial base, as a defining characteristic of economic development. Under ideal conditions the growth of the modern urban-industrial sector would create sufficient new employment opportunities to absorb redundant agrarian labour and bring about a net improvement in the productivity of the economy as a whole. Unfortunately, in many countries of the South which have embarked upon ambitious development programs, the growth of non-agricultural and urban employment has not matched either the total economic growth rate or the growth in surplus rural labour. Todaro has indicated that while many African economies have achieved output growth rates of five to eight percent per annum, the growth rate of non-agricultural employment has typically been negligible and, in many cases, negative.[81] Thus, the rural-urban population drift in the South might also be called a rural-urban unemployment drift.

Schumacher pointed out that the growth of rural-urban migration is a worldwide phenomenon, and that in the South this amounts largely to the growth of slums on the fringe of urban concentrations, beset by both poverty and unemployment. While developing his Intermediate Technology theory he noted that shanty towns of more than 100,000 inhabitants at the fringes of large cities concentrated twelve percent of the world's population and more than one third of the world's city population.[82] He considers mass urban migration to be more of a reinforcement of the South's severe problems of poverty and unemployment than a pathway for solving them.[83]

---

[80] S. Ominde and C. N. Ejiogu, eds., *Population Growth and Economic Development in Africa* (London: Heinemann, 1972); R. K. Som, "Some Demographic Indicators for Africa", in J. C. Caldwell and C. Okoanjo, eds., *The Population of Tropical Africa* (London: Longman, 1968), pp. 187-189.

[81] M. P. Todaro, "Income Expectations, Rural-Urban Migration and Employment in Africa", *International Labour Review*, 104 (1971), 387-413.

[82] Schumacher, *Economic Development*, p. 5.

[83] Schumacher's assessment of the nature of the urban migration problem was developed in various places; in particular, see: "No Future for Megalopolis"; *Resurgence*, 5, 6 (1974), 12-15; *Rural Industries*; "Patterns of Settlement"; "Social and Economic Problems"; *Development and Poverty*; "Intermediate Technology - its meaning and purpose"; *Small is Beautiful*, Part III. See also: G. McRobie, *Economic Growth* ; D. W. J. Miles, *Appropriate Technology for Rural Development: The ITDG 's Experience*, Occasional Paper #2 (London: Intermediate Technology Development Group, 1982).

Schumacher's analysis has been reinforced by the research of economic demographers, who have demonstrated that in the case of tropical Africa, for example, there are structural factors influencing poor rural dwellers to migrate to the cities for employment despite the low chances of actually obtaining a job. The major pressure appears to be the relatively high income levels of those who actually do obtain urban employment compared with those who remain in the rural areas. Despite the low probability of new rural immigrants actually finding employment, the fact that there is *some* chance of obtaining work with a relatively high income is enough to induce large numbers of people to endure long periods of privation as urban unemployed. The irony of this situation is that if substantial growth in urban employment opportunities occurs, rural-urban migration becomes a more attractive option to the rural poor, who then increase the urban population only to thereby reduce the probability of actually gaining employment. There is empirical evidence that policies to cure the urban unemployment problem in typical "underdeveloped" countries will generally result in the reverse of the intended effect if based upon attempts to increase urban labour demand - because their very "success" tends to stimulate rural-urban migration. Decisions which appear economically rational to the individual, in such circumstances, may be economically irrational for the society as a whole.[84]

In his 1965 article addressed not to economists but to the public at large, published in London's *Observer* newspaper, Schumacher summarizes the dilemma of rural-urban migration in the following way:[85]

> Unemployment and under-employment in developing countries are most acute in the areas outside a few metropolitan cities; so there is mass migration into these cities in a desperate search for a livelihood: and the cities themselves, in spite of 'rapid economic growth', become infested with ever-growing multitudes of destitute people. Any visitor who has ventured outside the opulent districts of these cities has seen their shanty towns and misery belts, which are growing 10 times as fast

---

[84] One useful survey of the salient trends at the time of the ITDG's initial development work in Africa is by J. C. Caldwell's *African Rural-Urban Migration* (New York: Colombia University Press, 1969). Cf.: M. P. Todaro, "The Urban Employment Problem in Less Developed Countries: An Analysis of Demand and Supply", *Yale Economic Essays*, 8 (1968), 331-402; M. P. Todaro, "A Model of Labor Migration and Urban Unemployment in Less Developed Countries", *American Economic Review*, 57 (1969), 138-148; J. R. Harris and M. P. Todaro, "Urban Unemployment in East Africa: An Economic Analysis of Policy Alternatives", *East African Economic Review*, 4 (1968), 17-36.

[85] Schumacher, "Help Themselves".

as the cities themselves. Current forecasts of the growth of metropolitan areas in India, and many other developing countries, conjure up a picture of towns with 20, 40 and even 60 million people - a prospect of 'immizeration' for a rootless and jobless mass of humanity that beggars the imagination.

In summary, one of the main influences behind Intermediate Technology was Schumacher's, at that time radical insight, that a process generally believed by economists to be a hallmark of economic and social improvement, namely rural-urban migration, was in fact turning out to be an obstacle to real economic improvement for the majority of people in the South.

International and Intranational Conflict

The four problems described above together formed the essential ingredients of the development problematique addressed by Schumacher in formulating the concept of Intermediate Technology. The final two, international political conflict and intranational political conflict, were of concern to him throughout his career and also formed part of his conception of the development problematique.[86] An analysis of political conflict was not, however, a substantial part of Schumacher's schema when evolving Intermediate Technology; conflict, of both a political and social kind, received greater attention at a later stage as part of a consideration of the propensity of technology to facilitate either violence or non-violence.[87] A fuller discussion of this subject will be left until later. Nevertheless, certain aspects of the subject may be raised here in a cursory manner.[88]

Firstly, while Schumacher did not believe that poverty in itself necessarily led to violence or political conflict, he did fear the social violence which could arise from the disparity of interests in a society where the majority of its citizens were faced with extreme poverty or even destitution, while a powerful minority lived in affluence.

---

[86] See, Wood, *Alias Papa*.

[87] See: Schumacher, *Small is Beautiful*, esp. Part I, (Ch.2), Part II(Ch.5) and Part IV; E. F. Schumacher, "The End of an Era" and "Toward a Human-Scale Technology"(Chapters 1 and 2) in Schumacher, *Good Work*.

[88] This precis of the political conflict issues behind Intermediate Technology is synthesized by the author from a large range of Schumacher's publications. Of particular relevance are: "Non-Violent Economics"; "Help Those Who Need it Most"; "Peace and Permanence", in *Small is Beautiful*, pp. 19-35; "Technology and Political Change", *Resurgence*, 7 (1976), 20-22; "Rural Industries".

Secondly, he feared that the growth of economic systems where the majority of people were dependent for their livelihood upon the economic decisions of a minority, would create a propensity for political instability.

Thirdly, he observed a global trend toward economic activities based upon massive and centralized production and the concomitant consumption of exotic supplies of largely non-renewable resources which, by their nature, were limited in supply. He felt that increasing international dependence upon and competition for such resources would of necessity lead to conflict of perceived economic interests and a greater likelihood of political and military friction. Attempts by countries to safeguard their economic interests would lead to the build up of military facilities, thereby diverting precious resources from uses more directly associated with alleviating extreme poverty.

Fourthly, Schumacher observed that economic growth strategies based upon massive levels of foreign investment have a tendency to increase the indebtedness of the investment-receiving country to the investors.[89] If such debts become too large in relation to the ability of the local economy to generate a surplus, the debtor country could be forced to adopt draconian economic policies in order to obtain foreign currency to repay its debts. Such policies could provoke civil and political conflict between advantaged and disadvantaged groups within the country. In situations where the debtor country might be forced to default on its repayments, the implications for the politics of international trade and finance could be immense.

In summary, Schumacher considered that trends in the economies of the South were creating a climate conducive to the development of political conflict, within societies and between them.

Intermediate Technology: Diagnosis of Problems

It has been argued above that six interdependent problems which together characterize the development problematique may be derived from Schumacher's writings. A schema will now be proposed which expresses diagnosis of this problematique which is also contained in Schumacher's writings.

---

[89] See *Small is Beautiful*, pp. 181-184; E. F. Schumacher, "Notes on Indian Development Problems", in *Roots of Economic Growth* (Varanasi: Gandhian Institute of Studies, 1962), pp. 47-48.

Dynamic Variables

The first important element which may be identified is Schumacher's focus on mass unemployment and mass urban migration as dynamic variables in the self-reinforcing system illustrated in Figure 4.1. Thus, while Schumacher's analysis portrays none of the six problems of the development problematique as discrete or independent variables, it does portray the "2nd order" problems of maldevelopment as being, in some sense, special. In other words he believes that they are of greater significance than the other variables for maintaining the vicious circle of poverty, maldevelopment and political conflict. While the problems of political conflict, starvation and extreme poverty are seen as causative influences on mass unemployment and mass urban migration, these latter problems are portrayed by Schumacher as variables with even greater scope for causative influences on the other problems. For this reason he points to the amelioration of unemployment and the reversal of mass urban migration as key, or dynamic, variables for breaking the vicious circle of underdevelopment. While Schumacher's starting point is with the "1st order" problems of poverty, his action focus for the solution of these problems is on the "2nd order" problems of maldevelopment.[90]

Schumacher's "action focus", on what he describes as the "twin evils" of mass unemployment and mass urban migration, differs from other approaches within economic planning which see these as either only "transitional" problems of economic development, as inevitable and unconquerable, or even as necessary for "sound growth".[91] In contrast Schumacher believes that the roots of sustainable economic growth for the South lie in tackling these "twin evils" directly rather than incidentally.

Mutual Poisoning of the Dual Economy

The second main element in Schumacher's diagnosis of the dynamic of underdevelopment is his recognition of the existence of a *dual economy*, or *dual society*, in most countries of the South.

Schumacher's writings describe an enormous gulf between the rich one-quarter of the world and the poor three-quarters, both globally and

---

[90] Schumacher, "Paths to Growth", esp. pp. 26-28.
[91] Schumacher: "Industrialization"; "Help Themselves".

within individual nations. This splitting of what "ought to be and used to be"[92] one world into two worlds, is considered by Schumacher to be a basic datum for effective economic analysis in the South. "Dual economy" implies two ways of life existing side by side with vastly different personal income levels. Schumacher refers to one as the *Westernized fringe economy*, the *metropolitan economy*, the *modern sector*, the *modern fringe economy* or *the urban industrialized sector*. The other he calls the *traditional sector*, the *rural economy*, the *rump economy*, the *hinterland* or the *great traditional body*, which typically includes up to eighty five percent of the population in an "underdeveloped" country. He claims that most of the professional knowledge of Western economists has evolved in Western economies which were not faced with the current structural problems of the South and that much of that know-how is therefore of little relevance to solving the difficulties of the dual economy.[93] He therefore speaks of the need for a two-fold approach to economic development whereby the dynamics of the two sectors are clearly recognized and where the distinctive problems of both the urban-industrialized economy and the traditional economy are directly confronted in their own terms.

Schumacher speaks of the process of *mutual poisoning* of both the urban-industrialized economy and the traditional economy. The advanced industries in metropolitan areas tend to kill off non-agricultural production in the rump economy, while a flight from the land into the metropolitan areas creates quite unmanageable problems in those areas, which increasing levels of urban industry seem incapable of mastering.

Large scale, capital intensive industry in urban areas, based upon modern technology, is often able to achieve lower unit production costs and higher profit levels than industries in the traditional, predominantly rural economy. In the absence of some form of protection, either natural or artificial, the modern industries tend to dominate the market, thereby forcing traditional industries out of business due to their lack of competitiveness. From a narrow economic perspective consumers in the rural sector are acting rationally when they purchase goods available at the cheaper prices offered by metropolitan industry. This rationality at the level of individual consumers reinforces the viability of metropolitan industry and weakens rural industries.

A result of this apparently rational behavior is that the income which traditionally would have accrued to members of the rural economy now accrues to the metropolitan economy, with the result that the

---

[92] E. F. Schumacher, *A Guide for the Perplexed* (London: Abacus, 1978), p. 9.

[93] Schumacher, "Paths to Growth".

ability of the rural population to purchase goods from either the relatively efficient metropolitan sector or the traditional sector, deteriorates. Consequently, the industry of the hinterland, which used to provide inexpensive employment opportunities for many, along with relatively expensive goods, is destroyed. Schumacher writes:[94]

> But is not this the essence of 'progress' - the substitution of superior methods of production for inferior ones. Does not the lower price benefit the villagers, raising their standard of living, enabling them to save and invest and finally to accomplish the 'take-off'? Many economists argue that way, but the truth is otherwise. Because their own production has stopped, the villagers are poorer than ever before; they may be unable to pay for any of the factory goods, except by getting into debt. It has happened even that the factory itself, having accomplished its frightful work of destruction in the villages, has had to close down for lack of a market.

Schumacher argues that this weakening of traditional industries in rural areas affects the ability of agriculture to meet essential food needs. He claims that agriculture alone is unable to sustain a fully human life and that it "thrives only when in contact with industrial crafts of all kinds and when vivified through cultural influences coming from thriving towns nearby."[95] It follows that a successful attack on the problem of starvation must not rely purely on the specific or exclusive development of agriculture. A "decent" standard of living for a sizeable community can only be produced by agriculture alone in very special circumstances which do not prevail in the hinterland of most underdeveloped countries. These special requirements are that agricultural output per person be exceptionally high, that there be a large market for agricultural products in the cities and therefore that the rural population be small in relation to the town population.

The attack on mass poverty in the South therefore requires, according to Schumacher's analysis, increased non-agricultural production in rural communities such that most basic needs can be covered without having to exchange food surpluses with the towns.

The process of the poisoning of the rural economy by the development of the Westernized fringe economy is exacerbated, according to Schumacher, by some of the very development programs introduced to assist the development of rural industry. Economic planners correctly

---

94 Schumacher, "Help to Those", p. 39.
95 Schumacher, "Indian Development", p. 43.

surmise that the kind of development which has brought wealth to the Western industrialized nations cannot proceed without some kind of infrastructure, such as a fast, efficient transport system. The introduction of such infrastructure in the hinterland of the South, however, tends to destroy the natural protection originally afforded the relatively "inefficient" rural industry by the existing slow, traditional transport.

Some theorists argue that the consequent deterioration experienced in the traditional economy is only temporary because development in the urban-industrialized economy will eventually spread over the whole nation revitalizing rural life. To Schumacher, experience has shown that this does not generally occur. He argues that the poisoning of the rural economy by the growth of the metropolitan economy actually sets in motion a reverse process which in turn poisons the metropolitan economy, ensuring its inability to rectify its influences on the hinterland; hence the phrase, "process of mutual poisoning".

The process of mutual poisoning is fed by the decay of rural industries, with the resultant increase in rural unemployment, the deterioration of agriculture and the shortage of food and wage-goods. A mass movement of the population into the cities follows. Because the cities are unable to satisfactorily absorb the mass migration of people and cater for their needs, however, the problems of the decayed hinterland become the problems of the metropolitan areas. Development centered in the cities often tends, by this process, to have an effect the exact opposite of what is intended. Generally, Third World cities thus affected become unmanageable and if they remain manageable, according to Schumacher, "they are only manageable because the great population in the shanty towns just become forgotten people, breeding vice and every kind of degradation".[96]

Schumacher observes that mutual poisoning between the economies of the metropolitan areas and the hinterland is also evident in the industrialized countries of the North. He points to the growth of "megalopolis" regions such as Boston/Washington, Chicago/Pittsburgh, and Los Angeles in the United States, and the conurbations around European cities such as London and Paris. A global trend may be observed whereby the stronger industrial regions in countries continue to grow at the expense of the weaker regions - for example, the north of Italy as compared with the south. Schumacher notes the contrast of urban congestion with vast emptiness in Australia, and a drift from smaller cities like Hobart to Melbourne and Sydney as the smaller Tasmanian industries are overtaken by the large scale indus-

---

[96] Schumacher, *Development and Poverty*, p. 6.

tries of the mainland.[97] He summarizes the dilemma of the dual economy with the phrases, "nothing succeeds like success" and "nothing
fails like failure."[98] In other words, he identifies a process at work in
most countries whereby the buoyancy of large cities and industrial areas makes it difficult for smaller and regional communities to compete -
with the latter becoming progressively weaker and the former becoming progressively more dominant. These tendencies are socially disruptive in both the North and South, but while funds may be made available in richer countries to ameliorate suffering, for poorer countries the
vicious circle of mutual poisoning of the dual economy leads to an extreme form of poverty as discussed earlier - more aptly called misery.

In summary:

- Schumacher saw the source of the dynamic of underdevelopment
  as the decay of economics and cultural life in the rural areas of
  the South, as expressed in mass unemployment and mass urban
  migration;
- he saw this decay as part of the process of mutual poisoning of
  the dual economy; and,
- he saw that the dominant approaches to development planning
  and aid failed to adequately address these phenomena and even
  tended to exacerbate them.

Schumacher, and colleagues such as George McRobie, therefore perceived that behind the failures of the various "five year plans" and
"development decades" of the so-called "developing countries" and the
international development community, lay a misdirected development
philosophy - the chief elements of which will now be described.

Misdirected Development Philosophy

Most countries of the North have some form of foreign aid program
directed ostensibly at solving problems of underdevelopment in the
South. Some critics of these aid programs claim that the amount of aid
is simply inadequate when compared with the enormity of the problems. Others claim that foreign aid itself is by nature inappropriate.
One school of thought even argues that some countries are beyond help,
and that available aid should be concentrated on those countries which

---

[97] E. F. Schumacher, An interview with Robin Williams, *Science Show*, Radio 2,
Australian Broadcasting Commission, Sydney, July 1977.

[98] Schumacher, "Industrialization".

promise the highest chance of success - this approach is often referred
to as "lifeboat ethics".[99]

Schumacher's approach involves two elements. Firstly he believes
that *foreign aid is able to play only a limited role in bringing about sustained economic development,* viz.[100]

> ... a country that makes development plans which utterly depend on
> the receipt of substantial foreign aid is doing such damage to the spirit
> of self-respect and self-reliance of its people that, even in the narrow-
> est economics terms, its loss is greater than its gains.

Notwithstanding this perspective, Schumacher holds that the
wealthier countries are honor bound to provide aid to poorer countries.[101] Thus the second element of his approach is that *the pattern of
aid is decisive.* He holds that aid of the wrong type is likely to induce
more harm than good. The widespread adoption of counterproductive
patterns of aid tends, in Schumacher's view, to be associated with misdirected philosophies of development in both the aid-receiving and
aid-giving countries.

Abandonment of Gradual Evolution. The first major misdirection in
dominant philosophies of development was, according to Schumacher,
the abandonment of an evolutionary approach in favour of an emphasis
on quantum leaps. He notes that the capacity for sophisticated industrial activity evolved in the North in a gradual manner from earlier
forms of industry, and that the *gradual transformation* in working patterns, innovative capacity, infrastructure requirements and organization of Western countries is a key to their relative industrial success.
Accordingly he holds that progress in industrial activity within poor
countries of the South will require the same gradual transition as in the
North and that short cuts to prosperity are virtually impossible:[102]

> Our scientists incessantly tell us with the utmost assurance that every-
> thing around us has evolved by small mutations sieved out through
> natural selection. Even the Almighty is not credited with having been
> able to create anything complex. Every complexity, we are told, is the
> result of evolution. Yet our development planners seem to think that

---

[99] G. Hardin, "Living on a Lifeboat", *Bio Science*, **24**, 6 (1974), 561-568; see also
Hardin's earlier paper, "The Tragedy of the Commons", *Science*, **162** (1968), 1243-1248.

[100] Schumacher, "Indian Development", p. 47,

[101] Schumacher, "Help to Those", p. 42.

[102] Schumacher, "Small is Beautiful", p. 155.

they can do better than the Almighty, that they can create the most complex things at one throw by a process called planning, letting Athene spring, not out of the head of Zeus, but out of nothingness, fully armed, resplendent, and viable.

With a note of sarcasm, Schumacher reflects that the dominant theories of economic development stand in stark contrast to almost every branch of modern thought, and to the empirical evidence of the West's own past, by emphasizing *creation* (i.e., quantum leaps) rather than evolution.

As mentioned earlier, Schumacher's conception of economic development emphasizes the actual people involved, rather than the formal abstractions of economic theory. He therefore argues that the introduction of new methods, capital and types of organization need to be *accessible* to the majority of people for whose benefit they are intended; and, that such "improvements" need to be introduced in small steps, so as to produce stimulation and not discouragement. Human beings and, in particular, broad groups of people who make up societies, are capable of effectively developing their personal capacities and changing their pattern of life only by means of a gradual, organic process - that is, if major trauma, neurosis, widespread disruption, or social anomie is to be avoided.[103] This is the main reason why Schumacher believes economic development must be a gradual process.

Schumacher acknowledges that there are exceptions to the general rule just stated, but warns that such exceptions do not annul the basic need for an evolutionary approach in development philosophy:[104]

Now, of course, extraordinary and unfitting things can occasionally be done ... It is always possible to create small ultra-modern islands in a pre-industrial society. But such islands will then have to be defended, like fortresses, and provisioned, as it were, by helicopter from far away, or they will be flooded by the surrounding sea.

He goes on to argue that whether or not such "islands" do well they tend to produce the dual economy because they are incapable of integrating into the surrounding society, and tend to destroy its cohesion.

---

[103] This has been argued at some length by the social psychologist, Erich Fromm; see, e.g.: "The Individual and Social Origins of Neurosis", *Am. Soc. Rev.*, **9**, 4 (1944), 380-384; *The Fear of Freedom* (1977 edition; London: Routledge and Kegan Paul, 1942); *The Sane Society* (1973 edition; London: Routledge and Kegan Paul, 1956).

[104] Schumacher, *Small is Beautiful*, p. 155.

With reference to the introduction of aid Schumacher avers, "There can be a process of stretching - never a process of jumping", implying that if new activities are introduced which depend upon special forms of education, organization and discipline, which are not inherent in the recipient society, the activity will hinder rather than promote healthy development.[105]

Neglect of the Human Dimension. The second major misdirection in dominant philosophies of development, identified by Schumacher, was a lack of serious attention given to the needs, skills, aspirations and dynamics of people. He considers the emphasis on ostensibly "non-human" factors by many economists to be counterproductive, even from the perspective of maximizing Gross National Product and other formal indicators. The following quote is indicative:

> Economic development is something much wider and deeper than economics, let alone econometrics. Its roots lie outside the economic sphere, in education, organization, discipline and, beyond that, in political independence and a national consciousness of self-reliance.

Two alternative points of departure may be identified for the theory and practice of economics - *goods* or *people*.[106] Schumacher argues that the dominant departure point for modern economics has been the question of how to maximize the total production of goods; this may be seen as easier than a focus on people because the production of goods is far more amenable to quantification and to being dealt with in strictly technical terms. "Development", according to Schumacher, has very little human meaning unless the development is actually experienced by the majority of the people who need it. Absolute growth in Gross National Product per capita ought not to be considered development if the majority of people in a country are excluded from participating in both the achievement and enjoyment of that growth.[107] An alternative focus to the question of how to produce more goods is the departure point: "How may people be made more productive?" Schumacher

---

[105] *Ibid.*, pp. 157-158.

[106] E. F. Schumacher, "Economics Should Begin with People, not Goods", *The Futurist* (December 1974), 274-275; also see E. F. Schumacher, "Two Million Villages", Part III, Ch. 13 in *Small is Beautiful* , by E. F. Schumacher (London: Blond and Briggs, 1973), pp. 178-191.

[107] This theme has been taken up at some length by the Commission on Churches' Participation in Development, of the World Council of Churches: Pascal de Pury, *Peoples' Technologies and Peoples' Participation* (Geneva: World Council of Churches, 1983).

claims that an attempt to answer this question evokes a different form of economics to that which has contributed to the growth of the dual economy and the dynamic of underdevelopment. Two of Schumacher's colleagues, George McRobie and Marilyn Carr, have produced a paper outlining the salient elements of such an approach, entitled *Mass Production or Production by the Masses?* The title, a phrase of Gandhi's, encapsulates Schumacher's contention that people are the most important resource and dynamic variable for economic development, and that other resources such as capital or physical resources, while important, are secondary.

In an international seminar on economic growth, during 1961, Schumacher posed a question, which undergirded his work thereafter, in the following terms:[108] "All the most decisive problems of development may be summed up, it seems to me, in the question: 'How can the impact of the West be canalized in such a way that it does not continue to throw the people into apathy and paralysis?' " In response Schumacher consistently argues that it is only valid to isolate technical economic factors from the rest of human life for the purpose of analysis, and that for fruitful action people need to be recognized as whole beings. He argues that if a "people centered" approach is not adopted and action is based solely on economic calculations, as laid down in elaborate central plans, coercion from the top becomes inevitable; and that if coercion succeeds it does so with the result that human freedom becomes "stultified by apathy and sullen disdain", and "the people sink ever deeper into misery."[109]

Schumacher's focus on people and, in particular, their education, organization and discipline, also explains the reasons why he advocates an evolutionary approach in development philosophy.[110] As mentioned above, he considers the capacity for humans to change to be the key limiting factor on economic change. This is because he considers that education, organization and discipline are the pivotal human factors in economic life, and that these three factors, of necessity, may only evolve slowly.

In summary, Schumacher holds that the starting point of economics ought to be people, rather than goods; and that such a change in the dominant focus of economics is necessary to break the vicious circle of underdevelopment as portrayed in Figure 4.1.

---

[108] Schumacher, "Help to Those", p. 42.

[109] *Ibid.*

[110] *Ibid.*, pp. 152-158.

Neglect of Rural Areas. The third major misdirection in dominant philosophies of development, according to Schumacher, is an overemphasis on the importance of the modernized metropolitan centers and a neglect of rural areas. From the very beginning of its work the activities of the Intermediate Technology Development Group were founded on the insight that the source and center of world poverty lies primarily in the rural areas of poor countries, which are largely bypassed by conventional aid and development programs.[111]

Schumacher claims that a failure to enhance the viability of non-agricultural production in the rural areas leads to the process of the "mutual poisoning of the dual economy". He writes, with the Indian economy in mind, as follows:[112]

> The crucial task ... is to make the development effort appropriate and thereby more effective, so that it will reach down to the heartland of world poverty, to two million villages. If the disintegration of rural life continues, there is no way out - no matter how much money is being spent. But if the rural people of the developing countries are helped to help themselves, I have no doubt that a genuine development will ensue, without vast shanty towns and misery belts around every big city and without the cruel frustrations of bloody revolution. The task is formidable indeed, but the resources that are waiting to be mobilised are also formidable.

He holds that the avoidance of world hunger is not simply a matter of raising agricultural yields, but a matter of raising the whole level of rural life.

As indicated by the earlier outline of the dynamics of the dual economy, it appears that mass unemployment, mass urban migration, mass poverty, and therefore famine, may not be eradicated in the South without the development of an agro-industrial culture which enables rural regions and communities to offer an attractive variety of occupations to their members.[113] Such a task requires an invigorated emphasis on the dynamics of self-reinforcing rural development in the philosophies of development economists. Subsequent to Schumacher's elucidation of these issues, research has revealed considerable eco-

---

111 Schumacher, "I.T. - Its Meaning and Purpose"; McRobie, *Small is Possible*, p. 33.
112 Schumacher, *Small is Beautiful*, p. 190.
113 *Ibid.*

nomic factors which should favour the development of decentralized capital-saving manufacturing in rural areas.[114]

Neglect of Meta-Economics. A fourth major misdirection in dominant development philosophies was, according to Schumacher, a tendency to assume that there is one mode of economics which is universally valid, and which if applied consistently throughout the South, would eventually lead to widespread prosperity.[115] In contrast Schumacher argues that many forms of economics are possible and that the best form for a given set of circumstances depends upon the nature of those circumstances.

He argues that economics is a "derived" science which accepts instructions from meta-economics. Meta-economics is the framework from which the basic assumptions used in economic modelling are derived; and, as such, it is beyond the analysis of economists in their capacity as economists. Meta-economics may be derived from the study of the fundamental characteristics of both people and nature, and embodies the normative context in which the conduct of positive economics may be pursued. Thus, different human needs or human cultural purposes, and different material conditions in different regions, may require the adoption of a different form of economics.

Schumacher argues that the circumstances of the South, combined with the growth of the dual economy, invoke a different meta-economic framework from that in which much modern economic theory has evolved:[116]

> I think economic analysis is an analysis of a given society, trying to find out small faults and to suggest small re-arrangements to maximize results, always on the assumption of other things remaining equal. This is really the science of the economist. Therefore he is a

---

114 D. J. Vail, *The Case for Rural Industry: Economic Factors Favouring Small Scale, Decentralized, Labour-Intensive Manufacturing*, Report of the Programme on Policies for Science and Technology in Developing Nations, Institute on Science, Technology and Development, Cornell University (Cornell University, July 1975); G. Jenkins, *Non-Agricultural Choice of Technique: An Annotated bibliography of Empirical Studies* (Oxford: The Institute of Commonwealth Studies, 1975); M. Carr, *Economically Appropriate Technologies for Developing Countries: An Annotated Bibliography* (London: Intermediate Technology Publication Ltd., 1976).

115 This theme occurs throughout Schumacher's writings, but is developed explicitly in his article, "The Role of Economics", Part I, Ch. 3 in *Small is Beautiful* (London: Blond and Briggs, 1973), pp. 36- 47; see also, E. F. Schumacher, "Does Economics Help? An Exploration of Meta-economics", in *After Keynes*, Papers presented to Section F (Economics) at the 1972 Annual Meeting of the British Association for the Advancement of Science, Edited by Joan Robinson (Oxford: Basil Blackwell, 1973), pp. 26-36.

116 Schumacher, "Paths to Growth", p. 27.

useful man when people don't want fundamentally to change the economy as it exists. Perhaps economists make a mistake when they think they can apply their ordinary economics to the problem of development, when the problem of development is not a matter of just developing (say) the British economy a little bit further, but of changing a large economy like India from a condition of misery into something more tolerable.

It follows that a redirection is required in development philosophy towards a more diversified approach to economic theory, towards an attempt to modify economics to match conditions in different regions and countries, and towards complementing the disciplinary study of economics with a study of meta economics.[117]

Neglect of Self-Reliance. Schumacher's insistence that economics should begin with people rather than goods, stresses that the development of people is the key to economic development. This is illustrated by his assertion that "aid can be considered successful only if it helps to mobilize the labour power of the masses in the receiving country and raises productivity without 'saving' labour".[118] In Schumacher's schema a high level of initiative taking, problem solving and effort is required by local people in a local region to enable the sort of innovation required for self-sustaining development. Thus a spirit of self-help is a necessary part of local economic development. He also holds that being productively employed is perhaps the most important means by which people in the hinterland of the South may gain the required capacity to innovate and solve problems. Thus, dependence by people in a local region on producers from outside their region for the provision of basic goods and services encourages longer term economic dependency of a debilitating kind, due to a decline in the human capacity for problem solving, innovation and self-initiated productive work. A lack of attention given to encouraging self-help economic activity is the fifth major misdirection identified by Schumacher in dominant development philosophies.[119]

The importance of self-help is confirmed by a consideration of the need for analyzing meta-economic factors as a part of development

---

[117] Schumacher makes such an appeal with acerbic wit when reviewing a book by development economist P. T. Bauer, entitled *Dissent on development: Studies and Debates in Development Economics* (London: Weidenfeld and Nicholson, 1972); see, E. F. Schumacher, "Don't Try to Help", *New Scientist* (13th April 1982), p. 95.

[118] Schumacher *Small is Beautiful*, p. 180.

[119] Cf., I. Low, "Humane economist", an interview with E.F. Schumacher, *New Scientist*, 63, 914 (1974), 656-657.

planning. The meta-economic context of development varies from culture to culture and region to region; this is due to both human factors and to variations in material resource endowments. Consequently, the adoption of the best economic package for a particular region requires a sound assessment of local conditions by people in a position to be aware of those conditions. Local people actually involved in the local economy would, for this reason, need to be a significant part of such assessment procedures. If economic decision making is normally done by those outside of a given region it is likely to have a dampening effect on the capacity of local people to develop their prowess as managers of the economy - through lack of practice.

As indicated earlier, Schumacher's schema requires that poverty and unemployment in the South be tackled at its source - in the rural areas. Given the enormity of the problem (e.g. typically three quarters or more of the population in "underdeveloped" countries lives in the hinterland) and the lack of available surpluses from the metropolitan areas to adequately address the problem, the need for a self-help approach in the rural areas becomes relatively self-evident. "Self-help" here needs to be distinguished from "absolute self-sufficiency": the latter is arguably impossible, while the former should be seen as an orientation in problem solving which does not exclude cooperation or trade with other regions.

In contrast to economic planners who focus on questions such as how to make more capital available for establishing new industries, or how to maximize return on capital so invested, Schumacher poses what he considers to be a more profound question:[120] "Why is it that the people are not helping themselves?" He argues that most peoples and all healthy societies have always known how to help themselves, and have discovered patterns of living which fitted their peculiar circumstance, and that societies and cultures have collapsed when they deserted their pattern or fell into decadence. He therefore considers that the chief role of economic planning ought to be the encouragement of the function of self-help by people in local economies.

Simplistic Approach to Technology. The sixth major misdirection identified by Schumacher in development philosophy was a propensity for an unrealistic or uncritical approach to the role of technology in economic development. This issue will be dealt with extensively further on, so cursory comments will suffice at this juncture. In short, while most economic planners acknowledge the crucial role of technology in development, Schumacher observes that the question of technology's

---

[120] Esp., Schumacher, "Help to Those", pp. 36-42, esp p. 37.

appropriateness vis-a-vis the hinterland of the South, was rarely adequately addressed. He holds that technology is not culturally neutral and that it should not be assumed that a technology designed for one economy will work effectively for another. He observes that widespread attempts to transfer production technologies from wealthier urban-industrialized countries to the South, without adequate assessment and discrimination, tended to exacerbate the mutual poisoning process of the dual economy.

In a 1961 report to the Gandhian Institute of Studies in India Schumacher concludes that:[121]

> ... it is erroneous to think that, for an underdeveloped country, the introduction of the highest level of technology is the best; that "high productivity", attained here and there through such an introduction, is better than nothing. It is in fact worse than nothing. The only hope, I should hold, lies in a broadly-based, decentralized crusade to support and improve the productive efforts of the people as they are struggling for their livelihood now. "Find out what they are doing and help them to do it better. Study their needs and help them to help themselves".

In accordance with the dictum, "economics should begin with people, not goods", Schumacher considers that the central purpose for introducing technology within a region beset by unemployment ought to be to empower people, who might otherwise be unemployed or underemployed, to become productively employed. In contrast, the main attitude in dominant philosophies of development appeared to be that technology's chief purpose should be the displacement of labour to enable improved labour productivity amongst those remaining in employment.

Summary of Schumacher's Diagnosis

In summary, Schumacher holds that the dynamic of underdevelopment in the South has developed and persists because a process of mutual poisoning has arisen between the urban-industrialized and traditional sectors of the dual economy. This is fuelled by the "twin evils" of mass rural unemployment and mass urban migration. These twin stimuli are in turn seen as arising from a general decay in rural life

---

[121] Schumacher, "Levels of Technology", p. 56.

linked to the decrepit state of non-agricultural industries in rural areas. While other factors are important, these are considered to be pivotal.

The dual economy and its self-reinforcing process of mutual poisoning is exacerbated and reinforced, according to Schumacher, by the philosophies of development which have predominated in the latter half of this century. The main features of those misdirected philosophies are: the abandonment of a concept of gradual evolution in economic affairs for one of creation or the quantum leap; a neglect of the human dimension; a neglect of rural areas; a neglect of meta-economics; a neglect of the role of self-reliance amongst the recipients of economic aid; and, a simplistic approach to the role of technology in economic development.

5

Schumacher and Intermediate Technology: Developments

Intermediate Technology: Proposed Solutions

While much public knowledge of Intermediate Technology has come from the publicized experiments of Intermediate Technology practitioners, it is important to realize that E. F. Schumacher's work was not just a matter of tinkering and experimenting with intermediate technologies: it was founded on deeply thought-out theory. The previous chapter dealt with the background of Intermediate Technology. This chapter will show Schumacher's analysis of problems was applied to the search for feasible solutions:

Enlightened Development Philosophy

> All history - as well as all current experience - points to the fact that it is man, not nature, who provides the primary resource: that the key factor of all economic development comes out of the mind of man.

In the passage from which the above quote is taken Schumacher argues that education is the greatest resource for achieving self-sustaining development - being the chief means for the cultivation of the human mind.[1] This attribute of education, however, also gives it an ambiguous status. If education, broadly conceived, produces a philosophy of development which is misdirected - as Schumacher claims has happened - then it may be viewed as an obstacle rather than a boon to sustainable and widespread development. The first requirement of a solution to the development problematique is, therefore, an enlightened development philosophy. Schumacher sees a need for education, in both the North and the South, to be oriented towards this end.[2]

An enlightened development philosophy would be one which avoided the pitfalls of "misdirected" development philosophies outlined above. Thus it would need to:

- emphasize *gradual evolution* rather than quantum leaps in economic development;
- take the needs and capacities of *people* rather than the production of goods as its departure point;
- focus on the integrated development of *rural areas* and the traditional sector, rather than on the promotion of "Westernized" industries in the metropolitan areas;
- incorporate the serious consideration of *meta-economic factors* as a precondition to effective economic analysis;
- stress *self-reliance* at all levels of the society as the dominant mode of economic problem solving; and,
- incorporate a more sophisticated appreciation of the role of technology in the dynamics of development and promote a serious interest in assessing the *appropriateness of technology.*

The most useful first step in evolving such an enlightened philosophy would be the acknowledgement of the self-reinforcing dynamic of underdevelopment illustrated in Figure 4.1. If such a description of the development problematique is accurate, as the weight of informed opinion increasingly indicates, it follows that it is essential for the vicious circle to be broken in some way. Schumacher's contention that the "2nd order" problems of mass unemployment and mass urban migration are dynamic variables, leads to the conclusion that eliminating these problems is the key to eliminating the others. It follows that providing massive numbers of employment opportunities and improving prospects for integrated development in rural areas ought to be *cardinal*

---

[1] E. F. Schumacher, *Small is Beautiful: A Study of Economics as if People Mattered* (London: Blond and Briggs, 1973), p. 180.

[2] See: E. F. Schumacher, "Education for Good Work", in *Good Work* , a collection of articles by Schumacher, ed. by P. Gillingham (New York: Harper Colophon, 1979), pp. 112-123; E. F. Schumacher, "The Greatest Resource - Education", in *Small is Beautiful*, pp. 70-92.

policy objectives for countries of the South. It is this conclusion which leads Schumacher to assert that people "must be given a chance to work [and] that nothing counts more than that".[3] In response to those who argue that "eliminating the human factor" is a necessary precondition for improving the economic performance of industry, Schumacher holds to the following position:[4]

> The greatest deprivation anyone can suffer is to have no chance of looking after himself and making a livelihood. There is no conflict between growth and employment. Not even a conflict between the present and the future. You will have to construct a very absurd example to demonstrate that by letting people work you create a conflict between the present and the future. No country that has developed has been able to develop without letting the people work. On the one hand, it is quite true to say that these things are difficult: on the other hand, let us never lose sight of the fact that we are talking about man's most elementary needs and that we must not be prevented by all these high-faluting and very difficult considerations from doing the most elementary and direct things.

This quote illustrates Schumacher's view that employment ought to be an end in itself, rather than simply a necessary means for the production of wealth. It is adduced here, however, to demonstrate his contention that the *social goal* of full employment may also be deemed as of paramount importance from the viewpoint of *economic rationality*.

The cardinal policy objectives mentioned above are expanded by Schumacher into the following four *action principles*: [5]

- Workplaces have to be created in the areas where the people are living now, and not primarily in metropolitan areas into which they tend to migrate;

---

[3] E. F. Schumacher, "Paths to Economic Growth", paper presented to International Seminar "Paths to Economic Growth", 21-28 January, 1961, Poona, India, published in *Roots of Economic Growth* by E. F. Schumacher (Varanasi: Gandhian Institute of Studies, 1962), p. 26.

[4] Schumacher, *Small is Beautiful*, pp. 204-205.

[5] *Ibid.*, p. 163. Also see, E. F. Schumacher: "How to Help Them Help Themselves", *Observer*, Weekend Review (29 August 1965); "Industrialization through Intermediate Technology", paper presented to the Conference on Rural Development, Cambridge University, 1964 (published in *Developing the Third World: The Experience of the 1960s*, ed. R. Robinson (Cambridge: Cambridge University Press, 1971).

- These workplaces must be, on average, cheap enough so that they can be created in large enough numbers without this calling for an unattainable level of savings and imports;
- The production methods employed must be relatively simple, so that the demands for high skills are minimized, not only in the production process itself but also in matters of organization, raw material supply, financing, marketing and so forth;
- Production should be largely from local materials for local use.

These action principles may only be implemented, argues Schumacher, by the adoption of a dual strategy of a *regional approach* to economic development and conscious effort to develop a *suitable technology* for the regional approach. These two elements of the strategy will now be discussed.

Regional Approach

The first element of Schumacher's proposed strategy for breaking the vicious cycle of underdevelopment is the adoption of an approach to economic planning and policy which takes *districts* or *regions* as the main focus of attention.[6]

The need for a regional approach is arguably self-evident, given the diagnosis outlined above. We have seen that underdevelopment is a problem of the *underdevelopment of actual regions or sectors* rather than something which may be adequately understood at the aggregate level. The notion of the dual economy indicates that "development" tends to be unevenly distributed under dominant modes of economic activity in the South; it is possible for a country as a whole to achieve economic growth, formally measured, while the actual experience of the majority of the population (excluded from the growth sector) is the opposite. The action principle that workplaces need to be created in rural areas rather than metropolitan areas is another way of saying that workplaces need to be created regionally. The following reasons for adopting a regional approach may be identified.

Firstly, as just intimated, the size of the gap between the metropolitan economy and the traditional economy of the hinterland is

---

[6] Schumacher develops this theme frequently throughout his writings; for an explicit discussion of this subject, see his chapter, "A Question of Size", in *Small is Beautiful* (pp. 57-68); see also: *ibid.*, pp. 164-165; "Industrialization"; "Notes on Indian Development Problems", report to the Gandhian Institute of Studies, written in April, 1961 (published in *Roots of Economic Growth*, by E. F. Schumacher [Varanasi: Gandhian Institute of Studies, 1962], pp. 43-48).

one of the principal causes of the dual economy's process of mutual poisoning. *A reduction in the size of this gap and of the destructive competition between the "efficient" industries of the large cities and the "inefficient" industries of the hinterland, appears necessary to enable non-agricultural production to grow in the rural areas* - to expand wealth creation in the rural areas and to service markets which are underdeveloped due to the vicious circle of poverty and unemployment. A regional focus is required, according to Schumacher, to provide the necessary economic structure to enable the bulk of the population in underdeveloped countries to participate productively in economic life.

He claims that the existence of the dual economy, at least in its extreme form, is a decisive feature of the countries of the South and therefore invokes the need for conscious policies of a type not required by Western Europe, or even Japan, during their periods of industrialization.[7] In a 1961 conference paper presented in India Schumacher speaks of the need for some kind of artificial protection for rural industries to compensate for the lack of natural protection of the kind experienced by Western industries during their development:[8]

> ... if what one might call organic protection of indigenous technological growth is withdrawn - and that is the actual situation because one is not developing the new techniques indigenously but importing them wholesale from another country and from another culture - then one must compensate for this by artificial means of protection. This is the specific situation of developing countries, which is unique.

Thus, Schumacher points to the need for some kind of "controlled isolation" (not complete separation) of districts within a large underdeveloped country to counteract the influence of modern infrastructure (e.g. efficient, fast transport systems) and modern technology on the traditional sector, so that local labor may be used primarily to cover local needs.

In addition to an altered planning focus he suggests such mechanisms as the use of local currencies as a means for protecting the economies of local regions from harmful competition from the "Westernized fringe economy" of the metropolitan areas.[9] This need

---

[7] Schumacher, "Paths to Economic Growth", pp. 14-20.

[8] *Ibid.*, pp. 20-21.

[9] E. F. Schumacher, "Help to Those Who Need it Most", paper presented to the International Seminar, "Paths to Economic Growth", 21-28 January 1961, Poona, India

for local currencies has been taken up recently by Jane Jacobs, who has argued forcefully that national currencies (in countries with diverse districts and regional economies) tend to transmit faulty messages to the market about the nature of local economies (within a country) to the detriment of those local economies. She shows how free trade mechanisms may work to the benefit of local or regional economies, but only if the currency mechanisms bear some direct relation to the actual production and trade activities within and between *local* economies.[10] Jacobs refers mostly to the problems of economic decline in cities of the geographically large industrialized nations; whereas, due to the differences between the North and the South, Schumacher applies similar arguments to the South's rural areas.

We may identify a second major justification for a regional approach by considering Schumacher's question, "How is it possible to make it clear to people that they work for one another, that the market is there, existing in their own poverty?"[11] The notion of the mutual poisoning of the dual economy emphasizes the following dilemma: an underdeveloped region in the South is not able to produce goods competitively enough for the metropolitan market, yet the decline in local incomes, due to local underemployment and unemployment, means the local market also does not provide enough effective demand for local production. Consequently, depressed regions are faced with simultaneous unmet human needs and underutilized human resources which could be used to meet those needs. The regional approach may be interpreted as a potential means for linking together these two problems to evoke a two-way solution.

A third justification for a regional approach may be adduced from Schumacher. He holds that political independence and economic self-reliance are mutually interdependent.[12] On this basis he argues that "a given political unit is not necessarily of the right size for economic development to benefit those who need it the greatest".[13] and that in the majority of cases in the South the political structure (i.e. the State) is too large to facilitate effective regional development. Schumacher's views imply that a significant degree of control over local politics and

---

(published in *Roots of Economic Growth*, by E. F. Schumacher [Varanasi: Gandhian Institute of Studies, 1962]), see esp. pp. 38–42.

[10] J. Jacobs, *Cities and the Wealth of Nations: Principles of Economic Life* (New York: Random House, 1984).

[11] Schumacher, "Paths to Growth", p. 28.

[12] E.g., Schumacher, "Small is Beautiful", p. 190.

[13] *Ibid.*, p. 164; cf. pp. 57–68.

the local economy is necessary to enable people in the hinterland of the South to mobilize their resources for local economic development. Thus he speaks of the need for some form of decentralized governance in rural districts with a substantial population.[14]

Fourthly, the need for a regional approach to economic development is directly implied by the need to prevent rural-urban population drift.

A fifth major justification for a regional approach is the essentially cultural origins of both development and underdevelopment. That Schumacher viewed the inclusion of cultural factors in economic planning as essential has already been discussed. His comment, as follows, confirms this unequivocally:[15] "It cannot be emphasized too strongly that the problem of agricultural poverty and rural misery is essentially and primarily a cultural problem".

This may be explained by considering Schumacher's comments on the appropriate size of a development region. He argues, for example, that India would be best served by being structured as a confederation of a few hundred *development districts*, each of about one to two million inhabitants and incorporating a thousand or so villages, a number of market towns, and a fairly substantial district center or capital. While the federal and state governments would have an important role to play (in the form of infrastructure provision and technical assistance etc.), the authorities of each district would have to be the principal authors and executors of their own development plans, basing their work on local materials and local methods, planning to meet local needs.[16] The essential point here, to use Schumacher's words, is that "the criterion for the choice of size must be a cultural, not an economic one, for the ultimate purpose of economic development is culture (in the comprehensive sense) and not mere economics".[17] Given the key role of education for innovation and development, Schumacher sees the necessity of an institution of higher learning in a development district and argues that the district ought to be large enough to support this. He also avers that it ought to be small enough to enable realistic participation by local people in economic planning and wealth creation within a given district:[18]

---

14 This was articulated clearly by Schumacher in "Help Themselves".
15 Schumacher, "Indian Development", p. 46.
16 *Ibid.*, passim., esp. pp. 46-47
17 *Ibid.*, p. 46.
18 Schumacher, "A Question of Size", p. 68.

What is the meaning of a democracy, freedom, human dignity, standard of living, self-realization, fulfillment? Is it a matter of goods, or of people? Of course it is a matter of people. But people can be themselves only in small comprehensible groups. Therefore we must learn to think in terms of an articulated structure that can cope with a multiplicity of small scale units.

In conclusion, Schumacher's principle, that the departure point of economics should be people rather than goods, might be no more than an impracticable ideal if it is not firmly rooted as part of a local or regional development strategy.

Finally, the sixth justification for Schumacher's regional approach to development is the need to provide structures which facilitate the maximum recirculation of locally created wealth within the local region, thereby enabling maximum local economic and employment multiplier effects.[19] Schumacher does not develop this theme systematically in his later writings, but he alludes to it often; and, it is logically connected to his insistence that industrial development strategies for poorer countries ought not to place too high a pressure on the need for obtaining foreign exchange. The existence of the dual economy within a country means that, in real terms, a depressed region in the hinterland may be thought of in relation to the metropolitan economy in the same way that a poor country of the South relates to a wealthy urban-industrialized economy of the North. Thus, a depressed economy of the rural hinterland may develop a "foreign exchange crisis" with the "foreign" of the metropolitan area, if it becomes too dependent upon unequal trading relationships with that more dominant economy.

The six areas considered above demonstrate the main factors in favour of a regional approach to development. These factors also indicate that *such an approach may not be viable unless self-sustaining means can be evolved to enable regional economies to achieve high enough levels of productive efficiency to meet basic needs and withstand the inevitable pressures emanating from the urban-industrialized economies.* It is for this reason that Schumacher emphasizes the importance of technology.

---

[19] Schumacher, *Small is Beautiful*, pp. 202-203.

Suitable Technology

Schumacher stresses repeatedly that the regional or district approach to development has no chance of success unless it is based on the employment of *suitable* technology.[20] In his 1962 report to the Indian Planning Commission he writes, "None of the developed countries has ever had to face the problems which are posed in India today and which arise from the existence and partial infiltration of a foreign technology which is at once vastly superior and vastly expensive."[21] He thus points to the adoption of *unsuitable technology* as a fundamental cause of the decay in economic and cultural life in the hinterland of the South and the consequent growth of the dual economy.

At the heart of the regional approach advocated by Schumacher are the cardinal policy objectives of generating massive employment opportunities and enabling integrated economic development in the rural areas. Neither of these objectives may be fulfilled without the use of technology because technology is necessary for achieving high enough productivity. Employment, in this context, must involve not only the human activity of work, but the production of goods and services at a sufficiently high level to raise the rural poor out of their extreme poverty. Under the conditions of the dual economy much primitive or traditional technology is unable to compete with the efficiency of modern technology and is therefore incapable of supporting financially viable workplaces. Yet, the apparently obvious solution of adopting the same modern technology in the rural areas is, in Schumacher's view, an unsuitable response. The reasons for this have been intimated above: the partial introduction of modern, highly efficient technology tends to produce the phenomenon of the dual economy, with the result that the economic decay of the hinterland robs it of the capacity to invest in such technology. In short, most of the technology of the Westernized fringe economy requires a level of resources beyond what is available in the traditional rural sector. Thus, attempts to create workplaces through the "advanced" technology lead to only

---

[20] The concept of Intermediate Technology occurs throughout Schumacher's writings from the early 1960s onwards (see references in Chapter Four). Cf., "Intermediate Technology", *The Center Magazine*, Center for the Study of Democratic Institutions, Santa Barbara, Calif., 8 (Jan/Feb 1975), 43-49; "On Appropriate Forms of Ownership and Action", in *Good Work* by E. F. Schumacher, ed. by P. Gillingham (New York: Harper Colophon, 1979), pp. 97-111.

[21] E. F. Schumacher, *Reflections on the Problem of Bringing Industry to Rural Areas* (New Delhi: Indian Planning Commission, 1962), p. 2.

limited success, or to failure. Schumacher summarizes the situation in these words:[22]

> A 'modern' workplace, moreover, can be really productive only within a modern environment, and for this reason alone is unlikely to fit into a 'district' consisting of rural areas and a few small towns. In every 'developing country' one can find industrial estates set up in rural areas where high-grade modern equipment is standing idle most of the time because of lack of organization, finance, raw material supplies, transport, marketing facilities, and the like. ... a lot of scarce capital resources - normally paid for from scarce foreign exchange - are virtually wasted.

The dilemma faced by rural economies in this situation is, then, that the "available" technology belongs to either of two extremes, neither of which is suitable for them under the conditions of the dual economy.

Schumacher labels this dilemma "the law of the disappearing middle":[23] technological change and the contingencies of technology transfer mean that either only the latest international "state of the art" technology or relatively primitive technology is available for communities to choose, while relatively efficient technologies from the recent past become obsolescent and unavailable. He therefore speaks of the need for a "middle way" in technology to bridge this gap.[24] The term "intermediate technology" is derived from this notion and embodies Schumacher's argument that in matters of development there is a problem of choosing the right "level of technology". He argues that there is a difficult *choice* required in technology because it cannot be assumed that what is best in conditions of affluence is necessarily the best in conditions of poverty.

He defines "intermediate technology" *symbolically* by arguing that the technologies most likely to be appropriate for development in

---

22 *Small is Beautiful* , pp. 165-166.

23 E. F. Schumacher: *Technology with a Human Face*, transcript of a lecture delivered at the University of Western Australia, July 1977 (Perth, Aust.: Campaign to Save Native Forests, 1977); "On Technology for a Democratic Society", in *Small is Possible*, by G. McRobie (London: Jonathan Cape, 1981), pp. 1-13, esp. p. 3.

24 Schumacher addressed this problem of the need to find a "middle option" in economic development and technology in a series of articles for the British journal *Frontier* in 1965 and 1966: "Buddhist Economics", *Frontier*, 8 (Autumn 1965), 201-205; "The Middle Way", *Frontier*, 9 (Winter 1965/1966), 297-301; "Crossing the Boundaries of Poverty", *Frontier*, 11, (Summer 1966), 115-118.

conditions of great poverty would be in some sense "intermediate" between "the hoe and the tractor, or the pange and the combine harvester".[25] In his paper to the 1964 Cambridge University Conference on Rural Development Schumacher acknowledges that certain sectors in the South are irrevocably committed to the employment of the "highest" levels of technology but stresses that this does not generally spread effectively to the service of the 85 percent of the population who typically make up the traditional sector. He writes:[26]

> The task is to re-establish a healthy basis of existence for these 85 percent by means of an "intermediate technology" which would be vastly superior in productivity to their traditional technology (in its present state of decay), while at the same time being vastly cheaper and simpler than the highly sophisticated and enormously capital-intensive technology of the West.

Intermediate Technology is therefore a special case of the broader concept of Appropriate Technology. It is a specific application of the general-principles approach of Appropriate Technology to specific circumstances. In short, it is technology which has been tailored to fit the psychosocial and biophysical context prevailing in the hinterland of the South (for the purpose of development within the framework of the dual economy).

Formal Definition of Intermediate Technology

The symbolic definition is useful for explanatory purposes but inadequate for the purposes of technology design and detailed policy formulation. Schumacher provides a formal and more precise outline of Intermediate Technology by focussing on the cost aspects of technology. Thus, an intermediate *level* of technology is one with an intermediate cost. Before a formal definition is provided this approach will be investigated in a little more detail.

For technology to be suitable to the conditions of the South's hinterland it must be *accessible* to the majority of the people and communities. From the point of view of cost, "accessible" means "affordable": machines and methods may not be generally accessible unless they

---

[25] E. F. Schumacher, "Intermediate Technology: Its Meaning and Purpose", mimeo (London: Intermediate Technology Development Group Ltd., n.d.), p. 1.

[26] "Industrialisation", p. 9.

stand in some definable relationship to the level of resources and wealth in the society in which they are to be used. In his 1962 report to the Indian Planning Commission, on rural industrialization, Schumacher makes this point in the following way:[27]

> If, therefore, it is intended to create millions of jobs in industry, and not just a few hundred thousands, a technology must be evolved which is cheap enough to be accessible to a larger sector of the community than the very rich and can be applied on a mass scale without making altogether excessive demands on the savings and foreign exchange resources of the country.

The *affordable level of expenditure* by a particular economy on technology is clearly not independent of the general level of technology already in use throughout that economy. This is because the general level of technology in a given economy (i.e., the degree of capitalization, defined as the amount of capital per workplace) is a chief factor which determines the income levels of people and, therefore, the levels of savings available for investment. Thus, the relatively high average income of people in the North is primarily dependent upon the relatively high level of capitalization in industry - but, and this appears to be a fundamental insight behind Schumacher's Intermediate Technology, the relatively high level of capitalization in the North's industry depends in turn upon the high income levels in the population. Consequently Schumacher states that *income per person* and *capital per workplace* stand in an organic relationship to each other, a relationship which can be "stretched" to some extent - for instance with the help of foreign aid - but cannot be disregarded.[28] While, strictly speaking, it is often possible to introduce technology which ignores this relationship, Schumacher argues that any society which does so for long will encounter "serious troubles" such as an undue concentration of power and wealth amongst the privileged, the growth of structural unemployment and the general malaise of the dual economy.[29] He points out that the economic development of the wealthier industrialized countries was in fact achieved without destroying the overall par-

---

[27] E. F. Schumacher, "Rural Industries", in India at Midpassage, ed., W. Clark (London: Overseas Development Institute, 1964), esp. pp. 32-33.

[28] Cf., "Industrialization", p. 8.

[29] Schumacher, *Small is Beautiful*, pp. 29-31.

ity between average income per person and average capital per work-place.[30]

The main content given by Schumacher to "suitable technology" is, in summary, that the *average per capita capital cost of a workplace ought to be affordable in terms of the average per capita income of people in a given development district.* Schumacher does not define "intermediate technology" in an analytically precise way, preferring instead to use either illustration and allegory, or to suggest "rule of thumb" cost levels suitable for particular circumstances. We may suggest then that intermediate technology is technology which leads to an average capital cost per workplace at an intermediate level between the low per capita income of the traditional economy and the high per capita income of the urban-industrialized economy. It is technology designed or chosen to exhibit cost characteristics to bridge the gap between the extreme capitalization levels which occur in the dual economy.

In accordance with the principle that technology ought to be suitable to the peculiar circumstances of a given development district, it follows that no universally valid figure may be quoted as an optimum "intermediate" level of capitalization: wealth and income levels vary between regions and over time, and other factors besides per capita income are influential. Schumacher, however, propounds a "rule of thumb" algorithm for application across different circumstances of time and place. Using the notion of "suitable technology" he argues that the upper limit for the average amount of capital investment per workplace is probably given by the annual earnings of an able and ambitious industrial worker.[31] His rationale for this is based upon an assessment of how long it would take a person to generate enough surplus wealth to invest in the plant required to establish his or her workplace. A ratio of roughly 1:1 between per capita investment and annual per capita income means that a workplace "costs" one person-year of work, or that a person would need to save one months' income for twelve years to be able to "afford" to "own" his or her job. Such a task does not seem unrealistic, whereas a ratio of 10:1 would require one months' income to be saved each year for 120 years - a forbidding prospect.[32]

---

[30] E.g., E. F. Schumacher: pp. 50-51 of "Levels of Technology", paper written in July 1961 for the Gandhian Institute of Studies (published in *Roots of Economic Growth*, by E. F. Schumacher [Varanasi: Gandhian Institute of Studies, 1962]; *Small is Beautiful*, pp. 167-168.

[31] *Small is Beautiful*, p. 31.

[32] Cf., *Ibid.*, pp. 167-168.

Schumacher uses "capital cost per workplace" to provide a further symbolic description of Intermediate Technology. This symbolism may be applied as in Table 5.1. The table illustrates the principle that, if a range of economically viable technologies are available with different capital-cost characteristics, then the most suitable choice for the purpose of encouraging regional development in a community with average annual per capita incomes in the vicinity of $100 to $1000, would be the $100 - $1000 technology.[33]

The $10 000 technology which, symbolically speaking, is more typical of the levels of capitalization involved in most official development programs in the South, would at best create one tenth of the number of direct jobs as the $100 - $1000 technology. At worst, which Schumacher points out is normally the case (because of the "negative demonstration effects" of the dual economy), the nett employment levels in the region may actually be reduced.

Table 5.1   *"Symbolic" Description of Intermediate Technology*

Technology type	Capital Cost per Workplace
Typical Indigenous Technology (South)	$10
Typical Industrial Technology (North)	$10,000
Intermediate Technology	$100 - $1,000

Source: Derived from various publications by E. F. Schumacher (e.g. "Levels of Technology".)
Figures included here are indicative only, and are chosen to illustrate a rough order of magnitude.

An example of how the introduction of apparently more "advanced" technologies to supplant "inefficient" indigenous means of production may have a nett adverse effect is illustrated by a widely quoted case study by the International Labour Office.[34] During the 1960s one country imported two plastic injection-moulding machines

---

[33] Figures chosen to illustrate rough order of magnitude only. These figures have been increased by a factor of 10 from those originally used by Schumacher, so as to create a more generally applicable symbolization of the options.

[34]   See K. Marsden, "Progressive Technologies for Developing Countries", *International Labour Review*, 101, 5 (1970), 475-502.

costing $100 000 with moulds. Working three shifts and with a total labor force of forty workers they produced 1.5 million pairs of plastic sandals and shoes a year. At $2 a pair these were better value (longer life) than cheap leather footwear at the same price. Thus 5 000 artisan shoemakers lost their livelihood; this in turn reduced the markets for the suppliers and makers of leather, hand tools, cotton thread, tacks, glues, wax and polish, eyelets, fabric linings, laces, wooden lasts and carton boxes, none of which was required for plastic footwear. As all the machinery and the material (PVC) for the plastic footwear had to be imported, while the leather footwear was based largely on indigenous materials and industries, the net result was a decline in both employment and real income within the country.

Despite wealth variations between regions and sectors, there are similarities between the hinterlands of the countries of the South. At the time when Schumacher was developing the Intermediate Technology concept the majority of the population in the South operated with an annual income per head of less than $200.[35] He therefore proposed a general guide for use by policy makers and business people in selecting intermediate technologies for rural based development - which, while not suitable as a universal design criterion, would indicate a level of cost normally appropriate. He argued that an intermediate technology should generally enable a workplace to be created at a per capita capital cost of not more than about $200.[36]

Schumacher's attempts to put a specific cost figure to intermediate technology for the South created an impression in the minds of a number of critics that he had adopted *inter alia* a static anti-efficiency or anti-development stance.[37] As the following quote indicates, however,

---

[35] *Ibid.*, 483.

[36] See "Industrialization", p. 9; in "Rural Industries" Schumacher advocates a capital expenditure per workplace for rural India (1962) of between Rs.1000 - Rs2000 (pp. 33-34).

[37] See, e.g.: A. Kestenbaum, "Criticisms of I.T.," in her booklet, *Technology for Development* (2nd ed.; London: Voluntary Committee on Overseas Aid and Development, 1977), p. 34; W. Rybczynski, chapter entitled "Millstone...", in *Paper Heroes: A Review of Appropriate Technology* (Dorchester: Prism, 1980), pp. 41-66; A. Agarwal, "Desai's difficulties with A.T.", *Nature*, 281 (1979), 172-174. In *Small is Beautiful* (p. 168), Schumacher labels such criticisms, "...the voice of those who are not in need, who can help themselves and want to be assisted in reaching a higher standard of living at once". He also summarized the issue during a public lecture in U.S.A. during 1977 in the following words: "Incredible amounts of money are being spent in trying to cope with the relentless growth of megalopolitan [sic] areas and in trying to infuse new life into 'development areas'. But if you say 'Spend a little bit of money on the creation of technologies that *fit the given conditions of development areas* ', people accuse you of wanting to take them back to the Middle Ages" ("Toward a Human-Scale Technology",

he was concerned essentially with promoting tangible resources for self-sustained growth:[38]

> Industrialisation on a mass basis is possible when there is a fairly high degree of self-sufficiency in equipment, so that the repair, maintenance, and replacement of equipment can be done largely from nearby resources. If in this sense the 'circle can be closed', then, and only then, can there be self-sustained growth. It is the structure of real things, not the appearance of symbols (like the rate of saving) which decides these matters.

Intermediate Technology, as it has been construed here, is a dynamic approach to economic progress rather than a reaction against economic progress.

The outline of Schumacher's ideas to this point has shown that Intermediate Technology derives from a *dynamic* approach to advancing an economy from a low level of capitalization to a much higher level of capitalization. Its distinguishing characteristic is not a rejection of the need for increases in capital productivity or labor productivity, but rather that increases in these factors ought to occur *gradually* so that the disparity between average capital cost per workplace and average annual income per worker does not become significantly large.[39]

Schumacher's emphasis on intermediate technologies as means for progressively attaining increased levels of industrialization, rather than as a means consistent with holding back economic development, has been recognized by a number of authorities. The International Labour Office, for example, has promoted the ideas under the rubric of "progressive technology".[40] The World Bank has incorporated the approach into some of its programs; the Bank has spoken of the need for technology users in "developing" countries to recognize that "in order to improve the lot of the vast majority of people they must, at least in the short run, accept standards of service and levels of

---

in *Good Work*, by E. F. Schumacher and edited by P. Gillingham [New York: Harper Colophon, 1979], p. 49).

[38] Schumacher, "Rural Industries", p. 34.

[39] N. Lockwood, in his study of Japanese industrialization, has confirmed the efficacy of this approach: "If Japan's experience teaches any single lesson regarding the process of economic development in Asia, it is the cumulative importance of myriads of relatively simple improvements in technology which do not depart radically from tradition or require large units of new investment" (*The Economic Development of Japan and Structural Change 1868-1938* [Princeton, N.J.: Princeton University Press, 1954], p. 198).

[40] Marsden, "Progressive Technologies".

'modernity' lower than those that might be found in more developed countries."[41] The language employed here clearly indicates the role of intermediate technologies as an economic "stepping stones".[42]

Other Aspects of Intermediate Technology

Intermediate Technology is a concept developed by Schumacher as part of a broad strategy for solving the problems of the vicious circle of underdevelopment in the South. It is essentially concerned with industrial methods which exhibit the capacity to generate employment opportunities at an intermediate level of capital cost. Despite its roots in a development philosophy which begins with people, rather than the production of goods, and which emphasizes meta-economic factors, Intermediate Technology is a formal economic concept defined in financial terms.

Nevertheless, as Schumacher and colleagues in the Intermediate Technology Development Group gained practical experience from the mid-1960s onwards applying Intermediate Technology, they evolved three other general criteria. In addition to the main criterion of *capital-cheapness*, they also came to emphasize *smallness, simplicity* and *non-violence*. They observed that it might not be possible for all four criteria to be concurrently satisfied in each case; but that any of the four criteria might lead to technology more suitable to integrated rural development, when taken seriously in engineering design work or as criteria for technology assessment and selection.

The reasons for adding the three further ("2nd order") criteria may be comprehended by reconsidering the background to Intermediate Technology. The concept holds that technology will serve development goals effectively only when it is tailored to fit the conditions of the development districts in which it is to be employed. The ("1st order") criterion of capital-cheapness derives from limitations in the availability of locally generated wealth in the South's hinterland for investment in technology. Most underdeveloped rural districts suffer, however, from shortages of other factors such as high-grade raw materials, low cost high grade energy (e.g. cheap grid-connected electricity), sophisticated technical skills and infrastructure for maintenance

---

[41] World Bank, *Appropriate Technology in World Bank Activities* (unpublished official document, Washington, D.C., July 19th, 1976), p. iv; cf., Appendix 1, p. 1.

[42] See, e.g., E. F. Schumacher, "Is Development Aid on the Wrong Road?", *Volunteers in Action* (March 1967), pp. 2, 3, 16, 17, 23.

and easy access to large, sophisticated markets. Consequently, a need may be identified for directly addressing these factors.

Schumacher points to smallness as a criterion of intermediate technology primarily because of the importance of the regional approach to development. He notes that large scale production processes are, in the main, only capable of maintaining economic efficiency if they maintain a high level of output, and that high output levels are only viable with access to large or dense markets. For technology to fit adequately into a rural development district, and therefore into a situation with a small effective market, it is therefore necessary for technology to operate competitively at a low output level. He therefore advocates that technology be designed to operate efficiently at a small scale. While small scale operations are not necessarily low or intermediate in terms of capital cost per person employed, relatively small scale technology is seen by Schumacher as potentially advantageous from a cost point of view. Self-reliant rural development requires the total capital cost of setting up a plant in a region to be affordable in terms of wealth generated in that region. Illustrative examples will be considered in the following section. In general, the smaller the scale of technology, the lower the total investment required to commence operations and the easier it is for a locally based group of people to own or establish a new enterprise. In the final analysis, however, Schumacher considers capital-cheapness (per workplace) to be a more important criterion.

Simplicity is stressed as another criterion of intermediate technology because, as Schumacher observes, much modern technology imported from abroad requires technical back-up, specialized maintenance and sophisticated management skills, of a kind which will most likely not be readily available in poverty stricken rural areas. Consequently, such technology has little chance of working effectively for long in such circumstances. Secondly, the development strategy advocated by Schumacher is radically dependent upon the gradual enhancement of the human capacities of *people* within a development district; in particular, it depends upon the development of their education, organization, discipline and entrepreneurship, through experience gained in productive economic activities.[43] In Schumacher's words:[44]

> It [economic development] cannot be 'produced' by skilful grafting operations carried out by foreign technicians or an indigenous elite

---

[43] Cf., "...education *en masse* can only be done through work- an education not just for a few people, who then will become alienated, but for the whole people" (E. F. Schumacher, *Economic Development and Poverty* [London: Africa Bureau, 1966], p. 12).

that has lost contact with the ordinary people. It can succeed only if it is carried forward as a broad, popular 'movement of reconstruction' with primary emphasis on the full utilization of the drive, enthusiasm, intelligence and labor power of everyone.

Thus, the complexity of technology employed in rural development projects needs to be manageable in terms of human skills available within the region. This feature is necessary for facilitating the gradual growth in the competence of local people in technical maintenance activities, product and process innovation and technology design.

The fourth criterion, non-violence, is more difficult than the others to define precisely and measure objectively, but is nevertheless considered by Schumacher to be of utmost importance - not just as a kind of wishful after thought, but as an active design parameter. Schumacher uses "non-violence" vis-a-vis Intermediate Technology to denote three things. Firstly, he holds that technology ought to be designed and operated in accordance with sound ecological principles, so as to efficiently utilize available supplies of *local* resources, and minimize local environmental damage. He believes that the violation of local ecological balance may in turn damage prospects for local economic development. Secondly, he holds that technology ought to be designed to avoid harmful impact on human beings as persons; this involves both avoiding damage to physical and mental health (from technology-related work patterns) and ensuring that work opportunities are created which allow for both human satisfaction and the cultivation of people's creativity and skills. Schumacher believes that this is important, not just for ethical or cultural reasons, but also because of economic necessity: the Intermediate Technology approach to development relies upon human beings as the cardinal economic resource. Thirdly, by "non-violence" Schumacher refers to the need for technology with a low propensity for unintended harmful impacts and side-effects; he advocates technology which is designed to prevent costly problems occurring, rather than technology which requires the introduction of further technology to compensate for its undesirable effects (for example, a factory with low pollutant emissions, rather than one with high emissions and the consequent need for further technology to counteract the resulting pollution).

In summary, Schumacher considers that all four of the criteria just listed should be adopted as guidelines for directing innovation and technology choice; he advocates them as directions in which to strive

---

[44] Schumacher, *Small is Beautiful*, p. 191.

rather than as rigid rules. For engineering design purposes he stresses the fundamental criterion of optimizing capital cost per workplace. The pivotal importance of this criterion has been recognized by the World Bank (at least in some of its work) and, as the following quote from one of the Bank's reports indicates, capital cheapness may be viewed as a key to making the broader objectives of Intermediate Technology viable:[45]

> The choice to be faced ... is whether to invest heavily in a few workers and in services for a few to increase their production and living standards substantially, leaving the rest unaffected by growth (or at best affected indirectly), or whether to make some gain in the productivity of many people by investments at lower cost per capita affecting the mass of the people in a country.

The report continues by arguing that the fundamental notions of equity point to the approach based on smaller investments per person and, furthermore, that it typically leads to faster growth and more widespread opportunity for individual advancement, as a greater number of people are employed productively with the limited capital. It concludes:[46]

> There is thus no choice but to develop technologies to raise productivity and to improve living standards at very low costs per beneficiary. These technologies must not require too great an input of scarce skills, and they must be simple enough to be extendable on a scale commensurate to the need.

The concept of Intermediate Technology may be seen as simultaneously based upon serious consideration of orthodox economic precepts and serious consideration of factors normally depicted by economists as outside the domain of professional economics.

Intermediate Technology: Some Practical Examples

Schumacher advocates the use of intermediate technology because, on the basis of his analysis, it is needed. He arrives at this stance in-

---

[45] World Bank, *Appropriate Technology*, p. 2.
[46] *Ibid.*

dependently of empirical evidence as to its actual technical feasibility. Work conducted by the I.T.D.G. and others since the mid 1960s, however, has added weight to Schumacher's claim that such technology is not only necessary, but also possible.[47] A 1979 conference of the International Economic Association was devoted to considering this issue. The conference noted that labor-intensive techniques, by definition, have lower labor productivity than capital-intensive ones, for the production of similar products; and, that to remain technically efficient and competitive in terms of total unit cost of production, labor-intensive alternatives must have a higher level of output per unit of investment. The following conclusion was reached:[48]

> The problem of inefficient labor-intensive technologies is that they waste scarce investment resources. However, there is no *a priori* reason why more labor-intensive alternatives should be inefficient in this sense. The question is empirical and there is a good deal of empirical evidence nowadays that efficient labor-intensive alternatives exist.

The weight of evidence now seems to demonstrate that Intermediate Technology may be justified on the grounds of both need and empirical evidence.

A considerable amount of detailed investigation has now taken place identifying actual technologies capable of being employed as part of the Intermediate Technology approach to industrialization in the South. The United Nations Industrial Development Organization, for example, has published twelve volumes documenting such technologies under the following industry classifications: low cost transport for rural areas; paper products and small pulp mills; agricultural machinery and implements; energy for rural requirements; textiles; food storage and processing; sugar; oils and fats; drugs and pharmaceuticals; light industries and rural workshops; construction and building materials; basic industries.[49]

Work in many countries of the South has now disproved the view that new, modern machines are necessarily indicative of efficient technological systems and old ones of inefficient systems; this fact is now

---

47   E. F. Schumacher, "Neue Technologien sind Notig und Moglich", *Nachrichtentechnische Zeitschrift*, 30, 4 (1977), 267-268.

48   C. Cooper, "A Summing up of the Conference", Proceedings of a Conference held by the International Economic Association, Teheran, Iran, *Appropriate Technologies for Third World Development*, ed. A. Robinson (London: MacMillan, 1979), p. 404.

49   United Nations Industrial Development Organization, *Monographs on Appropriate Industrial Technology* (13 vols.; New York: United Nations, 1979).

being recognized by economists from both within and without the field of Third World development economics.[50] The concept of Intermediate Technology transcends the issue of the relative modernity of technology; but the fact that efficiency and modernity are not necessarily correlated provides reinforcement for the assumptions concerning technology choice which underlie Intermediate Technology - e.g., that an intermediate "level" of technology is not necessarily intermediate vis-a-vis efficiency. A number of case examples will illustrate the potential for intermediate technologies to be economically competitive.

Efficiency and Scale in Production

It is a commonly held dogma that increases in the scale of production are necessary to achieve increases in total efficiency. In contrast, there is now a growing literature which indicates that such a principle has only limited validity and that the opposite is often true - particularly when transport factors are taken into account.[51] One British-based international study, of four different sized plants for producing animal feed, yielded the results shown in Table 5.2.

No significant variations in unit cost of production were found amongst the four technologies, ranging from the lowest total capital required of about 40 thousand pounds to the highest of about 250 thousand pounds. In fact the one scale of production which did present a marginal cost advantage was the "intermediate" technology, which exhibited a total capital requirement of about 100 thousand pounds.

---

[50] See, e.g., p. 28 of S. MacDonald, "Technology Beyond Machines" in *The Trouble with Technology*, ed. S. MacDonald, *et al.* (London: Frances Pinter, 1983).

[51] A review of the evidence for the efficacy of maintaining a modest scale to human institutions and therefore to technology (e.g. in the fields of housing, city planning, food production, waste disposal, transportation, health care, education and governance) has been conducted by K. Sale in *Human Scale* (New York: Coward, McCann and Geoghegan, 1980); an empirical study of the United States' economy, by B. Stein, has demonstrated that large industrial operations are not, by virtue of their size, inherently superior to smaller ones and that, within certain segments of the economy and above certain very modest limits on smallness, the opposite is more nearly the case (*Size, Efficiency and Community Enterprise* [Cambridge, Mass.: Center for Community Economic Development, 1974]); for a survey of more specific relevance to the South see S. Jackson, *Economically Appropriate Technologies for Developing Countries: A Survey*, #3 (Washington, D.C.: Overseas Development Council, 1972).

Table 5.2 *Efficiency and Scale in Production of Animal Feeds*

Technology	A	B	C	D
Scale (output - tons per annum)	2 400	6 000	10 500	16 800
Capital requirements (£)	42 096	105 213	168 236	254 518
Cost per ton (£)	37.3	36.0	37.2	36.5

Source: R. Palmer-Jones and D. Halliday, *The Small-Scale Manufacture of Compound Animal-Feed*, T.P.I. Report No. G. 67 (London: Tropical Products Institute, 1971)

Another example, a comparison of three different processes for the production of high grade salt for industrial use (based upon Sri Lankan data), is summarized in Table 5.3. In this case the "labor intensive" technological option turns out to be superior in terms of all six of the major economic parameters: it requires the lowest total amount of capital, it enables the largest absolute number of jobs, it averages the lowest unit output cost, it achieves the highest financial efficiency, the highest employment levels per unit of output and the lowest capital cost per workplace.[52]

It is interesting to note that the so-called "intermediate technology" option studied here (a mechanically "improved" version of the traditional method based upon solar evaporation), while less labor intensive than the traditional method, is also less efficient in financial terms. The study did not include a fourth highly labor-intensive method because of lack of reliable data and because the method only works effectively on a very small scale. It would appear then that the method labelled "labor-intensive" in Table 5.3 would be more aptly labelled "intermediate" than the "improved" method. The capital cost per workplace exhibited by this method ($5,600) is nevertheless higher than that suggested by Schumacher as appropriate for conditions of extreme poverty in rural areas (per capita incomes of only several hundred dollars); the $5,600-technology would therefore most likely be appropriate for a slightly higher level of industrialization,

---

[52] In his *Non-Agricultural Choice of Technique: An Annotated Bibliography of Empirical Studies* (Oxford: The Institute of Commonwealth Studies, 1975) Jenkins points out (p. 18) that if shadow prices for inputs were calculated in the study by Enos (*Production of Salt*) the labour-intensive technique would display considerably lower unit costs of production than the so-called "intermediate technique".

and some further innovation would most likely be necessary to make the technology accessible to an extremely poor community.

Table 5.3   *Efficiency and Scale in Production of Salt*

Technology	Partially Mechanized (labor-intensive)	Fully Mechanized ("intermediate")	Vacuum Pans (capital-intensive)
Fixed capital (U.S. $ million)	4.97	5.15	7.10
Employment (jobs)	893	364	250
Av. output cost (U.S. $ per ton)	8.41	8.55	18.50
I/O ratio ($/$ per year)	4.0	4.1	5.6
L/O ratio (jobs / 000 tons)	6.0	2.4	1.7
I/L ratio ($ / job)	5 600	14 100	28 400

Key: I = investment cost; L = employment; O = output.

Source: J. L. Enos, *Proposal for Regional Co-operation in the Production of Salt and its Derivatives*, Study #9 (Bangkok: Asian Industrial Survey for Regional Cooperation, 1973)

The above two examples demonstrate the principle that, to achieve economic viability, it is not always necessary to choose technology which exhibits high levels of either total capitalization or capitalization per person.

Schumacher and colleagues have followed three basic complementary approaches to the application of Intermediate Technology: upgrading traditional technologies; scaling down and redeveloping high cost technologies; and, designing new technologies (product and process).[53] Some examples from each of these categories will now be briefly considered.

---

[53] See, e.g., E. F. Schumacher and G. McRobie, "Intermediate Technology in Action", paper presented to the Third Intermediate-Congress of the Pacific Science Association, July 18-22, 1977, Bali, Indonesia, Sub-Theme 2, Appropriate Technology for Medium and Small Scale Industries (Jakarta: Indonesian Institute of Sciences, 1977), pp. 26-30; cf., G. McRobie and M. Carr, *Mass Production or Production by the Masses?* Occasional Paper #4,

Upgrading Traditional Technologies

The following two examples are taken from case studies conducted by the Small Industry Extension Training Institute (SIET Institute) in Hyderabad, India.[54]

Table 5.4 compares four techniques for manufacturing a hand-operated Japanese-style paddy weeder: an existing handicraft technique, an existing power technique, and an improved version of each. In each case the improved techniques have a higher capital-labor ratio (I/L) and lower unit costs. The hand techniques have a slightly lower unit cost than the power techniques for a small market (up to 2000 units per month).[55] Thus the improved techniques exemplify Intermediate Technology in that they make possible gradual increases in labor productivity without the need for inordinate leaps in the level of capitalization.

Table 5.5 compares a hand operated and a power-drive technique for manufacturing a cycle gear-case at a production rate of 3000 units per month. The hand operated version has a marginally cheaper unit cost of production and a significantly lower capital cost per workplace. Its capital productivity levels (not shown in the Table) are higher than those of the power-driven method, as is its level of financial surplus. In this case the "lower level" of technology is superior from the perspective of the main economic parameters. With the choice of technology limited to these two options there are no apparent productivity reasons for selecting the higher degree of capitalization. The figures would, however, point to the need for some upgrading of the hand-operated method for the purpose of achieving a higher productivity level at a similar level of capitalization.[56]

Intermediate Technology Development Group (London: Intermediate Technology Publications Ltd., 1981).

[54] *Appropriate Technologies for Indian Industry* (Hyderabad: SIET Institute, 1967).

[55] See comment on this study by M. Carr, *Economically Appropriate Technologies for Devloping Countries: An Annotated Bibliography* (London: Intermediate Technology Publications, Ltd., 1976), p. 22.

[56] At the time of the Institute's research the likely "optimum" level of average investment per workplace was, according to Schumacher's judgements, in the vicinity of Rs1000 to Rs2000 (Schumacher, "Rural Industries", p. 33).

Table 5.4   *Manufacture of Agricultural Equipment*

| Technology | Handicraft Technique (existing) | Power-driven Machinery | | Hand-operated Machinery (improved) |
		(existing)	(improved)	
Capital equipment (Rs)	150	25 000	31 000	7 200
Total employees	1	30	36	32
Production (units/month)	25	850	1 900	1 500
I/O ratio (Rs/units/month)	6.0	29.0	16.0	5.0
I/L ratio (Rs/job)	150.0	835.0	860.0	225.0
Cost per unit (Rs)	14.0	15.0	13.0	12.75

Source: SIET Institute, *Appropriate Technologies for Indian Industry* (Hyderabad: SIET Institute, 1964) reviewed by M. Carr, *Economically Appropriate Technologies for Developing Countries: An Annotated Bibliography* (London: I.T. Publications Ltd., 1976)

Table 5.5   *Manufacture of Cycle Gear-Case*

Technology	Hand-operated	Power-driven
Capital equipment (Rs)	35 200	80 000
Total employees	27	18
Cost per unit (Rs)	5.66	5.77
I/L Ratio (Rs/job)	1337	4444

Source: SIET Institute, *Appropriate Technology for Indian Industry* (Hyderabad: SIET Institute, 1964); reviewed by G. Jenkins, *Non-Agricultural Choice of Technique: An Annotated Bibliography of Empirical Studies* (Oxford: The Institute of Commonwealth Studies, 1975), p. 36.

McRobie and Carr, of the I.T.D.G., list a number of other projects where traditional techniques have been upgraded with modern knowledge and materials under the rubric of "intermediate technology" to increase the productivity of rural employment and to meet basic needs.[57] In many cases the mobilization of underutilized labor is a means for generating new capital which in turn may be used to enhance the effectiveness of local labor. These projects include: the development of efficient and affordable water storage tanks suitable for self-help

---

[57] McRobie and Carr, *Production by the Masses.*

construction; improved cotton-spinning techniques and weaving looms; alternative types of mortar to cement, based upon other locally available raw materials; better low-cost road construction methods; high efficiency animal-drawn agricultural equipment; and, upgraded rice mills.[58]

Scaling Down and Redeveloping High Cost Technologies

The following two examples arise from work conducted by the Intermediate Technology Development Group.

One of the Group's notable successes in making available intermediate technologies is the design and manufacture of a small paper-pulp moulding system (PPMS).[59] The Group identified a need in Zambia for the local manufacture of egg trays from waste paper - a need with which, as was subsequently discovered, a considerable number of countries were faced. At the time of the Group's investigations (early 1970s) the smallest available technology required a market of at least one million trays per month to operate economically; Zambia's total market was about one million trays per year.[60] This meant that with international "state of the art" technology Zambia was incapable of

[58] Intermediate Technology Industrial Services, *Rainwater Catchment Project Jamaica* (London: Intermediate Technology Publications Ltd., 1973); R. Bruce, *Proposal for a Project to Develop Small-Scale Mechanised Spinning*, I.T.D.G. mimeo (Eindhoven/London: Intermediate Technology Development Group Ltd., 1975); Technology Consultancy Centre, Annual Reports (1972-1973, 1973-1974), University of Science and Technology, Kumasi, Ghana; R. Spence, *Proposal for Research and Development on Small-Scale Pozzolana and Other Cements*, Mimeo (London: Intermediate Technology Development Group Ltd., 1975); McRobie, *Small is Possible*, p. 41; C. I. Ellis and J. D. Howe, "Simple Methods of Building Low Cost Roads", *Appropriate Technology*, 1, 2 (1974); Central Projects Staff of the World Bank, *Appropriate Technology and World Bank Assistance to the Poor* (Washington, D.C.: World Bank, 1979), pp. 36-44; G. W. Giles, *Toward a More Powerful Agriculture* (Lahore: Planning Cell, Agricultural Dept., Government of Pakistan, 1967); J. R. Arbolada, *Improvement in the Kisikisan Rice Mill*, Mimeo (Manila: International Rice Research Institute, 1975).

[59] Development Techniques Ltd., *Paper Pulp Packaging Unit*, Mimeo (London: Intermediate Technology Development Group Ltd., 1975); McRobie and Carr, *Production by the Masses*, p. 9; E. F. Schumacher and G. McRobie, "Intermediate Technology in Action", paper presented to the Third Inter-Congress of the Pacific Science Association, July 18-22, 1977, Bali, Indonesia, Sub-theme 2, Appropriate Technology for Medium and Small Scale Industries (Jakarta: Indonesian Institute of Sciences, 1977), p. 28; McRobie, *Small is Possible*, p. 64; Schumacher, "Intermediate Technology", pp. 46-47; E. F. Schumacher, "We Must Make Things Smaller and Simpler", interview, *The Futurist*, 8 (1974), 281-284.

[60] Schumacher's comments in "Intermediate Technology" (p. 47) indicate that an annual output in the order of 30 million trays was required to attain high efficiency.

entering production and was forced into the dilemma of either import-
ing egg trays (for which insufficient foreign exchange was available) or
doing without.

Through its Industrial Liaison Unit, based in the British Midlands,
the Group collaborated with the Royal College of Art, the University
of Reading and a small commercial manufacturer, and developed a
mini-machine and an improved egg tray design, suitable for Zambia's
conditions. The machine was capable of operating economically at
about two per cent of the market capacity of the previously available
technology and was capable of being scaled up as required. The capital
cost of the machine was approximately two percent of the smallest unit
previously available. In contradistinction to the pervasive opinion of
economists and engineers that large scale operations were necessary to
achieve adequate levels of capital productivity, this technology
demonstrated that if the rationale of Intermediate Technology was
adopted in the design process, capital-output figures could be as high
for small scale production as for large. The technology (PPMS) is now
being manufactured commercially in Britain, and sales have been
recorded in more than 30 developing countries.[61]

The Group has also identified a range of small-scale brick and tile
manufacturing units, making it easier to match the production system to
the size of local demand for products. Experiments in Ghana, Gambia,
Egypt, South Sudan and Tanzania demonstrated that cost savings were
possible through both air-drying of bricks before firing, and through
the reduction of transport requirements due to increased local production.[62]

A series of studies by the David Livingston Institute of Overseas
Development (University of Strathclyde) on the scope for technology
choice in the Third World, confirm that the sort of achievements illustrated by the above two examples are possible in a wide range of industry areas.[63] Detailed evidence exists for leather manufacturing, iron
founding, maize milling, brewing, footwear manufacture, metal-compo-
nents manufacturing, fertilizer production, farming equipment and tex-

---

[61] See *Organization Profile: Intermediate Technology* (Rugby: Intermediate
Technology Development Group Limited, 1989), p. 16.

[62] McRobie, *Small is Possible*, pp. 40-44; cf., J. P. M. Parry, "Intermediate
Technology Building", *Appropriate Technology*, 2, 3 (1975); J. Keddie and W. Cleghorn,
Brick Manufacture in Developing Countries (Edinburgh: Scottish Academic Press, 1980),
cited by G. McRobie, *Small is Possible* (London: Jonathan Cape, 1981), p. 282.

[63] Some of the Institute's work has been reported in a series of articles in *World
Development*, 5, 9/10 (1977), 773-882.

tile production.[64] The Institute's detailed empirical work demonstrates that there is a considerable range of choice, which does not disappear even when the quality of the product is specified in a fairly rigorous fashion; and that, despite the superiority of some capital intensive techniques, the least-cost technology is sometimes nearer the labor intensive than the capital intensive end of the spectrum. The studies also show that variations in profitability tend to be smaller across technologies than variations in levels of employment and capitalization.

Designing New Technologies

In addition to making available or improving traditional technologies, and to scaling down high-cost technologies, there is often a need for the development of new technologies and new products associated with those technologies.

Two examples may be cited here of new intermediate technologies developed for agricultural usage. The first is a power tiller developed by the International Rice Research Institute in Manila, Philippines; these are produced at half the cost of imported power tillers and have consequently diffused rapidly amongst farmers - due to their capacity to enhance agricultural labor productivity at an "affordable" level of investment. The tiller is designed to maximize use of locally popular standard machine parts and to simplify the required fabrication and assembly processes. It is now manufactured in at least six Asian countries - and slightly larger, more sophisticated versions have subse-

---

[64] M. M. Huq and H. Aragaw, "Technical Choice in Developing Countries: The Case of Leather Manufacturing", *World Development*, 5, 9/10 (1977), 777-789; B. A. Bhat and C. C. Prendergast, "Some Aspects of Technology Choice in the Iron Foundry Industry", *World Development*, 5, 9/10 (1977), 791-801; S. J. Uhlig and B. A. Bhat, "Capital Goods Manufacture and the Choice of Technique in the Maize-Milling Industry",*World Development*, 5, 9/10 (1977), 803-811; J. Keddie and W. Cleghorn, "Brewing Projects in Developing Countries: Technology, Costs and Employment", *World Development*, 5, 9/10 (1977), 813-828; N. S. McBain, "Developing Country Product Choice: Footwear in Ethiopia, *World Development*, 5, 9/10 (1977), 829-838; S. J. Uhlig and N. S. McBain, "The Choice of Technique in a Batch Production Industry: Some Lessons from a Study of Bolt and Nut Manufacture", *World Development*, 5, 9/10 (1977), 839-851; R. Disney and H. Aragaw, "The Choice of Technology in the Production of Fertilizer: A Case Study of Ammonia and Urea", *World Development*, 5, 9/10 (1977), 853-866; G. J. Gill, "Bottlenecks in a Single-Cropping System in Chilalo, Ethiopia: The Acceptance and Relevance of Improved Farming Equipment, *World Development*, 5, 9/10 (1977), 867-878; J. Pickett and R. Robson, "A Note on Operating Conditions and Technology in African Textile Production", *World Development*, 5, 9/10 (1977), 879-882.

quently been released.[65]   A second technology developed by the
National College of Agricultural Engineering in Britain, is a low cost
primary cultivator designed for manufacture in the country of use and to
be easy to use and maintain.[66]   This machine, known as the "Snail"
costs approximately £100 (early 1970s) and is extremely versatile. It is
a substitute for a tractor and consists of two parts: a self-propelled
winch, powered by a small engine, and a modified ox-tool farm implement which is attached by a cable to the winch unit. It is thus a useful
alternative to draft animals, which nevertheless avoids the pitfalls
which frequently accompany the introduction of large, costly tractors
into the South.[67]

The provision of useful energy for local applications from renewable sources, such as solar, wind and biomass, has provided a focus for
the development of much new intermediate technology.[68]   Examples
may be drawn from the fields of photovoltaic electricity generation,[69]
microhydro-electricity,[70] wood-based fuels[71] and other biomass energy
production,[72] and windpower (for either water-pumping or electricity

---

[65]   A. U. Khan, "Mechanization Technology for Tropical Agriculture", in
Appropriate Technology: Problems and Promises, ed. by N. Jéquier (Paris: Organization
for Economic Cooperation and Development, 1976), pp. 213-230.

[66] *The Other Way*, film (London: British Broadcasting Commission, 1974);   T. B.
Muckle, et al., *Low Cost Primary Cultivation: A Proposed System for Developing
Countries*, Occasional Paper #1 (Silso, U.K.:   National College of Agricultural
Engineering, 1973);   cf., Carr, *Economically Appropriate Technologies*, p. 20.

[67]   For a discussion of the difficulties of adopting tractors in underdeveloped rural
areas see: G. S. Aurora and W. Morehouse, "Dilemma of Technological Choice: The Case
of the Small Tractor", *Economic and Political Weekly*, 7, 31-33 (1972), 1633-1644;   A.E.
Deutsch, "Tractor Dilemma for Developing Countries", *World Crops*, 24, 5 (1972).

[68]   National Academy of Arts and Sciences, *Energy for Rural Development:
Renewable Resources and Alternative Technologies for Developing Countries*
(Washington, D.C.: National Academy of Arts and Sciences, 1976).

[69]   C. Flavin, *Electricity from Sunlight: The Future of Photovoltaics*, Worldwatch
Paper #52 (Washington, D.C.: Worldwatch Institute, 1982), esp. pp. 16-20;   S. Hogan, *The
Future of Photovoltaic Energy Conversion in Developing Countries* (Golden, Col.: Solar
Energy Research Institute, 1980).

[70]   D. Deudney, *Rivers of Energy: The Hydropower Potential*, Worldwatch Paper
#44 (Washington, D.C.: Worldwatch Institute, 1981), esp. pp. 21-25.

[71]   N. Smith, *Wood: An Ancient Fuel with a New Future*, Worldwatch Paper #42
(Washington, D.C.: Worldwatch Institute, 1981),.

[72]   W. Palz, P. Chartier and D. Hall, eds., *Energy from Biomass* (London: Applied
Science Publishers, 1981);   C. R. Prasad, K. K. Prasad and A. K. N. Reddy, "Bio-gas
Plants: Prospects, Problems and Tasks", *Economic and Political Weekly* , special edition,
9, 32-34 (1974).

generation).[73] Some of these technologies (such as photovoltaic systems) often require further innovation to become affordable vis-a-vis Intermediate Technology, but there is now plenty of evidence of low cost "intermediate" renewable energy technologies.[74]

Other new technologies-cum-products developed under the rubric of "intermediate technology" include the use of ferro-cement for roofing, water and crop storage and the construction of boats. The I.T.D.G. has developed new low cost methods of manufacturing corrugated roofing sheets and other building components, using locally available natural fibres and other materials. These products can be manufactured in small, self-contained workshops thereby enabling new manufacturing industries in rural areas with small markets.[75]

Summary Note

In this section on practical examples a selection of cases has been extracted from a much larger range. There is a growing number of publications which catalogue actual examples of intermediate technologies.[76] A number of data bases and documentation services are now in operation to provide a continually updated source of information on actual choices vis-a-vis intermediate technology.[77] The purpose of the present study is not to examine such evidence in detail but to obtain an

---

[73] C. Flavin, *Wind Power: A Turning Point*, Worldwatch Paper #45 (Washington, D.C.: Worldwatch Institute, 1981).

[74] C. D. Ouwens, "Renewable Energy Sources for Third World Countries", in *Appropriate Technology for Developing Countries*, ed. by W. Riedijk (Delft: Delft University Press, 1984), pp. 301-318; E. U. von Weizsäcker, J. Boltersdorf and P. Miller, "New Energy Technologies Suitable for Upgrading Traditional Technologies: A Selective Annotated Bibliography", in *New Frontiers in Technology Application: Integration of Emerging and Traditional Technologies*, ed. by E. U. von Weizsäcker, M. S. Swamanithan and A. Lemma (Dublin: Tycooly, 1983), pp. 128-136.

[75] See National Academy of Sciences, *Ferro-cement: Application in Developing Countries* (Washington, D.C.: National Academy of Sciences, 1973); McRobie, *Small is Possible*, pp. 43-44.

[76] E.g.: United Nations Industrial Development Organization, *Monographs on Appropriate Industrial Technology*; I.T.D.G., *Appropriate Technology* (journal).

[77] E.g.: SATIS (an international network based in the Netherlands), VITA (consulting and information services group, Virginia, U.S.A.), VIA Appropriate Technology Microfiche Project (Palo Alto, California) etc. Recent initiatives include a program of the International Labour Organisation for the establishment of an Asian regional clearing house for information on improved technology for cottage industries (Asian Employment Program [ARTEP], *Progress Report 1988* [New Delhi: International Labour Organisation, ARTEP, 1988]).

impression of what intermediate technologies entail in practice. The following conclusions are justified:

- Intermediate Technology is not just an idea - there are many actual technologies (either in wide usage, or at an earlier stage) which exemplify the concept;
- there is ample empirical evidence to justify the view that real choice is possible between different technologies;
- it is possible for technologies which are capital-saving, small-scale and relatively simple to be economically viable;
- it is possible for an expanded range of technology choices to be made available to communities wishing to embark upon self-reliant economic development - if the rationale of Intermediate Technology is consciously adopted in the processes of designing, developing, disseminating and diffusing technology.

The feasibility of Intermediate Technology, both theoretically and as demonstrated by practical applications, was not in itself sufficient to ensure the uptake of the approach on a large scale. Schumacher believed that there were special organizational requirements linked to Intermediate Technology. The first major step he and his colleagues took was to establish a suitable organizational framework for promoting Intermediate Technology.

Intermediate Technology: Organizational Requirements

Establishment of Independent Body

The first major step for Schumacher and colleagues in their attempts to promote Intermediate Technology in a practical way was to set up the Intermediate Technology Development Group Limited. Schumacher recounts that step in these terms:[78]

> So, for a number of years I have been talking and thinking and lecturing about intermediate technology. Then comes the awful moment - with some people this never comes - when you ask yourself, or your friends say, "Are we only talkers or are we doers?" But what can one do? Talking and giving lectures is not illegitimate, *but if one wants to do something one sets up an organization.*

---

[78] Schumacher, "Intermediate Technology", p. 44 [emphasis added].

We may deduce that the need for special Intermediate Technology organizations arises from the need for networks of people capable of mounting innovative programs independently of pressures to follow "misdirected" development philosophies. The self-help and "grass roots" orientation of Intermediate Technology also emphasizes the need for direct rather than indirect action by those involved in promoting the approach.

Filling the Knowledge Gap

The cardinal objective of the I.T.D.G. was identified as *filling the knowledge gap* about technologies suitable for a self-help approach to development in the rural areas of the South.[79] This objective was based upon the observation that a large gap existed in the availability of knowledge about technology choices, particularly in the South. The Group began its experiments at mobilizing information to fill the technology gap by aiming to conduct or facilitate four main activities.

Reviewing "State of the Art" Technology. The foundation work of the I.T.D.G. was based on the view that much of the technological knowledge required for intermediate technologies was already in existence somewhere. Thus, conducting systematic surveys of intermediate technologies which were either in use somewhere but not widely known, or which had been widely employed in the past but were no longer used, was seen as a valuable information resource - to be assembled in the form of catalogues, bibliographies, drawings, design specifications, photographs, industrial profiles and other materials. The first tangible program of the I.T.D.G. was the production of a catalogue of this type, which became much sought-after internationally.[80] It became the first of a series of technical catalogues.[81] The need for regular updates on intermediate technologies was eventually met by the Group's quarterly journal *Appropriate Technology*.[82] This need is

---

[79] This discussion is based inter alia on: E. F. Schumacher, "The Work of the Intermediate Technology Development Group in Africa", *International Labour Review*, 106, 1 (1972); Schumacher and McRobie, "Intermediate Technology in Action", pp. 1-3; D. W. J. Miles, *Appropriate Technology for Rural Development: The ITDG Experience*, ITDG Occasional Paper #2 (London: Intermediate Technology Publications Ltd., 1982).

[80] Intermediate Technology Development Group, *Tools for Progress: A Guide to Small-scale Equipment for Rural Development* (London: Intermediate Technology Development Group, 1967).

[81] Intermediate Technology Development Group, *Tools for Agriculture* (London: Intermediate Technology Development Publications, Ltd. 1976).

[82] London: Intermediate Technology Publications Ltd.

also now addressed by a number of international intermediate technology data services (mentioned above) which have evolved from the initial stimulus by the I.T.D.G.

Research and Development. A reason for conducting surveys of currently existing technical knowledge was that actual gaps in the spectrum of technological knowledge could be more readily identified. Thus, when the parameters of needs in target communities or development districts were known, it was possible to target research and development activities to fill the technological gaps in the most productive way. The I.T.D.G. approached this function primarily by mobilizing or collaborating with established research and development institutions.

Communication. Given the I.T.D.G.'s focus on knowledge, one of the group's main activities became the communication of the information it accumulated through its review procedures and efforts at facilitating the development of new technical knowledge. In addition to the use of publications, the Group pursued its communication function in a number of other ways involving participation in development programs and cooperation with partner groups in the South. People themselves were viewed as very important means for the effective dissemination of Intermediate Technology knowledge.

Technical Consulting. The Group's officers, most with technical competence and experience, took an active role in practical field demonstrations, testing, and technical training in various intermediate technologies and their associated management skills. This practical field approach has been an essential part of the information system of the I.T.D.G., and of crucial importance for ensuring that intermediate technologies introduced to meet a need in a particular place really were suitable for those unique circumstances. These factors are illustrated by the principle adopted very early by the Group in its methodology for introducing new agricultural technologies:[83]

> ... innovation must be based on detailed knowledge of the farmer's problems and needs, and on an accurate picture of the farming calendar and cropping sequence in a locality, taking into account communal customs and obligations and the economic pattern on which the present farming system is based.

While Intermediate Technology is essentially an economic concept, and stems from supra-technical human concerns, it nevertheless directly

---

[83] Schumacher, "Work of ITDG in Africa", 84.

invokes technical considerations. Technical problem solving is an essential activity at the heart of Intermediate Technology.

Evolving Structure for Mobilizing Knowledge

Initially the I.T.D.G. worked through a series of panels comprised of professionals and experts in the relevant fields, who served voluntarily. The panels were found to work most effectively when made up of a balanced mixture of people from industry, the professions, academia, administration (including government) and experienced practitioners, and when served by the employment of full-time staff. The panels covered such fields as: agriculture; building and building materials; chemistry and chemical engineering; co-operatives; ferrocement; fertilizers; forestry and forest products; homestead technology; nutrition; power (wind, solar, methane and hydro-electric); rural health; low-cost printing techniques; transportation and simple vehicles; and, water. An Industry Liaison Unit and a University Liaison Unit was also operated.

Over time the I.T.D.G. has changed its structure and has devoted much time to institutional challenges.[84]

The panel system was eventually found to be less attractive than initially thought, and the Group's work is now divided into three main divisions: operations activities, dissemination activities, and institution building activities. The operations activities are subdivided into the four general fields of: agriculture and fisheries; agro-processing; mineral industries and shelter; and, manufacturing. The dissemination activities are subdivided into: establishing permanent representatives (preferably local people) in the main countries of the Group's work; training programs; a technical enquiry unit; an information office; a development education program; a publications company; a consulting company; and a commercial technology licensing company. The institution building activities involve: advising governments and international agencies on the institutional aspects of Intermediate Technology; assisting in the establishment of various "development" institutions and Appropriate technology organizations both in the South and in the U.K.; and, establishing various companies to deal with specialized aspects of the Group's work (such as in energy technology or transport technology).

---

[84] See *Organization. Profile: Intermediate Technology* (Rugby: Intermediate Technology Development Group Limited, 1989).

The evolution of the Intermediate Technology Development Group has reflected the basic tenet embodied in the concept of Appropriate Technology as expounded in this book: that particular technologies form part of technology-practice, and therefore that the development of new technologies or the improvement of old technologies must involve much more than just technical-empirical factors.

Review

We may conclude this section by making the following observations about the principles embodied in the I.T.D.G.'s organizational response to implementing Intermediate Technology.

The first principle is that structures and forms of assistance are required in Third World development programs which free poor rural communities (and poor urban communities) from dependency upon aid from wealthier communities. For this to occur, means of production need to become more available which make *self-reliance* and *self help economic activity* possible. It follows that organizations established for the promotion of Intermediate Technology ought not to encourage, either deliberately or by default, a self-reinforcing relationship of dependency upon them by the poor communities for whose advancement such organizations are established.

The second principle may be derived from the first: *the most suitable form of assistance is knowledge.* If the major focus of Intermediate Technology aid programmes becomes the transfer of "off the shelf" intermediate technologies from the technologically strong communities to the technologically weak, then most of the experience and development of "intermediate" technological ingenuity will accrue to the donor community. Because the integrated regional development advocated by Schumacher centers around the development of people, rather than the development of things, it follows that the transfer of technological expertise - *of the appropriate kind* - is an important ingredient for the evolution of a sustainable economic development pathway (in accordance with the principles of Intermediate Technology). Thus the prime task for Intermediate Technology organizations (if we take the approach of Schumacher and colleagues as a bench mark) is to transfer knowledge on self-help technologies in such a manner as to encourage the capacity in the receivers of such knowledge to conduct technological innovation of their own, in conjunction with local institutions.

The third principle is that the *countries of the North do have an important role to play in the development of Intermediate technology in the South.* At first glance it might appear that the failure of past development aid from the North, combined with the need for self-reliance in the South, points to no useful role for the North - apart from careful planning to avoid harmful economic impacts on the South as a consequence of the North's activities. The I.T.D.G.'s experience demonstrates, however, according to Schumacher, that for Intermediate Technology work to be fruitful "a considerable part of it must of necessity be done in the developed countries".[85] Because knowledge of technological options would appear to be a prerequisite to effective technological choice, it may be argued that those countries which already possess the experience and resources to store, process and mobilize a diverse range of technical information, are in the best position to provide the initial stimulus to awaken an effective demand for intermediate technologies within target communities. Such a level of technical information handling capacity is by no means an exclusive attribute of the North; because the majority of the world's systematic science and technology research and development occurs in the North, however, it would follow that the North has a pivotal role to play. It is nevertheless important that this be done in cooperation with scientific and technological institutions in the South, so as to facilitate the development of indigenous technological capacity.[86]

Having surveyed and partly analyzed its origins in the development of Intermediate Technology and the work of Schumacher and colleagues we may now proceed to survey the broader Appropriate Technology movement.

International Response to Intermediate Technology

Following the pioneering work of Schumacher and colleagues the Appropriate Technology movement became firmly established. By the early 1980s, as indicated earlier, over one thousand organizations throughout the world were involved in work directly related to

---

85 *Ibid.,* p. 79.

86 Interesting examples of where the I.T.D.G. has cooperated with science and technology institutions in the South for the purpose of developing indigenous Intermediate Technology programs, may be found, for example, in the cases of Kumasi University of Science and Technology in Ghana (S. Holtermann, *Intermediate technology in Ghana: The Experience of Kumasi University's Technology Consultancy Centre* [Rugby: Intermediate Technology Industrial Services, 1979]) and the Birla Institute of Technology in India (M. Carr, *Developing Small-scale Industries in India: An Integrated Approach* [London: Intermediate Technology Publications Ltd., 1981]).

Appropriate Technology. While by no means the only cause for the immense growth of the movement, the activities of Schumacher and colleagues have been the most prominent stimuli for the widespread debate and polemic which has ensued under the rubric of "Appropriate Technology". Internationally, Intermediate Technology came to symbolize the possibility that people, no matter how powerful or wealthy, could actually do something to help eradicate underdevelopment and to control the direction of technological change. This gave the concept considerable emotive power. Intermediate Technology, though proposed by Schumacher as a very specific solution to a very specific set of problems, stemmed from and in turn reinforced a more general concept - Appropriate Technology.

Having established the notion of Intermediate Technology, Schumacher increasingly concentrated his efforts on articulating the broader rationale of Appropriate Technology. Given his concern and active involvement with Intermediate Technology, and given that work on its development and dissemination was the main tangible and internationally identifiable example of the deliberate implementation of its parent notion, Intermediate Technology was generally held up as the exemplar of Appropriate Technology. Quite reasonably, a large number of Intermediate Technology artefacts and programs came to be known as "appropriate technology", or as "appropriate technologies". In due course "Intermediate Technology" and "Appropriate Technology" became interchangeable in popular discussion and also in professional or academic literature. In addition to the interchanging of terms the conceptual distinction between Intermediate Technology and Appropriate Technology eventually became blurred in published material. This is reflected in the differences between the general-principles and specific-characteristics approaches to defining Appropriate Technology discussed in the Chapter Two.

These inconsistencies of language and ideas did not, on the surface level at least, do much damage to the growth of interest in Appropriate Technology. The unity of rationale between Intermediate Technology and Appropriate Technology means that most people active in the movement have been able to concur on certain fundamental ideas and strategies despite differences in perspective stemming from either a North, South or global orientation. For example, an international workshop on Appropriate Technology attended by representatives from 24 different Appropriate Technology organizations, during September 1979, showed unanimity in its preference for the term "appropriate technology".[87] Despite the fact that many participants in effect presented papers on Intermediate Technology, the general conclusions of these papers bore remarkable resemblance to the conclusions of those which addressed problems of the wealthier industrialized nations. Nevertheless, semantic ambiguities undergird much of the actual debate which has occurred in the above and other forums on Appropriate Technology.

Despite the ability of people involved in the Appropriate Technology movement to concur on some fundamental aspects of the movement's concerns, the semantic and conceptual ambiguity mentioned above has led to a wasteful diversion of energy into the repeated resolution of conflicting viewpoints which, upon closer examination, may have been no more than different formulations of essentially similar ideas.[88] Much of the voluminous literature on Appropriate Technology reflects the fact that the movement has been concerned principally with practical action (including the promotion of its ideas) rather than with the rigorous implementation of detailed abstract theory. This disposition has meant that the movement has had to formulate much of its theory "on the run". Consequently, in addition to the quasi-debate just mentioned, the literature in the decade following the mid-1970s reveals genuine differences of perspective which may not have been immediately apparent. This is not surprising for at least two reasons.

The first reason is the inordinate influence exerted by Schumacher on the movement, combined with the fact that Schumacher's own writings do not fully articulate the theory which underlay his action and polemic. While he was an erudite man with considerable academic experience, Schumacher himself stressed that he was primarily a practitioner rather than a scholar.[89] Referring to *Small is Beautiful* he states:[90]

---

[87] The workshop was conducted at the Centre for Appropriate Technology, Delft University of Technology, 4th-7th September, 1979; proceedings of the workshop, edited by J. de Schutter and G. Bemer, are published as *Fundamental Aspects of Appropriate Technology* (Delft: Delft University Press and Sijthoff and Noordhoff International Publishers, 1980).

[88] A similar observation has been made by H. Brooks ("A Critique of the Concept of Appropriate Technology", in *Appropriate Technology and Social Values: A Critical Appraisal*, ed. by F. A. Long and A. Oleson [Cambridge, Mass.: Ballinger, 1980], pp. 53-78) who notes that exponents of different points of view within the "AT" movement "often talk past one another and do not address the same issues" (p. 53).

[89] See, e.g., "Alternative Technology: Gordon Laing, Dorothy Emmet and Anthony Appiah talk to Fritz Schumacher, Founder of the Intermediate Technology Development Group", transcript of interview, *Theoria to Theory*, 9, 1 (1975), 5-22.

[90] *Ibid.*, p. 20.

If the book was just speculation it would be worthless. Every chapter
has grown out of practical activity. It was written as a cry of agony.

He often interpreted pedantic questions as veiled excuses by theo-
reticians for inaction, and frequently refused to engage in theoretical
debate with such people.[91] In an attempt to mobilize popular support
for the work of Appropriate Technology Schumacher would often speak
directly of activities in the areas of building construction for the poor,
agricultural equipment or rural energy supplies, for example, rather
than spend time dealing with what he viewed as tangential arguments.[92] Consequently many proponents of Appropriate Technology
who considered themselves supporters of Schumacher's ideas could
easily differ with each other on the detailed make-up of those ideas.
This was compounded by there being no single publication of
Schumacher's which fulfils the role of a *systematic*, comprehensive
and authoritative statement of his theory, and by the fact that
Schumacher died in 1977 at about the time when debate over his ideas
was reaching its peak (thereby preventing him from ever producing
such an authoritative publication).

The second reason is that the Appropriate Technology movement
has attracted the support of people from a wide array of backgrounds,
disciplines, political interests and philosophical views.[93] It is not
surprising that unanimity of theory has not been achieved despite the
fact that the movement has achieved a sufficient degree of consensus to
qualify as a social movement.[94] This tension between consensus and

---

[91] This gained a reputation for Schumacher, at some international conferences, as an
arrogant man (cf., R. Williams, An Interview with E. F. Schumacher, *The Science Show* ,
Radio 2, [Sydney: Australian Broadcasting Commission, 1977]).

[92] See, "Alternative Technology", *Theoria to theory*. Note: In private
conversations with people seriously concerned with the practical work of Appropriate
Technology, however, Schumacher was more prepared to discuss freely his broader
theoretical foundations, which included an extensive knowledge of Western philosophy
and political-economy (Personal communication with George McRobie, November 1982);
a study by the current author of over 80 of Schumacher's publications provides evidence of
a systematic theoretical framework behind most of the popular ideas for which
Schumacher is widely known. (K. W. Willoughby, *Technology and Traditional Wisdom:
A Study of the Thought and Work of Ernst Friedrich Schumacher*, unpublished Honours
thesis, Murdoch University, Perth, Western Australia, 1979).

[93] This is noted by C. Baron in "Appropriate Technology Comes of Age: A Review of
Some Recent Literature and Aid Policy Statements", *International Labour Review*, 117, 5
(1978), pp. 625-634.

[94] The consensus of views in the Appropriate Technology movement is clear enough
to be articulated in a manageable way, as evidenced for example by the chart provided

diversity is typical of social movements, as opposed to dogmatic or tightly proscribed academic schools of thought, and need not be seen as reflecting fundamental flaws in the theory behind the popular statements of the leaders of social movements.[95]

In addition to actual conflicting ideas which do occur within the Appropriate Technology movement, there is a considerable diversity of perspectives which derive from the intrinsic nature of the Appropriate Technology concept - a diversity which has nothing to do with theoretical conflict.[96] Appropriate Technology implies that the appropriateness of technology will vary with the diversity of psychosocial and biophysical contexts. The confluence of a huge variety of people dealing with different problems in different contexts, in different parts of the world, inevitably leads to a diversity of concerns and perspectives.

The literature of the Appropriate Technology movement is in large part a response to the growth of activities and publications which ensued under the rubric of "intermediate technology" during the mid 1960s. It is also an expression of a much broader and diffuse set of concerns which had little or nothing to do with Schumacher and colleagues. The next three chapters will provide a précis of these different concerns, from the contexts of both the North and the South, along with a list of some of the antecedents and complementary schools of thought to the Appropriate Technology movement.

In Part Three a systematic theoretical statement of Appropriate Technology will be developed. Articulating a general theory of Appropriate Technology is made very difficult, however, by the fact that the movement comprises a heterogeneous set of activities and ideas. This difficulty is compounded by the inconsistency in the terminology used by the movement's protagonists and observers, which creates obstacles to precise comparisons of the movement's different streams. The movement is still evolving, furthermore, with new themes emerging in company with those already established. For the time being, therefore, the diversity of streams and foci which comprise Appropriate Technology will be described.

---

by D. G. Lodwick and D. E. Morrison in "Research Issues in Appropriate Technology", paper presented to the Rural Sociological Society, Cornell University, Ithaca, New York, 20-23 August, 1980 (Michigan Agricultural Experiment Station Journal Article No. 9649).

[95] Cf., L. Killian, "Social Movements", in *Handbook of Modern Sociology*, ed. by R. E. L. Faris (New York: William Morrow and Co., 1964), pp. 426-455.

[96] This is illustrated in the compendium of readings compiled and edited by M. Carr (*The AT Reader: Theory and Practice in Appropriate Technology* [London: Intermediate Technology Publications, Ltd., 1985]).

The subdivisions into which Chapters Six and Seven are categorized represent actual foci which have emerged in the Appropriate Technology movement, but they are not all conceptually equivalent. Some represent actual sub-movements within the whole movement while others represent environmental pressures, socio-political forces or special themes and activities which are found across various sub-movements within the whole movement. The subdivisions should therefore be treated simply as convenient categories for analysis of a social movement which is otherwise quite fluid and difficult to categorize.

6

Appropriate Technology
in the South

Intermediate Technology was formulated as a specific response to the problems of the South, and many of the problems of technological dependency, technology transfer, technology choice, technological underdevelopment and technological non-competitiveness etc., as discussed in the burgeoning technology studies literature, are most severe in countries of the South. This has meant that the bulk of the Appropriate Technology literature is predominantly concerned with the South - a fact which appears neither surprising nor unreasonable given the close connections between Appropriate Technology and Intermediate Technology. Because of the sheer volume of relevant publications, it is beyond the scope of this study to comment on all the literature in detail. Only the main features of Appropriate Technology in the South will therefore be covered, with particular cases being discussed for illustration where useful.[1]

Small-scale Industries

A considerable amount of Appropriate Technology work has appeared within the framework of developing small-scale industries. This is in keeping with Schumacher's emphasis on the need for non-agricultural industries in rural areas in the South as a means of building

---

[1] For a sample of both the general literature on the technology-related problems of the South, and of the Appropriate Technology literature in particular, see the appendix at the end of this chapter.

up the diversity and vigor of economic life in depressed rural communities, for the longer term goal of breaking the vicious circle of underdevelopment.[2]

The reputed advantages of small-scale industry include: the ability to operate when only a small market is available; the low level of total capitalization required to establish an industry in a local area; the increased stability and security which a local economy may experience as a result of the increased diversity which is made possible by reliance upon a series of small industries rather than one or two large industries; the reduced size of organizational and administrative obstacles to business activity; and the enhanced opportunities for participation in industry by local entrepreneurs and investors.[3] Small-scale industries are also frequently claimed to have greater scope than large-scale industries for labor intensive forms of employment.[4] In summary, because rural communities in the hinterland of the South are by nature small in scale, and are normally excluded from the advantages of ready access to large urban markets for non-agricultural produce, it follows that small-scale industries lend themselves more closely to the resource levels and environments of rural communities.[5]

---

[2] See, M. Carr, et al., "Technology for Development: Manufacturing, Mining and Recycling", in *The AT Reader: Theory and Practice in Appropriate Technology*, ed. by M. Carr (London: Intermediate Technology Publications, Ltd., 1985), pp. 287-318.

[3] The general references on the role of small-scale industries and technology in the South's economic development include: H. W. Singer, *International Development: Growth and Change* (New York: McGraw Hill, 1964); E. Staley and R. Morse, *Modern Small Industry for Developing Countries* (New York: McGraw Hill, 1965); Organization for Economic Cooperation and Development, *Transfer of Technology for Small Industries* (Paris: Organization for Economic Cooperation and Development, 1974).

[4] While small-scale industries are not necessarily the most labour intensive, a series of studies has demonstrated that in general they do tend to be labour intensive, e.g.: P. N. Dhar and H. F. Lydall, *The Role of Small Enterprises in Indian Economic Growth*, (New York: Asia Publishing House, 1961); B. V. Mehta, "Size and Capital Intensity in Indian Industry", *Bulletin of the Oxford University Institute of Economics and Statistics*, 31, 3 (1969), pp. 189-204; K. A. Di Tullio, *The Role of Small Industries in the Political Economy of Pakistan*, unpublished doctoral dissertation, Syracuse University, 1972; R. K. Vepa, *Small Industry in the Seventies* (London: Vikas, 1971); S. Watanabe, "Reflections on Current Policies for Promoting Small Enterprises and Subcontracting", *International Labour Review*, 110, 5 (1974), 405-422; L. J. White, "Appropriate Factor Proportions for Manufacturing in Less Developed Countries: A Survey of the Evidence", in *Appropriate Technologies for Third World Development*, ed. by A. Robinson (London: MacMillan, 1979), pp. 300-341.

[5] Cf.: D. H. Perkins, ed., *Rural Small-scale Industry in the People's Republic of China* (Berkeley: University of California Press, 1977); D. H. Perkins, "China's Experience with Rural Small-scale Industry", in *Appropriate Technology and Social Values: A Critical Appraisal*, ed. by F. A. Long and A. Oleson, (Cambridge, Mass.: Ballinger, 1980); M.Carr, *Appropriate Technology for Rural Industrialization*,

Small industry may be seen as of relevance to urban areas as well as rural areas.[6] Given that a significant proportion of the poor in urban areas are bypassed by the economic growth which does occur in urban areas, and given that most governments in the South do not have enough resources to adequately compensate the poor through massive welfare programs, it follows that the urban poor and their communities may be faced with the need for *self-generated* business activities. It would appear that small-industries are far more within the grasp of the economically non-dominant population in the South than are large ones. This is not to deny the potentially legitimate role of some large-scale industries, which may often operate in a symbiotic fashion with small industries. Small industries fit more easily, however, with self-help and "bottom-up" approaches to employment generation and wealth creation.[7]

There is an important connection between technology and the scale of industry.[8] Unless technology can be made available which enables a small-scale industry to become efficient and competitive, that industry will have little chance of becoming firmly established. If economically efficient industrial activity is only possible on the basis of large-scale production systems, then hope for widespread employment generation through small industry development is vacuous. On the conviction that an expanded role for small-scale industries in the South would enhance prospects for equitable and employment-expanding economic development, many groups have evolved programs to produce,

Intermediate Technology Development Group Occasional Paper #1 (London: Intermediate Technology Publications, 1981).

[6] Three sources which illustrate this are: J. W. Powell, "University Involvement in Appropriate Technology", *Appropriate Technology*, 6, 4 (1980), 12-14; D. Murphy, "Intermediate Technology and Industry", *Appropriate Technology*, 7, 1 (1980), 27-28; M. Davies, "IT Suburban Style in Africa", *Appropriate Technology*, 3, 1 (9176), 22-24.

[7] Cf., R. K. Diwan and D. Livingston, *Alternative Development Strategies and Appropriate Technology: Science Policy for an Equitable World Order* (New York: Pergamon, 1979), esp. pp. 71-120.

[8] These matters have been addressed in: ECAFE, *Appropriate Technology for Small Manufacturing Plants* (Bangkok: ECAFE, 1969); A. S. Bhalla, "Small Industry, Technology Transfer and Labour Absorption", in *Transfer of Technology for Small Industries*, ed. by Organization for Economic Cooperation and Development (Paris: Organization for Economic Cooperation and Development, 1974), pp. 107-120; C. Riskin, "Intermediate Technology in China's Rural Industries", in *Appropriate Technologies for Third World Development*, ed. by A. Robinson (London: MacMillan, 1979), pp. 52-74; C. Norman, "Technology for the Small Producer", in *The God that Limps: Science and Technology in the Eighties*, by C. Norman (New York: W.W. Norton, 1981), pp. 154-163; Stewart, *Technology and Underdevelopment*.

adapt or transfer technology suitable for small-scale industries.[9] Intermediate Technology is normally viewed as both scaled for small industry usage and as, in turn, most practicable at the level of small industries.

The concept of Intermediate Technology in Schumacher's terms, refers principally to an intermediate level of capital expenditure *per workplace*. However, the concept is also relevant to the *total* level of fixed capital required to establish an enterprise or industry. It may well be that the minimum capital-cost-per-workplace for a given industry may only be attainable at a scale of production beyond what is possible for a given local community. Consequently a criterion of affordable cost-per-enterprise would appear to be a necessary adjunct to that of affordable cost-per-job, for the purpose of industry development in local regions or by poorer people.

Empirical evidence indicates that, with judicious choice of technology, small industry may be equally or more competitive than large industry.[10] Suitable technology alone is no guarantee , however, of the effectiveness of small-scale industries; business management skills, technology-use skills, resource supplies, regional and international economic climate, degree of government involvement, degree of monopolization of industry areas, market volatility, or local cultural patterns - amongst other variables - play crucial roles in determining the success of small industrial enterprises.[11] When combined with good economic

[9] A. K. Basy, "SIRTDO: The Experiences of a Technology Institute in Appropriate Technology", *Appropriate Technology*, 6, 2 (1979), 15-17; B. Behari, "Technological Adaptation in Small-scale Industries: An Indian Experience", *Options Méditerranéenes*, 27 (1975), 81-85; W. M. Floor, "Activities of the U.N. System on Appropriate Technology", and A. K. N. Reddy, "National and Regional Technology Groups and Institutions: An Assessment" (both of these articles are published in *Towards Global Action for Appropriate Technbology*, ed. by A. S. Bhalla [Oxford: Pergamon, 1979], pp. 138-163 and pp. 63-137, respectively).

[10] See, "Part II, Empirical Case Studies", in *Technology and Employment in Industry*, ed. by A. S. Bhalla (Geneva: International Labour Office, 1975), pp. 85-324; cf., chapters 8-10 in Stewart, *Technology and Underdevelopment*, (Stewart concludes [pp. 196-197]: "Most micro case studies of choice of technique have concluded that, given certain assumptions and within certain limits, there is a technically efficient range of techniques of differing labour and investment intensity, in the industries studied, thus rejecting the technological determinist view and providing support for the neo-classical assumption ... The industries for which this conclusion has been reached include salt and sugar production, rice and maize milling, textiles (spinning and weaving) and carpet weaving, brewing and gari processing, the production of metal cans, and metal working industries, shoe production, road building and cement block manufacture").

[11] Cf.: P Bourrières, "Adaptation of Technologies to Available Resources", in *Appropriate Technologies for Third World Development*, ed. by A. Robinson (London: MacMillan, 1979), pp. 1-11; N. Jéquier, "Building-Up New Industries", in *Appropriate Technology: Problems and Promises*, ed. by N. Jéquier (Paris: Organization for Economic

planning and management, however, and when especially chosen for their suitability for specific small industries, the positive value of intermediate level industrial technologies has been demonstrated.

The development of small-scale industries in the South through the use of Intermediate Technology has been most successful when an *integrated* approach has been adopted, involving both technological considerations and non-technological considerations.[12] Particular success has been achieved in regions lacking in economic infrastructure and where entrepreneurship has not traditionally flourished, when innovative public or community institutions have been established with a specific charter to facilitate the development of small industries using the integrated approach.[13] The most successful of these, such as the Birla Institute of Technology in Ranchi, India, involve cooperation between tertiary research institutions, community organizations, financial institutions (e.g. public or private banks) and government departments, and have been maintained by the vigorous support of a core group of people imbued with both professional skills and a personal commitment to the work.[14]

Basic Needs and Alternative Development Strategies

During the 1950s and 1960s an array of ambitious economic development programs were promulgated by national governments and international agencies under the rubrics of the "five-year plan" and the "development decade", based upon the large-scale injection of western

Cooperation and Development, 1976), pp. 85-97; M. Harper and T. Phiam Soon, *Small Enterprises in Developing Countries: Case Studies and Conclusions*, (London: Intermediate Technology Publications Ltd., 1979); V. K. Chebbi, "Management of Appropriate Technology", in *Conceptual and Policy Framework for Appropriate Industrial Technology*, Monographs in Appropriate Industrial Technology #1, ed. by United Nations Industrial Development Organization (New York: United Nations, 1979), pp. 122-133.

[12] D. H. Frost, "Appropriate Industrial Technology: An Integrated Approach", in *Conceptual and Policy Framework for Appropriate Industrial Technology*, Monographs in Appropriate Industrial Technology #1, ed. by United Nations Industrial Development Organization (New York: United Nations, 1979), pp. 72-87.

[13] The Botswana Technology Centre (Botswana), the Technology Consultancy Centre, (Ghana) and the Appropriate Technology Development Association (India) are three examples surveyed by A. Sinclair, as part of a larger survey of 36 such bodies, in *A Guide to Appropriate Technology Institutions* (London: Intermediate Technology Publications, 1984)

[14] M. Carr, *Developing Small-Scale Industries in India: An Integrated Approach* (London: Intermediate Technology Publications, 1981); this publication is based upon a 1980 study of the small industry scheme of the Birla Institute of Technology.

capital and technology into the nations of the South. The pronouncements of officials responsible for such programs were extremely sanguine. The following quote from a 1962 speech by a United Nations official is typical:[15]

> This deliberate effort to improve the economic and social conditions in the less-developed countries of the world evidences a significant step in the history of human relations. There have been economic successes. Few countries or areas have not benefitted in some measure by some aspect of this peculiar phenomenon of our century. ... The development of the emerging countries will not follow the pattern of the older industrial societies. The international resources of capital and technical assistance provide a new element in the process; more important is the availability of new technologies for accelerating growth. New methods permit these countries to leapfrog certain periods and programs that were necessary but costly in time and resources in the development of the industrial countries.

After a couple of decades of experimenting with development strategies which emphasize the growth of gross national product *per se*, based upon the expansion of capital-intensive industries and foreign technology, however, we find the Director-General of the United Nations' International Labour Office declaring of the South:[16]

> ... the industrialization process [predominantly followed in the decades leading up to the mid 1970s] contributes little to the alleviation of poverty, whether in urban or rural areas. Generally, the benefits of modern sector growth accrue to the modern sector alone, and in many developing countries industrialization remains largely irrelevant to the really pressing problems of our time.

He goes on to argue for redefining the place of industrialization and industrial technology in the development process, and for carefully examining alternative policies for technology, not only in industry but also in agriculture, construction and rural development. The above comments represent a growing realization amongst professionals in eco-

---

15 A. Goldschmidt, "Technology in Emerging Countries", paper presented to the Encyclopædia Britannica Conference on the Technological Order, March 1962, Santa Barbara, California (*The Technological Order*, ed. by C. F. Stover [Detroit: Wayne State University Press, 1963], pp. 197, 204).

16 F. Blanchard, "Foreword", *Technologies for Basic Needs*, by H. Singer, (Geneva: International Labour Office, 1977), p. v.

nomic development planning that, despite some identifiable enclaves of success during the last few decades, the basic needs of the majority of the poor in the South are still not being addressed.

This realization has led a number of organizations and commentators to speak of the need for moving away from reliance upon conventional development strategies, towards an integrated socioeconomic and political strategy which places special weight on the fulfillment of material and non-material basic needs in the shortest possible time span - and which is specifically oriented towards the elimination of poverty and unemployment. This is generally labelled as the *basic needs* strategy.[17] The basic needs strategy rejects the notion that the putative benefits to future generations from the present accumulation of capital, as part of either private industry or socialist reconstruction, may justly be given higher importance than the provision of benefits to currently impoverished people. It aims at leading to the satisfaction of the basic needs of future generations via the pathway of finding means to satisfy the basic needs of the current generation.

Sometimes referred to as a *basic human needs* strategy, this approach seeks to provide: basic consumer goods (e.g. food, clothing, housing, etc.); universal access to basic services (primary and adult education, clean water, health, adequate and sound habitat, communications, etc.); the right to productive employment (including self-employment); an infrastructure capable of producing goods and services required, generating a surplus to basic communal services, and providing investment capacity to improve productive forces; and, mass participation in decision-making and implementation of projects.[18]

Given the important role of technology in development efforts, some authorities have advocated the promotion of technologies especially suited to the basic needs strategies.[19] The practical characteristics of technologies for basic needs, as promulgated by these authorities, are similar to those proposed by Schumacher for Intermediate Technology. In addition to the general requirement of

---

[17] The notion of a "basic needs strategy" was first publicly articulated at the 1976 World Employment Conference in Geneva, by the International Labour Office; see, International Labour Office, *Employment, Growth and Basic Needs: A One-World Problem* (Geneva: International Labour Office, 1977).

[18] This list is based on the comments of R. H. Green in the *I.D.S. Bulletin*, 1978 (reprinted in *The AT Reader: Theory and Practice in Appropriate Technology*, ed. by M. Carr [London: Intermediate Technology Publications Ltd., 1985], pp. 33-36).

[19] H. Singer, *Technologies for Basic Needs* (Geneva: International Labour Office, 1977); A. S. Bhalla, "Technologies for Basic Needs", in *Towards Global Action for Appropriate Technology*, ed. by A. S. Bhalla (Oxford: Pergamon, 1979), pp. 23-61.

Appropriate Technology that technology be tailored to fit the circumstances of a particular region or place, the basic needs approach stresses that technologies ought to be tailored to underpin the *redistribution of incomes* towards the poor in the South. Functionally the most important characteristics of technologies appropriate to the basic needs strategy are that they build up technological self-reliance in regions and sub-regions in the South and that they assist in better integrating sectors of the economy in a region so that the productivity and technology gaps between different sectors are reduced.

Singer argues that if access to scarce capital or other scarce resources (e.g., foreign exchange or planning) is liberally given on a privileged basis to a certain sector or group of projects, these sectors or groups of projects will form a modern or "formal" sector which will be counterbalanced by a capital-short or "informal" sector.[20] As part of the basic needs strategy Singer claims that it is necessary for governments to intervene in the economy through fiscal or other policies to prevent a non-integrated or dualistic structure to the economy from growing, with inherent inequalities of income distribution and a likelihood that those outside the modern/formal sector will fail to satisfy their basic needs. He claims that reducing capital intensity of technologies across the economy is a key to the basic needs strategy:[21]

> If a relatively labor-intensive technology can be chosen for all sectors and all projects, this will make for a more unified or integrated economy, with a fair degree of equality of labor productivity in different sectors creating a favorable situation for a reasonable equality of income distribution.

Other technology-related aspects of the basic needs strategy, outlined by Singer, are reductions in the scale of technology, "unpacking" the production process into components to identify greater scope for technology choice, varying products to enable more technology choice, simplifying production processes, and "labor-addition" (by multiple-shift working of equipment, or by retaining people to maintain old equipment rather than installing new equipment which will require less maintenance).[22]

Closely related to the notion of a basic needs strategy is that of the *alternative development strategy*. Some commentators who prefer the

---

[20] Singer, *Technologies.*

[21] *Ibid.*, p.1.

[22] *Ibid.*, Chapters 2 - 6, passim.

latter criticize proponents of the former on the grounds that it is not re-ally possible to adopt a basic needs strategy while maintaining a clear commitment to economic growth (as does the International Labour Office) and that the basic needs strategy may not be viable unless it is part of a broader political movement on the national and international level.[23] The first criticism is not backed-up by substantive evidence, whereas there is substantive evidence from elsewhere that economic equity and economic growth in the South may be compatible.[24] The second criticism is contestable as it appears to be based upon a misrepresentation of the views of advocates of the basic needs strategy; Singer, Bhalla and others also refer to the need for broader political and insti-tutional reforms on both a national and an international level.[25]

Village Technology

Another emphasis within the Appropriate Technology movement is on the provision of technical services tailored specifically to fit the conditions and needs of rural villages in the South.[26] Despite the ex-odus of the South's rural population to the city, the villages and small towns still contain over three quarters of the population - up to 80 per-cent or 90 percent.[27] It would therefore seem counterproductive for technology development programs ostensibly aimed at improving the economic prospects of the poor to ignore the distinctive dynamics, re-sources and problems of village life.[28] There is a strong tradition, in India, for example, of scholarship and political-cum-organizational efforts in support of villagers themselves bringing about the develop-

---

[23] Diwan and Livingston, *Alternative Development Strategies*, pp. 83-85.

[24] International Labour Office, *Employment, Incomes and Equality: A Strategy for Increasing Productive Employment In Kenya* (Geneva: International Labour Office, 1972).

[25] E.g., Bhalla, *Technologies for Basic Needs*, pp. 57-61.

[26] Schumacher himself also stressed the importance of the approach in his essay "Two Million Villages", first published in *Britain and the World in the Seventies*, ed. by G. Cunningham (London: Weidenfeld and Nicolson, 1970) and subsequently published in Schumacher's *Small is Beautiful*, pp. 178-191.

[27] Rural Communications Service, *Village Technology Handbook* (Geneva: Lutheran World Service, 1977), introduction, sheet 1.

[28] Cf., A. W. Browne, "Appropriate Technology and The Dynamics of Village Industry: A Case Study of Pottery in Ghana", *Transactions of the Institute of British Geographers*, 6, 3 (1981), 279-292.

ment of their own local economies.[29]   There is also considerable evidence that this is possible. In the late nineteen seventies, for example, Indian village industries established within this tradition employed (directly) about ten million people in handloom weaving, 1.34 million potters, 230 000 *ghani* oil workers, 80 000 leather processors, and 385 000 cereal processors, with this sector continually expanding.[30]

The potential for economically viable employment generation on a massive scale through village industries will be extremely limited, however, unless villagers may gain access to technologies which are both appropriate to village conditions and yet which enable village production to survive in the competitive climate created by modernized industries elsewhere. Serious attempts are now being made to mobilize science to assist in this regard, although progress is limited at this stage.[31]   Nevertheless several concerted attempts have been made to provide systematic sources of technical information to villagers and village development workers on technologies amenable to appropriation by villagers, and which may provide low cost means for self-help problem solving in villages.[32]   While not adequate for dealing with all the requirements of an economy of the South in the modern environment, village industry and village technology are obviously essential to self-help economic improvement and may be seen as specific examples of the application of the broader Appropriate Technology rationale.   One American organization, Volunteers in Technical Assistance Inc., (VITA), was established in 1959 in recognition of this fact.[33]   By the late 1970s VITA, the North American equivalent to Schumacher's London-based I.T.D.G. Ltd., was mobilizing and coordinating the work of over 7 000 volunteer professionals representing 96 countries and 2 000

---

[29] E.g.: S. C. Biswas, ed., *Gandhi: Theory and Practice, Social Impact and Contemporary Relevance* (Simla: Indian Institute of Advanced Studies, 1969);   B. Kumarappa, *Capitalism, Socialism or Villagism?* (Varanasi: Sarva Seva Sangh Prakashan, 1965; written in 1944 and first published in 1946).

[30] Harrison, *Third World Tomorrow*, pp. 171-177.

[31] The journal *Science for Villages*, edited by D. Kumar and published in New Delhi, is representative of the small initiatives occurring in this field.

[32]   E.g.:   Appropriate Technology Development Association, *Appropriate Technology Directory,Vol. 1* (Lucknow, India: Appropriate Technology Development Association, 1977);   Darrow, Keller and Pam, *A.T. Sourcebook*;   Rural Communications Service, *Village Technology*;   South Pacific Appropriate Technology Foundation, *Liklik Buk. A Rural Development Handbook: Catalogue for Papua New Guinea* (Boroko, P.N.G.: Wantok Publications, 1977).   Other publications such as by the Canadian Hunger Foundation (*Handbook*), while being broader reference books on appropriate technology also have technical catalogues relevant to village self-help activity.

[33] Volunteers in Technical Assistance,*Village Technology*, p. vii.

corporations, universities and other institutions, for village technology development work. It had responded by that stage to over 25 000 requests for assistance. While village technology may appear less "serious", in popular thinking within the urban-industrialized context and in comparison with typical modern industrial technology, the level of activity of groups such as VITA indicates the contrary.

International Equity

Another stream within the Appropriate Technology movement flows from concern about the role of technology in either reinforcing or abetting international economic and political inequities.

There is growing debate on the relations between rich and poor nations, involving more than professional specialists in international relations and political economy.[34] This debate places increasing attention on relationships of dependency by the poor upon the rich, and on the development of center-periphery power relationships between rich and poor nations.[35] Recognition of this situation leads to calls, largely unheeded, for massive increases in international transfers of resources from the rich to the poor.[36]

It is now widely recognized that technology provides a medium for the development and maintenance of such dependency relationships and that viewing technology as "neutral" in terms of political economy is unjustifiable in view of its tendency to reinforce or evoke international inequity.[37]

---

[34] J. Merson, ed., "Rich Against Poor Nations: Jan Tinbergen, E.F. Schumacher, Charles Birch and Bruce McFarlane", *Transcripts on the Political Economy of Development from A.B.C. Radio Programs "Lateline" and "Investigations"* (Sydney: Australian Broadcasting Commission, 1977), pp. 219-264.

[35] See, G. K. Helleiner, *A World Divided: The Less Developed Countries in the International Economy* (Cambridge: Cambridge University Press, 1976).

[36] R. S. McNamara, "A Pittance for International Aid", *Bulletin of the Atomic Scientists* (November 1977), 36-38.

[37] G. K. Helleiner, "International Technology Issues: Southern Needs and Northern Responses", in *The New International Order: The North-South Debate*, ed. by J. N. Bhagwati (Cambridge, Mass.: The M.I.T. Press, 1977), pp. 295-316; A. Jamal, "The Consequences of Technological Dependence in the Least-Developed Countries", in *Mobilizing Technology for World Development*, ed. by J. Kamesh and C. Weiss, Jr. (New York: Praeger, 1979), pp. 58-65; K. R. Roby, "Science and Technology for Human Development: The Global Context", in *Challenges for Einstein's Children: Keith Roby's Vision of Science in Community Life*, ed. by I. Barns (Perth: Murdoch University, 1984), pp. 55-63; H. Johnston, *Technology and Economic Interdependence* (London: St. Martin's Press, 1975).

    The role of multinational business institutions as instigators or cat-
alysts for technology-related economic inequity is debated with some
intensity,[38] with calls ensuing for increased technological independence
by the poorer nations, or at least mutual interdependence.[39] Increased
technological independence is seen as a means for building up the
bargaining power of countries of the South in dialogues over the redistribution of international economic power.[40] Ambitious science and
technology development programs have been planned and experi-
mented with by the South in an attempt to build up the desired independence.[41] Given the currently existing situation of dependency,
however, the prospects of poorer nations succeeding in "catching up" *using conventional development strategies*, sufficiently to break free of
the situation of dependency, appear daunting.[42] It is for this reason
that commentators now point to the need for new directions in science
and technology innovation.[43] Some call for limited withdrawal from

---

[38] Positions in the debate range from those who view multinational enterprises as
"culprits" (e.g., R. J. Barnet and R. E. Müller, *Global Reach: The Power of Multinational
Corporations* [New York: Simon and Schuster, 1974]), through those who seriously
question multinationals (e.g., M. Lubis, "Some Second Thoughts About Foreign
Investments", *Appropriate Technology*, 1, 2 [1974], 19) to those who see their role as
primarily beneficial (e.g., R. L. Meier, "Multinationals as Agents of Social
Development", *Bulletin of the Atomic Scientists* [November 1977], 30-35).   Some
empirical studies on the role of multinational enterprises vis-a-vis technology are now
emerging; one such study by the International Labour Office points to the need for more
self-reliance by poor countries in technological innovation, but finds no unequivocal
evidence that singular blame for inappropriate technology choice need be laid on
multinationals (N. Jequier, et al., *Technology Choice and Employment Generation by
Multinational Enterprises in Developing Countries* [Geneva: International Labour Office,
1984]).

[39] L. A. Yuzon, ed., *Towards an Asian Sense of Science and Technology* (Singapore:
Christian Conference of Asia, 1984); Ramesh and Weiss, *Mobilizing Technology*.

[40] Galtung, *Development, Environment*; Bhalla, *Technologies*; Goulet, *Uncertain
Promise*.

[41] Cf.: A. Ahmed, "Science and Technology in India", *Bulletin of the Atomic
Scientists* (November 1980), 38-41; A. Parthasarathi, "India's Efforts to Build an
Autonomous Capacity in Science and Technology for Development", *Development
Dialogue*, 1 (1979), 46-59; A. Rahman and S. Hill, *Castasia II: Science, Technology and
Development in Asia and the Pacific* (SC.82/CASTASIA II/Ref.1; Paris/Manila:
UNESCO, 1982). Nominal support for such approaches are also found from countries such
as Australia (cf., K. L. Sutherland, "Toward an Australian Attitude to Science and
Technology for Development of Third World Countries", *Search*, 9, 11 [1978], 396-400).

[42] An interesting exposé of the Brazilian situation is provided by J. L. Lopes in
"Technology Transfer and Dependence", in *Faith, Science and the Future*, preparatory
readings for the conference of the World Council of Churches, Massachusetts Institute of
Technology, July, 1979, ed. by P. Abrecht, et al. (Geneva: World Council of Churches,
1978), pp. 185-192.

open interaction in the international technology arena (as a transitional strategy).[44] A growing body of opinion points to work of the style which occurs under the rubric of Appropriate Technology as providing the best opportunities for countries of the South to begin to break out of their vicious cycles of technological dependency.[45]

Intranational Equity and Self- reliance

The dynamics of inequality and dependence which occur at the international level have their equivalents at the intranational level, and many of the commentators just considered apply their analyses to both levels. Centers of economic and technological power evolve in certain parts of a country (mostly in the metropolitan regions) and other regions (mostly rural) fall increasingly into the periphery of the centers of power and into self-perpetuating situations of dependency. Centers of power within a country tend to benefit from relations with international centers of power, while communities in the periphery often do not. It is often also true that certain classes of people within a country obtain fuller participation in the benefits of the "center" than others. Technology may also play a role in exacerbating and reinforcing such inequities of power.[46]

Concern for the development of Appropriate Technology as part of a strategy to address these problems has come from at least two sources.

---

[43] Dag Hammarskjøld Foundation, *Another Development* (Uppsala: Dag Hammarskjøld Foundation, 1975); K. R. Roby, "Science and Tecnology for Human Development: New Directions", in *Challenges for Einstein's Children: Keith Roby's Vision of Science in Community Life*, ed. by I. Barns (Perth: Murdoch University, 1984), pp. 64-75.

[44] W. Morehouse, "Third World Disengagement and Collaboration: A Neglected Transitional Option", in *Mobilizing Technology for World Development*, ed. by J. Ramesh and C. Weiss, Jr. (New York: Praeger, 1979), pp. 74-81.

[45] For a survey of this school of thought see: Diwan and Livingston, *Alternative Development Strategies*.

[46] A "futures studies" research programme, which places more emphasis on social and political factors than the Club of Rome's famous study (D. Meadows, et al., *The Limits to Growth* [New York: Universe Books, 1972]) has examined technology and resource issues in the context of the interaction between international and intranational centre-periphery relationships (A. O. Herrera, et al., *Catastrophe or New Society? A Latin American World Model* [Ottawa: Fundacion Bariloche (Argentina) and International Development Research Centre (Canada), 1976]). This problem is discussed explicitly by D. Chabrol, from G.R.E.T., a leading international Appropriate Technology organization based in France ("Vent du Nord, Soleil du Sud: La Nouvelle Alliance?" *Autrement*, 27 [1980], pp. 18-32).

Firstly, there is a growing interest in the importance of self-reliance at the level of local communities and on the importance of analyzing economic parameters at this level. This "local" focus tends to reveal more clearly conflicts of interest between regions and social groups which do not readily show up in national level aggregate data.[47] Secondly, there have been concerted efforts to address problems of technological and economic dependency from an ethical point of view. The concept of the Just, Participatory and Sustainable Society, promulgated by the World Council of Churches, is a prominent example of a guiding principle which has emerged within this approach.[48] These two sources combined have led to heightened calls for science and technology programs to reverse forces of inequity between geographical regions and between social groups, on the grounds of justice.[49]

One concerted attempt to develop a serious technological strategy as an application of an ethically-based approach to self-reliance has been conducted by Ignacy Sachs and colleagues at the *Centre Internationale de Recherche Sur l'Environnement et le Developpement* in Paris, under the rubric of "ecodevelopment". Sachs' approach combines both international and intranational equity perspectives, with the insights of human and natural ecology. For example:[50]

---

[47] E.g.: B. Stokes, *Helping Ourselves: Local Solutions to Global Problems* (New York: W.W. Norton, 1981); M. A. Max-Neif, "The Tiradentes Project: Revitalization of Small Cities for Self-Reliance", *Development Dialogue*, #1 (1981), pp. 115-137; S. Christie, "Community-Based Societies: An Alternative Structure for the Development of Just, Participatory and Sustainable Societies", mimeo (Perth, W.A.: Community Aid Abroad, 1981). Most serious scholars or official reports on this subject make a clear distinction between self-reliance and autarchy, expressing preference for the former, e.g.: I. Sachs, "Autonomie, oui, Autarcie, non", *Autrement*, 27 (1980), 34-39; Dag Hammarskjøld Foundation, *What Now?*, Report on Development and International Cooperation, prepared on the occasion of the Seventh Special Session of the United Nations General Assembly (*Development Dialogue*, 1/2) (Uppsala: Dag Hammarskjøld Foundation, 1975).

[48] P. Abrecht, et al., *Faith, Science and the Future*, preparatory readings for the 1979 Conference at Massachusetts Institute of Technology on the Contribution of Faith, Science and Technology in the Struggle for a Just, Participatory and Sustainable Society (Geneva: World Council of Churches, 1978); E. F. Schumacher, "Growth - Yes, but Who For and How Fast? Three Horns of a Dilemma" (World Council of Churches Document SE/50), *Study Encounter*, 9, 4 (1973), 1-4.

[49] Commission on the Churches Participation in Development, *On Appropriate Technology* #2 (Geneva: World Council of Churches, 1979); S. L. Parmar, "The Quest for Appropriate Technology", in *Faith Science and the Future*, ed. by P. Abrecht, et al. (Geneva: World Council of Churches, 1978), pp. 193-200; de Pury, *People's Technologies*.

[50] I. Sachs, "Technology of Self-Reliance: Self-Reliance in Technology", *Human Futures* (Spring 1979), 16.

The ... search for technologies of self-reliance departs radically from the traditional approach: instead of starting with ready-made technologies and then adapting at great pains the economy and environment to these transferred technologies, it is suggested to do it the other way round, making the best possible use of the specific natural and cultural resources of each ecosystem (the - *ecodevelopment approach*).

Sachs goes on to argue that rejecting transfer of technology and imitative growth as development paradigms is not tantamount to severing links with the external world. Self-reliant development, according to Sachs, calls for *selectivity* in foreign contacts and not their banning. He holds that the strategy of self-reliance reinforces the need for countries to avail themselves of the opportunities created by international cooperation, so long as this is properly handled.

The political goal of enhanced equity within nations has been an entry point for many into the field of Appropriate Technology.

Food Production and Health

The Appropriate Technology movement in the South has by no means been concerned exclusively with the manufacturing sector. Intermediate Technology, as first propounded by Schumacher, was concerned primarily with non-agricultural production in rural areas.[51] Others soon applied the general Intermediate Technology approach to agriculture and other fields as well, however, as did Schumacher and the I.T.D.G.[52] Famines experiences by poor communities in the South have placed importance on the role of small communities and peasants themselves growing their own food.[53] Relatively efficient technology

---

[51] In considering this and the following topics of relevance to the South we will limit our survey to cursory comments, due to restrictions in space. Most of the Appropriate Technology work conducted in the following streams has been structured along lines already discussed. Our purpose here is to illustrate the diversity of streams within the Appropriate Technology movement, and not to study each stream in detail.

[52] Many reports and articles in the general field of agriculture, food and health.have been published in the I.T.D.G.'s journal *Appropriate Technology*.

[53] This theme has been developed at some length by F. Moore-Lappé and J. Collins, who argue that all countries and groups of malnourished peasants in the South do have the capacity to feed themselves if allowed the freedom to do so. Moore-Lappé and Collins have shown that maximum increases in the rate of food *production* will not solve

for this purpose, which is both accessible to rural peasants and which is compatible with local (often stringent) environmental and resource conditions, is necessary to enhance the prospects for food self-sufficiency in the South.[54] Such technology is of even greater importance if the rural poor are to be able to create sufficient agricultural surplus to rise above a subsistence lifestyle.

Some side-effects and practical limitations of the "green revolution" have become apparent thus intensifying the search for viable technological alternatives for Third World agriculture.[55] Considerable research and development activity has demonstrated the sound prospects for increasing productivity and community self-reliance in agriculture through technologies which fit the ambience of the South's rural hinterland.[56] Problems *inter alia* of culture, politics, the dissemination of technologies and the cultivation of indigenous expertise, rather than technical obstacles as such, appear to be the main limitations.[57]

Extensive Appropriate Technology work has occurred in such areas as food storage and processing, water supply and irrigation, fertilizer production, livestock usage, fishing/aquaculture and post-harvest sys-

---

hunger problems and that policies based upon this approach have in fact often compounded the problems. They argue that providing assistance to poor communities to produce their own food is the best strategy to solve hunger problems (*Food First: Beyond the Myth of Scarcity* [Boston: Houghton Mifflin, 1977]). See also, S. George, *How the Other Half Dies: the Real Reasons for World Hunger* (Harmondsworth: Penguin, 1974).

54   Cf., S. Biggs, *Appropriate Agricultural Technology in Bangladesh: Issues, Needs and Suggestions* (Dacca: Ford Foundation, 1975).

55   L. Surendra, "Seeds of Disaster, Germs of Hope", *Far Eastern Economic Review*, 22nd March, 1984, pp. 64-65; N. Wade, "Green Revolution (I): A Just Technology, Often Unjust in Use", *Science*, 186, 4169 (1974), 1093-1096; N. Wade, "Green Revolution (II): Problems of Adapting a Western Technology", *Science*, 186, 4170 (1974), 1186-1192; J. P. Grant, "An International Challenge: Science and Technology for Managing the World Food Problem", in *Mobilizing Technology for World Development*, ed. by J. Ramesh and C. Weiss, Jr. (New York: Praeger, 1979), pp. 204-211.

56   C. H. Gotsch and N. B. McEachron, "Technology Choice and Technological Change in Third World Agriclture: Concepts, Empirical Observations and Research Issues", in *Technology Choice and Change in Developing Countries: Internal and External Constraints*, ed. by B. G. Lucas and S. Freedman (Dublin: Tycooly, 1983), pp. 29-62; D. A. G. Green, *Ethiopia: An Economic Analysis of Technological Change in Four Agricultural Production Systems* (Michigan State University, Institute of International Agriculture, Monograph #2, 1974); J. Worgan, "Possible Solutions to the World Food Problem", *Appropriate Technology*, 2, 2 (1975), 27-28.

57   F. Moore-Lappé and J. Collins, "More Food Means More Hunger", *The Futurist*, April 1977, pp. 90-95; L. R. Brown and E. C. Wolf, *Reversing Africa's Decline*, Worldwatch Paper #65 (Washington, D.C.: Worldwatch Institute, 1985).

tems.[58] There is now also a notable array of programs aimed at providing simple affordable technologies for health care.[59]

Transport and Building

Two further fields of interest which have contributed to the development of the Appropriate Technology movement in the South are transport and building. In both cases the major concern has been to identify, improve and promote technologies and products which are affordable by the poor and by those without ready access to extensive resources.

In the field of transport, attempts have been made to evolve simple, low-cost, labor-absorbing methods of constructing roads and bridges, without the need for using exotic supplies of materials.[60] Low-cost, locally producible vehicles have also featured prominently.[61]

In the field of building construction emphasis has been placed on techniques, machines, materials and designs conducive to owner-builder approaches.[62] A second emphasis has been on ensuring buildings are

---

[58] See: Carr, *AT Reader*, pp. 133-186; Dunn, *Appropriate Technology*, pp. 55-84; McRobie, *Small is Possible*, pp. 46-49. Note: The crucial importance of the *right choices* being made in water sanitation technologies, as a key to solving other problems in food and health is now well recognized internationally (cf.: A. Charnock, "Ten Years to Slake a Global Thirst", *New Scientist*, 10th March, 1983, pp. 661-665; S. B. Watt, "The Social Context for Choosing Water Technologies", in *Introduction to Appropriate Technology: Toward a Simpler Life-Style*, ed. by R. J. Congdon (Emmaus, PA.: Rodale Press, 1977), pp. 17-33).

[59] For general surveys of these topics see: Carr, *AT Reader*, pp. 259-286; Dunn, *Appropriate Technology*, pp. 136-145; McRobie, *Small is Possible*, pp. 40-44, 49-51.

[60] Ellis and Howe, "Simple Methods"; M. Muller, "A Design for a Medium-Span Wooden Bridge in Kenya", *Appropriate Technology*, 1, 4 (1974); World Bank, "Highway Construction", in *Appropriate Technology and World Bank Assistance to the Poor*, ed. by World Bank (Updated edition [1979]; Washington, D.C.: World Bank 1978), pp. 36-44.

[61] R. Keatley, "Ferrocement Boatbuilding in a Chinese Commune", *Appropriate Technology*, 2, 1 (9175), pp. 4-6; MacAlister Elliot and Partners, "The Introduction of Ferrocement Fishing Boats to Lake Malawi", *Appropriate Technology*, 7, 2 (1980), 10-12; S. S. Wilson, "The Oxtrike", *Appropriate Technology*, 3, 4 (1976), 21-22.

[62] J. Aitken, ed., "Homes for the Poor: Building on What They Have", special issue of *Appropriate Technology*, 11, 3 (1984); J. P. M. Parry, "Intermediate Technology Building", in *Introduction to Appropriate Technology: Toward a Simpler Life-Style*, ed. by R. J. Congdon (Emmaus, PA: Rodale Press, 1977), pp. 53-62; J. P. M. Parry, "The Growth in the Use of Intermediate Building Technologies", *Appropriate Technology*, 11, 3 (1984), 30-31; J. F. C. Turner, *Housing by People: Towards Autonomy in Building Environments* (London: Marion Boyars, 1976); United Nations Centre for Human Settlements (Habitat), *Small-Scale Building Materials Production in the Context of the Informal Economy* (Nairobi: United Nations Centre for Human Settlements, 1984).

adequately adapted to local conditions from the points of view of lifestyle, safety and environmental comfort.[63]

Energy

A major part of Schumacher's professional career was spent as an energy economist, with the British National Coal Board. This gave him insights into the crucial role which energy plays in economic activity and led him to pronounce that, "energy is for the mechanical world what consciousness is for the human world. If energy fails, everything fails".[64] The use of energy is essential to all activity and all material economic processes inevitably involve the nett degradation of useful energy into waste heat.[65] Energy may be viewed as a universal medium for conducting and comparing tangible processes.[66] Myths about the place of energy in civilization began to proliferate with the growth of coal usage in the nineteenth century.[67] Since the rapid rise in energy prices in the early 1970s debate over and interest in energy has soared, with a deepening awareness ensuing of our radical dependence upon energy.[68] The prospects for the South and its economy in particular, are linked incontrovertibly with the energy issue. It is in this context that a stream within the South's Appropriate Technology movement has evolved, concerned with the technology-related aspects of energy scenarios and problems.

It is beyond the scope of this study to canvass all the relevant issues in the field of energy which pertain to either the South or the North. The main point to stress here is that much serious energy-related engi-

---

[63] J. Herklots, "Ferrigloo Houses in Papua New Guinea", *Appropriate Technology*, 7, 3 (1980), 11-12; S. N. Sibtain, *To Build a Village: Earthquake-resistant Rural Architecture - A Technical Handbook* (Sydney: Australian Council of Churches, 1982); Note: greater attention appears to be paid to thermodynamic considerations in the Appropriate Technology building literature of the North more than the South.

[64] Schumacher, *Small is Beautiful*, p. 112.

[65] N. Georgescu-Roegen, *The Entropy Law and the Economic Process* (Cambridge, Mass.: Harvard University Press, 1971).

[66] I. Prigogine and I. Steugers, "The New Alliance: From Dynamics to Thermodynamics: Physics, the Gradual Opening Towards the World of Natural Processes", *Scientia*, 112 (1977), 319-332.

[67] G. Basalla, "Some Persistent Energy Myths", in *Energy and Transport: Historical Perspectives on Policy Issues*, ed. by G. H. Daniels and M. H. Rose (Beverly Hills: Sage, 1982), pp. 26-38.

[68] Cf., H. T. Odum and E. Odum, *Energy Basis for Man and Nature* (New York: McGraw Hill, 1976).

neering and technical design work is conducted under the rubric "Appropriate Technology", and is based upon an analysis of the broader socioeconomic context of the South.[69] In other words, a conscious attempt is made to ensure that technical activities in the field of energy are not artificially divorced from the meta-technical analysis which undergirds the Appropriate Technology movement.

In accordance with now firmly established schools of thought in the field of energy analysis, the Appropriate Technology movement places emphasis on moving away from reliance upon non-renewable energy sources (fossil fuels in particular) towards increased conservation and use of renewable energy resources; particular regard is also paid to the interdependence of the South and the North.[70] The Appropriate Technology approach stresses that the circumstances of the South are, however, different to those of the North and require a different response. While adopting the view that it is neither just nor practicable for total global primary energy consumption to continue to increase at the rate which occurred in the industrialized nations after the second World War, the Appropriate Technology movement does not adopt the view that there should be a universal moratorium on expanded energy use in the South.

There are vast differences in per capita energy consumption throughout the world.[71] During 1983 the per capita consumption in the United States was over 32 times the level which prevailed in Asia (excluding Japan and China). While the myth that higher levels of economic growth always require higher levels of energy use has now been debunked, it does not follow that growth in energy use is not required by many poor communities as a means of enhancing their situation.[72] The dilemma for energy policy is how the energy use of the poor may be increased where appropriate, without causing undue pressures on global consumption and without creating too much dependence by the

---

[69] A good example is the work of British engineer based at the University of Reading, Professor Peter Dunn, who pioneered the programmes of the I.T.D.G. in the energy area - as Chairman of its Power Panel; cf., P. D. Dunn, "Energy in Rural Areas: An Intermediate Technology Approach", in *Introduction to Appropriate Technology: Toward a Simpler Life-Style*, ed. by R. J. Congdon (Emmaus, PA: Rodale Press, 1977), pp. 63-79.

[70] E.g., K. R. Roby, "The Argument for Conservation and the Fullest Possible Use of Renewable Energy Sources", in *Faith, Science and the Future*, ed. by P. A. Abrecht, et al. (Geneva: World Council of Churches, 1978).

[71] C. Flavin, *World Oil: Coping with the Dangers of Success*, Worldwatch Paper #66 (Washington, D.C.: Worldwatch Institute, 1985), p. 52.

[72] R. Grossman and G. Daneker, *Energy, Jobs and the Economy* (Boston: Alyson, 1979); W.U. Chandler, *Energy Productivity: Key to Environmental Protection and Economic Progress*, Worldwatch Paper #63 (Washington, D.C.: Worldwatch Institute, 1985).

poor on essential resources over which they have little control and perhaps little capacity to afford.

It is towards resolving this dilemma that the Appropriate Technology movement makes an invaluable contribution. Extensive research and development work has now been conducted on this subject, by authorities which are by no means outside of the "mainstream".[73] There is now clear evidence that renewable energy technologies are available which both conform with the Intermediate Technology rationale or with the village development approach.[74] Particular interest has been generated in the prospects for using and enhancing systems for digesting organic waste for the production of both methane gas and fertilizer.[75] Some controversy has arisen over the use of such technology. This controversy has reinforced a basic tenet of the Appropriate Technology movement that technology ought to be tailored to social and political goals and circumstances rather than the other way round.[76] Experiments with the "biogas" digestors have shown that if wrongly designed (from the point of view of scale) or if introduced without regard for social relations, they may in fact increase inequality and poverty. Nevertheless, despite these risks and the need for some technical improvements, the systems have been adopted extensively.[77]

Recent research has revealed that if less effort is placed on accelerating the pace of expansion of power systems and more effort is put into carefully ensuring that such systems actually promote local devel-

---

[73] E.g.: J. Drevon and D. Thery, *Ecodevelopment and Industrialization: Renewability and New Uses of Biomass*, Ecodevelopment Study #9 (Paris: International Research Centre on Environment and Development, 1977); R. Merchert, ed., *Using Renewable Energy Resources in Developing Countries* (West Berlin: German Foundation for International Development, 1980); National Academy of Sciences, *Energy for Rural Development: Renewable Resources and Alternative Technologies for Developing Countries*, Report of a Panel of the Advisory Committee on Technology Innovation (Washington, D.C.: National Academy of Sciences, 1976).

[74] V. Smil, "Intermediate Energy Technology in China", *Bulletin of the Atomic Scientists* (February 1977), 25-31; N. L. Brown and J. W. Howe, "Solar Energy for Village Development", *Science*, 199 (1978), 651-657.

[75] H. Kaufman, "Biogas in China: Small Scale, Decentralized Anærobic Digestion", *Alternative Sources of Energy*, 51 (1982), 24-27; Prasad, Prasad and Reddy, "Biogas Plants".

[76] J. B. Tucker, "Biogas Systems in India: Is the Technology Appropriate?" *Environment*, 24, 8 (1982), 12-20, 39.

[77] By August 1976, for example, it was reported that more than 2.8 million peasant families in China's Szechwan province alone were producing their own biogas using these systems (Smil, "Intermediate Energy", p. 30); Smil estimates that approximately three quarters of China's rural population live in areas where biogas production is viable (*ibid.*, p. 31).

opment, through increased efficiency and through paying careful attention to the special energy needs of villagers, then new ways of meeting Third World power needs through decentralized electricity technologies become feasible.[78]

The encouraging prospects for intermediate-level and renewable energy technologies in the South have prompted calls for the North to assist the South in attaining self-reliance in energy technology and energy supply.[79] The World Bank, on the basis of extensive feasibility studies, developed an ambitious international proposal to finance the development of renewable energy projects in the oil-poor countries of the South. It involved the stimulation of projects to the value of about $100 billion, from the initial capital outlay of about $1.0 billion to $1.5 billion from the Bank's affiliates (member governments). The plan was effectively vetoed by the Reagan Administration in favour of increased development of oil resources for the South (for a budget saving to the U.S. Government of about $250 million).[80] Despite obstacles such as this to the successful promotion of alternative energy programs in the South, much work continues directly through the programs of independent agencies.[81]

Despite the lack of energy self-reliance in poorer regions and communities of the South, awareness has now been raised of the importance of such work, and sufficient evidence has now been accumulated to demonstrate the technical feasibility of such an approach. This work makes up a significant stream within the Appropriate Technology movement.[82]

---

[78] C. Flavin, *Electricity for a Developing World: New Directions*, Worldwatch Paper #70 (Washington, D. C.: Worldwatch Institute, 1986).

[79] E.g.: L. Gordon, "Energy Development: Crisis and Transition", *Bulletin of the Atomic Scientists*, 37, 4 (1981), 24-29.

[80] C. Norman, "U.S. Derails Energy Plan for Third World", *Science*, 212, 4490 (1981), 21-24.

[81] Imaginative schemes for independent action exist at the University training level (e.g., P. D. Dunn, *Alternative Energy for Developing Countries*, Brochure for an MSc/Diploma Course in the Department of Engineering, University of Reading, United Kingdom [1983]) and through individuals or community organizations (e.g., K. R. Roby, P. A. Abrecht and L. Maciel, *Energy for My Neighbour* (Geneva: World Council of Churches, 1978).

[82] See Appendix 4.4 for a sample list of reports and articles in the I.T.D.G.'s journal *Appropriate Technology* in the general area of appropriate technologies for energy in the South. General overviews may also be found in: Carr, *A.T. Reader*, pp. 222-258; McRobie, *Small is Possible*, pp. 51-55; and Norman, *Soft Technologies*, pp. 23-30.

Environmentally Sound Development

The predominant stimuli in the South for Appropriate Technology have been social and economic concerns, and, as explained in Chapter Four, Intermediate Technology (as originally espoused) was essentially an economic concept. Environmental or ecological motivations have, in general, been marginal in the South in comparison with the North. In view of this, many commentators have criticized environmental approaches to technology and development as class-based and luxurious diversions by wealthy people from the true economic and political problems confronting the poor.[83]

The main proponents of Appropriate Technology and Intermediate Technology for the South have always tended to exhibit environmental commitments, however, despite their ostensibly narrower focus on "technology" and "economics" issues.[84] Since the promotion of the concept of "another development" in 1975 by the Dag Hammarskjøld Foundation these commitments have become more explicitly articulated and more widely accepted.[85] "Another development" is a strategy for social and economic development based upon three elements: the satisfaction of needs, beginning with the eradication of poverty; endogenous and self-reliant activities and programs; and, harmony with the environment.[86] These principles have also been taken up, refined and augmented by authorities such as the Centre International de Recherche sur l'Environnement et le Développement (C.I.R.E.D.) under the rubric of "ecodevelopment", as mentioned earlier.[87]

---

[83] Cf., T. O'Riordan "Environmental Ideologies", *Environment and Planning A*, 9 (1977), 3-14.

[84] Schumacher's environmentally-oriented ethics, for example, are abundantly obvious to even the casual reader of *Small is Beautiful* (e.g., the chapter entitled "The Proper Use of Land", pp. 93-107); cf., Morrison, *Soft, Cutting Edge*.

[85] Dag Hammarskjøld Foundation, *What Now?*

[86] *Ibid.*, p. 28.

[87] See "Development, Environment and Technology Assessment", by the Director of CIRED, I. Sachs (*International Social Science Journal*, 25, 3 [1973], 273-283). It appears that Sachs' systematic work on this topic predates that of the Dag Hammarskjøld Foundation. Sachs' concept of ecodevelopment has more recently been applied specifically to the underdeveloped economy by others such as R. Riddell (*Ecodevelopment: Economics, Ecology and Development - An Alternative to Growth Imperative Models* [New York: St. Martin's Press, 1981]).

Under the auspices of the United Nations Environment Programme, Professor A. K. N. Reddy has promulgated the phrase *"environmentally sound and appropriate technology"* in an attempt to synthesize environmental concerns with the more dominant economic and social concerns of the Appropriate Technology movement in the South.[88] As employed here, the word "appropriate" appears to give Reddy's phrase the semantic problems associated with specific-characteristics definitions (as discussed in Chapter Two). Nevertheless Reddy's phrase has served to highlight the need for an economic strategy aimed at satisfying the "inner limits" of fundamental human needs while not violating the "outer limits" of the Earth's physical and ecological integrity.[89] This theme has been prominent in U.N.E.P.'s journal *Mazingira*.[90]

The potential for fusing environmental concerns in the South with economic development objectives has been reflected by increasing interest in the recycling of waste materials.[91] The World Council of Churches, through the programs of its Commission on Church and Society, has emphasized the overall unity of environmental and socioeconomic concerns vis-a-vis the South, but does so within the context of predominantly global concerns, rather than with a specific focus on the South.[92] Finally, while compatible with environmental concerns, and while it is moving increasingly to embrace environmental concerns, Appropriate Technology in the South has, in the main, emerged separately from environmentalism.

Political and Cultural Freedom

There is a further stream within the Appropriate Technology movement in the South concerned pre-eminently with what might be

---

[88] A. K. N. Reddy, *Technology, Development and the Environment: A Re-Appraisal* (Nairobi: United Nations Environment Programme, 1979). Reddy is Professor and Convenor, ASTRA (Cell for the Application of Science and Technology to Rural Areas), Indian Institute of Science, Bangalore.

[89] This notion of "inner limits" and "outer limits" was first promoted at a symposium in Cocoyoc, Mexico, October 1974, chaired by Barbara Ward (sponsored by U.N.E.P. and U.N.C.T.A.D.). The symposium's statement, known as *The Cocoyoc Declaration*, was published in the *Bulletin of the Atomic Scientists*, March 1975, 6-10.

[90] Nairobi: United Nations Environment Programme (Periodical).

[91] J. Vogler, *Work from Waste: Recycling Wastes to create Employment* (London: Intermediate Technology Publications Ltd. and Oxfam 1981).

[92] See, Abrecht, et. al, *Faith Science and the Future*.

called "political and cultural freedom". It is an emphasis which has been inherent in Appropriate Technology from its beginnings, but which has been more explicitly addressed by some. The concern with politics here goes beyond the political desire for equity in economic life discussed earlier.

This approach has become most widely known through the writings of Ivan Illich and, more specifically, through his notion of *convivial tools*.[93] By "convivial tools" Illich wishes to denotes technologies which serve politically interdependent individuals within a *convivial society*.[94] By "convivial society" he denotes a society characterized by autonomous and creative intercourse among persons and between persons and their environment. This contrasts with a situation of conditioned responses by persons to the demands made upon them by others and the human-made environment. Illich states: "I believe that, in any society, as conviviality is reduced below a certain level, no amount of industrial productivity can effectively satisfy the needs it creates among society's members."[95] Illich's approach is not concerned exclusively with the South, but is included here because of its connections with certain analyses of the need for liberation from conditions of dependency in the South. These connections have been described elegantly by Roby in the following terms:[96]

> Science and technology must respect and build upon traditional sources of knowledge and wisdom, and respect the people's culture. With such roots in the community, science and technology may be a liberating force - liberating people from hopelessness, drudgery and dependency, and liberating their inherent ingenuity and inventiveness. The general requirements are highlighted now in the developing countries, but their relevance to the industrialized countries is rapidly being realized. ... The conversion required is from a center-periphery scientific-technological relationship to a center-center relationship of mutual respect, co-operative diversity and learning from one another.

---

93 Illich, *Tools for Conviviality*. Note: An extensive bibliography (858 items) of materials in keeping with this concept has been prepared by Valentina Borremans, the Founder and Director of Centro Intercultural de Documentación (CIDOC) in Cuernavaca, Mexico (*Guide to Convivial Tools*, Library Journal Special Report #13 [New York: R. R. Bowker, 1979])

94 *Ibid.*, p. xii.

95 *Ibid.*, p. 11. Note: a similar point is made by Johan Galtung ("The Technology that Can Alienate", *Development Forum*, 4, 6 (1976), 6).

The broader cultural dimension behind Illich's approach has been developed in a critical-cum-philosophical way by the Latin American scholar Paulo Freire. Freire stresses a dynamic approach to education as the most important ingredient of the process of liberation towards a "convivial" society.[97] Something of this approach has been taken up by the United Nations University, based in Tokyo, through its projects on "Socio-cultural Development Alternatives in a Changing World"and "Goals, Processes and Indicators of Development".[98] Specific attention was given to the issues surrounding Appropriate Technology, using this perspective, at a major international conference sponsored by the United Nations University on "Science and Technology in the Transformation of the World"; the conference attempted to define realistic strategies by which the countries of the South could overcome the present unequal distribution of power over the world's technological resources and develop their own scientific and technological creativity rather than rely on transferred technology; the theme of "technology for liberation" was prominent.[99]

In summary, there is a stream of the Appropriate Technology movement which has as its point of departure the psychic, social and cultural needs of human beings and of those who are oppressed, in particular. This stream largely adopts a political mode of analysis, in a very broad sense, and tends to view technology as both a product of human cultures and an influence upon cultures. Its focus of attention is upon prospects for evolving technologies and technical innovation capacities, within dependent cultures, conducive to the development of greater vitality, independence and freedom in those cultures. Such concerns are considered to be of greater importance than the growth of economic productivity as such.

---

[96] Roby, "New Directions", p. 75.

[97] P. Freire, *Pedagogy of the Oppressed* (Harmondsworth: Penguin, 1972)

[98] J. W. Golebiowski, *Social Values and the Development of Technology*, Working Paper HSDRSCA-78/UNUP-304 (Tokyo: United Nations University, 1982); R. Preiswerk, *Cultural Identity, Self-Reliance, and Basic Needs*, Working Paper HSDRGPID-8/UNUP-60 (Tokyo: United Nations University, 1979); O. Kreye, *Perspectives for Development through Industrialization in the 1980's: An Independent Viewpoint on Dependency*, Working Paper HSDRGPID-37/UNUP-151 (Tokyo: United Nations University, 1980).

[99] A. Abdel-Malek, G. Blue and M. Pecujlic, eds., *Science and Technology in the Transformation of the World* (Tokyo: United Nations University, 1982).

Appendix to Chapter Six
Sample of General Literature on Technology and the South
and Appropriate Technology and the South

General publications on the technology-related problems of the South include: J. Baranson,*Technology for Underdeveloped Areas: An Annotated Bibliography* (Elmsford, N.Y.: Pergamon, 1967); C. Cooper, *Science Technology and Development: The Political Economy of Technical Assistance in Underdeveloped Countries* (London: Frank Cassirer, 1973); D. Ernst, ed., *The New International Division of Labour, Technology and Underdevelopment* (Frankfurt, a.M.: Campus Verlag, 1981); D. Goulet, *The Uncertain Promise: Value Conflicts in Technology Transfer* (New York: IDOC/North America, 1977; published in cooperation with the Overseas Development Council, Washington, D.C.); Organization for Economic Cooperation and Development, *Choice and Adaptation of Technology in Developing Countries* (Paris: Organization for Economic Cooperation and Development, 1974); Research Policy Program, *Technology Transformation of Developing Countries* (Lund: University of Lund, 1978); D.L. Spenser, *Technology Gap in Perspective* (New York: Spartan Books, 1970); Study Group of the N.G.O. Committee on Science and Technology for Development, *N.G.O. Report on the United Nations Conference on Science and Technology for Development* (New York: United Nations Economic and Social Council and N.G.O. Forum on Science and Technology for Development, 1979); J. Tinbergen, *Reshaping the International Order* (New York: E.P. Dutton 1976); United Nations Conference on Trade and Development, *Transfer of Technology: Its Implications for Development and Environment* (New York: United Nations, 1978).

General publications on Appropriate Technology and the South include: M. J. Betz, P. McGowan and R. T. Wigand, eds., *Appropriate Technology Choice and Development* (Durham, North Carolina: Duke Press Policy Studies, 1984); A.S. Bhalla and C.G. Baron, "Appropriate Technology, Poverty and Unemployment: The I.L.O. Project", *Appropriate Technology*, 1, 4 (1974), 20-22; B. van Bronckharst, *Development Problems in the Perspective of Technology* (Eindhoven: Technische Hogeschool Eindhoven, 1984); M. Carr, *Appropriate Technology for African Women*, (Addis Ababa: African Training and Research Centre for Women, United Nations Economic Commission for Africa, 1978); K. J. Charles, "Intermediate Technology: Can It Work?", *Appropriate Technology*, 1, 4 (1974), 14-15; K. Darrow, K. Keller and R. Pam. *Appropriate Technology Sourcebook* (Compilation of 2 vols.; Stanford, Cal.: Volunteers in Asia, 1981); D.D. Evans and L.N. Adler, eds., *Appropriate Technology for Development: A Discussion and Case Histories* (Boulder: Westview Press, 1979); R.S. Eckaus, *Appropriate Technologies for Developing Countries* (Washington, D.C.: National Academy of Sciences, 1977); M.K. Garg, "Problems of Developing Appropriate Technology in India", *Appropriate Technology*, 1, 1 (1974), 16-17; Insititution of Civil Engineers (Great Britain), *Appropriate Technology in Civil Engineering*, proceedings of the Conference held by the Institution of Civil Engineers, 14-16 April, 1980 (London: T. Telford, 1981); A.D. Jedlicka, *Organization for Rural Development: Risk Taking and Appropriate Technology* (New York: Praeger Publishers, 1977); S.F. Johnston, "Intermediate Technology: Appropriate Design for

Developing Countries", *Search*, 7 (1976), 27-33; A Kestenbaum, *Technology for Development* (London: Voluntary Committee on Overseas Aid and Development, 1977); J. McDowell, ed., *Appropriate Technology for the Rural Family* (Nairobi: UNICEF Eastern Africa Regional Office, 1976); I.N. Mazonde, *Science and Technology for Development in Botswana: A Critical Appraisal with Emphasis on Appropriate Technology for Rural Development* (Gaberone: University College of Botswana and Swaziland, 1978); R.J. Mitchell, ed., *Experiences in Appropriate Technology* (Ottawa: Canadian Hunger Foundation, 1980); A. Noyes, *The Poor Man's Wisdom: Technology and the Very Poor* (Oxford: Oxfam, 1979); National Technical Information Service, *Appropriate Technology Information for Developing Countries: Selected Abstracts from the NTIS Data File* (3rd ed., Washington, D.C.: U.S. Department of Commerce, 1981); H. Pack, "Policies to Encourage the Use of Intermediate Technology", paper prepared for the Committee on Intermediate Technology of the Agency for International Development, Washington, D.C. April 1976; J. Pickett, et al., "The Work of the Livingston Institute on 'Appropriate' Technology", *World Development*, 5, 9/10 (1977), 773-882; J. Ramesh and C. Weiss, Jr., *Mobilizing Technology for World Development* (New York: Praeger, 1979); A.K.N. Reddy, "Alternative Technology: A Viewpoint from India", *Social Studies of Science*, 5 (1975), 331-342; H.V. Singh, *Appropriate Technology*, M.Phil. Thesis in Economics, University of Oxford, Christ Church, Oxford, 1981; United Nations Department of Economic and Social Affairs, *Appropriate Technology and Research for Industrial Development: Report of the Advisory Committee on the Application of Science and Technology to Development on Two Aspects of Industrial Growth* (#E.72.II.A.3; New York: United Nations, 1972); J.J. de Veen, *The Rural Access Roads Programme: Appropriate Technology in Kenya* (Geneva: International Labour Office, 1980); C. Weiss, Jr., "Mobilizing Technology for Developing Countries", *Science*, 16, (March, 1979), 1083-1089; World Bank, *Appropriate Technology and World Bank Assistance to the Poor* (Updated ed. [1979]; Washington, D.C.: World Bank, 1978).

7

Appropriate Technology in the North

Categorizing the Movement

It is extraordinarily difficult to tightly and consistently categorize the Appropriate Technology movement, largely because it is in fact a dynamic, fluid social movement. This difficulty is pronounced in the North.

Despite this difficulty an attempt has been made by one of the veteran protagonists of the overall movement, Godfrey Boyle, to categorize the different streams (using "alternative technology" as the all-embracing rubric); He portrays them on a simple one-dimensional political spectrum, ranging from those which are supposedly conservative, centralist or elitist (on the "right"), and those which tend towards socialism, decentralization and egalitarianism (on the "left"). Boyle places what he terms the "radical technology", "liberatory technology", and "human-centered technology" streams on the furthest left, ranging through "utopian technology" and "convivial technology" to "community technology" in the center, through "Intermediate Technology", "soft technology" and "Appropriate Technology" to "low-impact technology" on the furthest right.[1]

Boyle's schema is rather awkward, however, as none of the different streams fit easily within a left-right continuum - although most of the streams have a "left" element in the sense that they embody a cri-

---

[1] G. Boyle, "A. T. is Dead - Long Live E. T.!" paper presented to *A. T. in the 80s* Conference, London, 16th June 1984, pp. 3-4, 10. Note: Boyle advocates the term "ecotechnics" as a replacement for the now ubiquitous but ambiguous "alternative technology" or "appropriate technology".

tique of the existing society and its technology along with proposals for reform. The inadequacy of a left-right continuum for categorizing Appropriate Technology has been stressed by Morrison who observes that, while it might be described as left-leaning and while it embodies certain elements of Fabian, communalistic and agrarian socialism, the society envisaged by the movement is qualitatively different than that of the centralist and expansionist images of socialism and does not reject some elements dear to capitalism.[2] Morrison writes:[3]

> This means that there is both "creative tension" in the A.T. movement as well as some genuine conflicts of emphasis within A.T. and between A.T. and other movements as the strange and not-so-strange bedfellows ... try to determine whether it is possible to get comfortable under a common cover which is, in fact, an ideological quilt.

Taken together, the comments of both Boyle and Morrison reflect the fact that the Appropriate Technology movement is neither simple nor uniform.

Despite the profusion of *terms* within the rhetoric of the movement an attempt will not be made here to document the ideas and activities which correspond with each term; the differences in meaning associated with various terms are often either superficial or ambiguous. Instead, the various *themes* which typify the movement in the North will be summarized, noting that many of these themes may be denoted by any one of a number of rubrics.

A further difficulty with discussing the Appropriate Technology movement in the North is that the associated organizations mostly do not deal exclusively with the concerns of the North, notwithstanding the fact that certain themes do stem especially from the North. An O.E.C.D. survey has discovered that less than two-fifths of the world-wide activities of Appropriate Technology organizations based in the North are focussed on the North per se.[4] This chapter will therefore primarily consider Appropriate Technology *themes* vis-a-vis the North rather than the activities of *organizations*.

---

[2]   D. E. Morrison, "Energy, Appropriate Technology and International Interdependence", paper presented to the Society for the Study of Social Problems, San Francisco, September 1978., pp. 5-24.

[3] *Ibid.*, p. 17.

[4]..N. Jéquier and G. Blanc, *The World of Appropriate Technology* (Paris: Organization for Economic Cooperation and Development, 1983), p. 34.

There is a greater volume of literature dealing explicitly with Appropriate Technology for the South than for the North. There is nevertheless a substantial range of compendia, readers, books and other publications which attempt a general overview of the latter. Examples may be found in the work of Bender, Coe, Davis, Dorf and Hunter, Frahn and Buttel, Henderson, Kohr, McRobie, de Moll and Coe, Pausacker and Andrews, Roby, Sachs, Sale, Todd, VanDerRyn and Willoughby.[5] This material demonstrates that the topic is being seriously addressed. There is, however, little consensus at this stage on systematic theory for the application of Appropriate Technology specifically for the North.

Responses to the Technological Society

The bulk of the impetus for Appropriate Technology in the North appears to have come from responses to the perceived growth of the

[5] T. Bender, *Sharing Smaller Pies* (Portland, Or.: Rain, 1975); T. Bender, ed., *Rainbook*(Portland, Or.: Rain, 1977); Coe, *Present Value*; J. Davis, "Appropriate Technology for a Crowded World", *New Universities Quarterly*, 32, 1 (1977), 25-36; J. Davis, *New Directions for Technology*, the Tawney Lecture, 1981 (London: Intermediate Technology Publications, Ltd., 1982); J. Davis, *Technology for a Changing World* (London: Intermediate Technology Publications Ltd., 1978); Dorf and Hunter, *Appropriate Visions*; A. Frahn and F. H. Buttel, "Appropriate Technology: Current Debate and Future Possibilities", *Humboldt Journal of Social Relations*, 9, 2 (1982), 11-37; H. Henderson, *Creating Alternative Futures: The End of Economics* (New York: Perigree, 1978); H. Henderson, "Coming of the Solar Age", in *The Schumacher Lectures*, ed. by S. Kumar (London: Blond and Briggs, 1980), pp. 165-181; L. Kohr, "Appropriate Technology", in *The Schumacher Lectures*, ed. by S. Kumar (London: Blond and Briggs, 1980), pp. 182-192; L. Kohr, *The Overdeveloped Nations: The Direconomies of Scale* (Reprint; New York: Schocken, 1978); G. McRobie, "Technology Choice in Rich Countries", in *Small is Possible*, by G. McRobie (London: Jonathan Cape, 1981), pp. 73-179; L. de Moll and G. Coe, eds., *Stepping Stones: Appropriate Technology and Beyond* (London: Marion Boyars, 1979); I. Pausacker and J. Andrews, *Living Better with Less* (Ringwood, Aust.: Penguin, 1981); K. Roby, "The Prospects for Appropriate Technology", in *Challenges for Einstein's Children: Keith Roby's Vision of Science in Community Life*, ed. by I. Barns (Perth: Murdoch University, 1984), pp. 126-137; I. Sachs, "Conditions of Development: Whither Industrial Societies?", *ITCC Review*, 30 (July 1979), 4-10; I. Sachs, "How Do We Get There? Transition Strategies Towards Another Development in the North", *IFDA Dossier*, 3 (January 1979), 1-12; K. Sale, *Human Scale* (New York: Coward, McCann and Geoghegan, 1980); N. J. Todd, ed., *The Book of the New Alchemists* (New York: E. P. Dutton, 1977); S. VanDerRyn, *Appropriate Technology and State Government* (Sacramento, Cal.: Office of Appropriate Technology, 1975); K. W. Willoughby, "Appropriate Technology for Australia", report of a workshop at Melbourne University, December, 1983, in *1984 and Beyond*, ed. by C. Ledger (Sydney: Australian Council of Churches, 1984), pp. 80-82.

technological society. These responses may be described briefly in the following terms.

Technology or technicity appear to have played an increasingly dominant role in Western civilization (and now almost all cultures) with the progression of time; and, with the increasingly technological nature of society has come increasing contention over the status, social value and tangible impacts of technology. While technology has always played a role in human society and culture, responses-to and perceptions-of technology now appear more vehement.

The majority of the innovations which make up the international effusion of modern technology-practice originate from the North and were developed to suit the dominant interests of the North. Not surprisingly, such technology has brought undeniable benefits to people in the North, to a degree that has been far from matched in the South. Until recently the majority response of people in the North to its technology has been one of optimism and acceptance. Technology has been viewed as a relatively benign provider of continually increasing benefits. Beginning with the criticisms of a small minority as the middle of the twentieth century drew near, however, a groundswell of dissent from faith in technology erupted in the North, gathering momentum in the early 1960s.

A major stimulus for this dissent was the explosion of atomic weapons at the end of the Second World War. The development of nuclear fission technology represents one of the greatest feats of task-oriented cooperation in modern science and technology; it involved the successful translation of front-line science into technology (making it an example of "high technology" *par excellence*). Realization that one of humankind's supposedly greatest achievements had unleashed such terrible destruction, however, led many people to question the supposed benevolence of technology and to doubt whether technological progress could necessarily be acquainted with human progress. This new mood of incredulity also extended to doubts about the validity of the notion of progress *per se*.[6]

With the release in the early 1960s of Rachel Carson's seminal and popular book, *Silent Spring* the growing skepticism spread to widespread concerns about technology's systemic harm to the environ-

---

[6] Critical reviews of the emerging disquiet over the status of science and technology include: B. Gendron, *Technology and the Human Condition* (New York: St. Martin's Press, 1977); A. Mazur, *The Dynamics of Technical Controversy* (Washington, D. C.: Communications Press, 1981); D. Nelkin, ed., *Controversy: Politics of Technical Decisions*, 2d ed. (Beverly Hills: Sage, 1984); S. Yearly, *Science, Technology and Social Change* (London: Unwin Hyman, 1988).

ment.[7] The rapid growth of environmentalism was linked to the realization that increasing use of technology had led to insidious effects on human health. Attention had been drawn to this as far back as the early 1950s in the work of Murray Bookchin (alias Lewis Herber).[8] This became more widespread as the effects of pollution became more obvious by the 1960s.

The 1960s were also marked by a growing awareness that technology's role in modern industrialism could be associated with other risks. The work of Ralph Nader and others indicated that harmful impacts could arise not only from the unforeseen side-effects of technology but also from the normal use of industrial products by consumers.[9] The ruthless use of science and technology for mass destruction in the Vietnam War added further vigor to the growing dissent.

The publication of Jacques Ellul's book *The Technological Society* in the United States in 1964 fueled the view that modern industrial society exhibited intrinsic characteristics which were inimical to traditional human values and needs, and that this was somehow related to technology. Ellul's writings encapsulated the view that technology could no longer be viewed simply as neutral tools to be used by human beings as means towards autonomous human ends; because so many human activities had become dependent upon technology, and because so much technology in turn depended upon other technology for its effectiveness, society itself had become technological.[10]

While technology is a human creation and is in principle subject to human control, a growing number of writers began to point out that certain dynamics and imperatives of technological development transcended the control of individual people and groups, thereby confound-

---

[7] R. Carson, *Silent Spring* (Boston: Houghton Mifflin, 1962).

[8] L. Herber [M. Bookchin], "The Problem of Chemicals in Food", *Contemporary Issues*, 3, 12 (1952), 206-241. Bookchin, under the pseudonym of Lewis Herber, later developed a systematic and more general treatment of the environmental aspects of the technological society in his book *Our Synthetic Environment* (New York: Alfred A. Knopf, 1963); this was later republished under his real name (New York: Harper and Row, 1984). Note: Schumacher indicates that he had been influenced by this book (see *Small is Beautiful*, pp. 104-105, 285).

[9] E.g., R. Nader, *Unsafe at Any Speed* (New York: Grossman, 1965).

[10] Ellul's book was originally published in France a decade earlier as *La Technique ou l'Enjeu du Siècle* (Paris: Librarie Armand Colin, 1954); the title means, literally, "La Technique - the Stake of the Century". Other publications which express this perspective include: G. F. Jünger, *The Failure of Technology* (Chicago: Regnery, 1956); Meyer, *Der Technissierung*; Galbraith, *New Industrial State*; J. Habermas, *Toward a Rational Society*, trans. by J. Shapiro (London: Heinemann, 1971); P. L. Berger, B. Berger and H. Kellner, *The Homeless Mind: Modernization and Consciousness* (New York: Random House, 1973).

ing human control of technological change. At the aggregate level (the level of sociological analysis) technology was seen as being subject to laws of progress independent, or at least semi-independent, of social judgements as to what was desirable. This problem had been put into words by Oswald Spengler in the early 1930s: "The lord of the World is becoming the slave of the Machine."[11] Spengler's observation (elaborated and extended by Ellul) is arguably consistent with claims about the forces of production made during the previous century by Marx and Engels:[12]

> This consolidation of what we ourselves produce, which turns into an objective power above us, growing out of our control thwarting our expectations, bringing to naught our calculations, is one of the chief factors in historical development up to now.

Ellul himself claimed to be merely *describing* the trends of the technological society, rather than *evaluating* them.[13] A popular movement appeared from the early 1960s onwards, however, which held that: modern industrial society was becoming inimical to human health in both psychosocial and biophysical terms; that technology was at the center of this process; that the emerging technological society was insidiously eroding people's control over their own affairs; and, that some decisive response to "technology" (often vaguely defined) was required for the maintenance of human and environmental health. Concerns were being raised about technology *per se*, and not just about individual technologies.

In the general literature on the technological society two broad schools of thought may be discerned. One school tends to view the growth of technology in civilization as a threat to human health and to the cultivation of distinctively human attributes (such as subjectivity, self-consciousness, personal creativity and individual freedom). Another school tends to portray the increasing technological nature of civilization as an authentic expression of human-ness and as a central characteristic of human progress. Both schools acknowledge the existence of human and environmental problems; the former sees the

---

[11] O. Spengler, *Man and Technics*, trans. by C. F. Atkinson (New York: Alfred A. Knopf, 1932), p. 90. Spengler had also raised such a theme in his earlier volume, *The Decline of the West*, trans. by C. F. Atkinson (New York: Alfred A. Knopf, 1928).

[12] K. Marx and F. Engels, *The German Ideology*, ed. with an introduction by R. Pascal (New York: International Publishers Inc., 1939), p. 23.

[13] Ellul, *Technological Society*, pp. xxvii-xxxiii.

growth of technology as a cause of such problems, however, whereas the latter sees an intensification of the technological society as a precondition for their solution.

The Appropriate Technology movement in the North may be seen as part of an emerging third school of thought (with roots in earlier occidental traditions) which avoids both of the above two extremes. Whereas proponents of the former school tend to advocate a reduction in the scope for technology and proponents of the latter advocate an increase in the scope for technology, Appropriate Technology is biased towards neither option. Proponents of Appropriate Technology tend to be concerned about the *direction* of technology-practice and the *nature* and *control* of technology. Appropriate Technology accords with the views of those who reprove the technological society (in so far as technology has tended to harm people and the environment) and also of those who laud the technological society (in so far as technology may act as a means to human fulfillment and ecologically sound management of the environment).

"Humanized Technology"

Many critics of the technological society counterpoise technology or technicity, on one hand, and humanity, on the other hand. Humanity is often typified, for example, by: spontaneity, transcendence, subjectivity, freedom, individuality, diversity, personal autonomy, organic processes, open-endedness, communicative and symbolic interaction, self-consciousness, spirituality, normative evaluation, or personal creativity. In contrast, technology or technicity are often typified as embodying, for example: order, uni-dimensionality (leaving no room for transcendence), objectivity (to the exclusion of subjectivity), determinism, uniformity, automatism, mechanism, closed and linear processes, unconscious information processing and transfer, instrumental action, purposive-rational action, positive action (i.e. free from normative evaluation), cause-and-effect reactions and predictability.

The inconsistency of the literature's terminology means that it is often difficult to work out exactly what it is that is being counterpoised, and whether the contrasts made are ontological or accidental-cum-historical. In any case, for proponents of this viewpoint, the *historical reality* of this generation is that technology-practice has become largely "inhuman" or "antihuman" (irrespective of the ontological meaning of "humanity" and "technology"). Herbert Marcuse became a popular spokesman for this critique of technology during the 1960s with his view that human life had become "one-dimensional" in the technological society amidst "a comfortable, smooth, reasonable, democratic unfreedom".[14] Marcuse states: "In the medium of technology, culture, politics and the economy merge into an omnipresent system, which swallows up or repulses all alternatives."[15].

In response to the perceived "dehumanization" of modern society a number of thinkers have, rather than advocate rejection of technology *per se*, called for the "humanization" of technology.[16]

Environmentalism

There has been a close link between environmentalism in the North and the spread of the Appropriate Technology movement.[17] There is a long tradition in European thought and North American thought of viewing the processes of nature and the principles of technology as being somehow in conflict.[18] In the mid 1960s a spate of scholarly and popular work began to appear which lay direct blame for modern ecological problems on Western technology.[19] It was not until the growth

---

[14] H. Marcuse, *One Dimensional Man* (Boston: Beacon Press, 1964), p. 1.

[15] *Ibid.*, p. xvi.

[16] E.g.: Fromm, *Revolution of Hope*; P. Goodman, "Can Technology be Humane?" in *New Reformation: Notes of a Neolothic Conservative*, by P. Goodman (New York: Random House, 1969) (reprinted in *Western Man and Environmental Ethics*, ed. by I. G. Barbour [Reading, Mass.: Addison-Wesley, 1973], pp. 225-242); W. Schirmacher, "From the Phenomenon to the Event of Technology", in *Philosophy and Technology*, ed. by P. T. Durbin and F. Rapp (Dordrecht: D. Reidel, 1983), pp. 275-289. Cf.: F. J. Dy, "Technology to Make Work More Human", *International Labour Review*, 117, 5 (1978), 543-555; K. K. Murthy, "Man, Technology and the Natural Holism", *Impact of Science on Society*, 30, 2 (1980), 81-85; J. Rodman, "On the Human Question", *Inquiry*, 18 (1975), 127-166.

[17] A survey of the development of environmentalism in the North and of the relations to various political and social factors has been conducted by S. Cotgrove (*Catastrophe or Cornucopia: The Environment, Politics and the Future* [Chichester: John Wiley and Sons, 1982]); Cotgrove demonstrates that environmentalism, though unified by a concern with environmental issues, is a heterogeneous movement which is in fact a medium for the expression of profound discontents over the fundamental nature of industrial society (viz., "society" is of as much concern within environmentalism as "nature").

[18] E.g., Meyer, *Technisierung* (European). A survey of the American tradition has been conducted by L. Marx (*The Machine in the Garden: Technology and the Pastoral Idea in America* [Oxford: Oxford University Press, 1964]).

[19] A highly influential article which sparked off a complex debate on this topic was published in the journal of the American Association for the Advancement of Science

of the Appropriate Technology movement, however, that the different literatures concerned with the human impacts of technological society, on one hand, and environmental impacts of technological society, on the other hand, began to merge.[20] Appropriate Technology accorded with the growing tendency to view environmental problems and socio-cultural problems as intertwined.[21]

Notwithstanding the fact that the Appropriate Technology movement has tended to advocate certain types of technology-practice on *both* human *and* environmental grounds, much of the literature exhibits a distinctly environmental focus. A number of compendia and readers make this focus explicit[22] and it is a pervasive influence on most of the Appropriate Technology journals and magazines from the North.[23] In addition to an abiding concern for energy issues, most of the Appropriate Technology groups dealing with urban-industrialized cultures place much attention on horticulture, aquaculture and other innovative aspects of agriculture.[24] The movement draws upon both

in 1967 (L.White, Jr., "The Historical Roots of Our Ecologic Crisis", *Science*, 155, 3767 (1967), 1203-1207).

[20] This trend is reflected in the writings of T. Roszak (*Where the Wasteland Ends: Politics and Transcendence in the Post-Industrial Society* [New York: Anchor, 1973]; *Person/Planet: The Creative Disintegration of Industrial Society* [New York: Anchor Doubleday, 1978]) and B. Commoner (*The Closing Circle* [New York: Alfred A. Knopf, 1971]; *Alternative Technologies for Power Production*, ed. with H. Boksenbaum and M. Corr [New York: MacMillan Information, 1975]; *The Poverty of Power* [New York: Alfred A. Knopf, 1976]; "Freedom and the Ecological Imperative: Beyond the Poverty of Power", in *Appropriate Visions: Technology the Environment and the Individual*, ed. by R. C. Dorf and Y. L. Hunter [San Francisco: Boyd and Fraser, 1978], pp. 11-49). Roszak's knowledge-of and commitment-to Appropriate Technology was made explicit in his foreword to a paperback edition of Schumacher's *Small is Beautiful* for the North American audience (New York: Harper Colophon, 1975).

[21] See: M. North and S. Kumar, *Time Running Out? Best of Resurgence* (Dorchester: Prism Press, 1976); cf., E. Goldsmith, et al. [The Ecologist], *A Blueprint for Survival* (Harmondsworth: Penguin, 1982).

[22] E.g.: B. McCallum *Environmentally Appropriate Technology: Renewable Energy and Other Developing Technologies for a Conserver Society in Canada* (4th ed; Ottawa: Department of Fisheries and Environment, 1977); Coe, *Present Value*; Bender, *Rainbook*; J. Baldwin and S. Brand, *Soft-Tech* (Harmondsworth: Penguin, 1978); S. Brand, ed., *The Updated Last Whole Earth Catalog: Access to Tools* (New York: Point/Random House, 1974); S. Brand, ed.,*The Whole Earth Epilog* (Baltimore: Point/Penguin, 1974).

[23] E.g.: *Acorn* (Governors State University, Park Forest South, Illinois); *A.T. Times* (Butte, Montana); *De Kleine Aarde* (Munsel, Netherlands); *Journal of the New Alchemists* (Woods Hole, Massachusetts); *Mother Earth News* (Hendersonville, U.S.A.); *New Roots* (Amherst, Massachusetts); *Rain* (Portland, Oregon); *Self-Reliance* (Washington, D.C.); *The CoEvolution Quarterly* (Sausalito, California); *The Ecologist* (Wadebridge, Cornwall); *Undercurrents* (London).

[24] G. Friend, "Nurturing a Responsible Agriculture", in *Stepping Stones: Appropriate Technology and Beyond*, ed. by L. de Moll and G. Coe (London: Marion

scientific knowledge from the discipline of ecology and from the emerging field of environmental ethics[25]. These influences normally lead to an active interest in methods of recycling waste materials[26] and in food production methods which avoid the use of synthetic fertilizers, pesticides and herbicides.[27] A substantial body of empirical evidence and international opinion is now available as to the economic and technological viability of ecological agriculture (as it is often denoted).[28] Experiments with operating agricultural activities in urban

Boyars, 1979), pp. 146-157; T. Cashman, "The New Alchemy Institute: Small-Scale Ecosystem Farming", *Appropriate Technology*, 2, 2 (1975), 20-22; M. Morrissey, "Agricultural Self-Sufficiency: The Recent History of an Idea", *Studies in Comparative International Development*, 17, 1 (1982), 73-95; R. Merrill, ed., *Radical Agriculture* (New York: Harper Colophon, 1976).

    25 See: D. E. Marietta, "The Interrelationship of Ecological Science and Environmental Ethics", *Environmental Ethics*, 1 (1979), 195-207; E. Katz and S. Barbash, "Environmental Ethics and Consumer Choice: A Conceptual Case Study", *Humboldt Journal of Social Relations*, 9, 2 (1982), 142-160; I. G. Barbour, ed., *Western Man and Environmental Ethics: Attitudes Towards Nature and Technology* (Reading, Mass.: Addison-Wesley, 1973).

    26 A. H. Purcell, *The Waste Watchers* (Garden City, N.Y.: Anchor/Doubleday, 1970); N. Seldman, *Garbage in America: Approaches to Recycling* (Washington, D.C.: Institute for Local Self-Reliance, 1975); Institute for Local Self-Reliance, "Community-Based Recycling: Three Successful Models" (Part I published in *Self-Reliance*, 21 (1980), 4-5, 10-11; Part II published in *Self-Reliance*, 22 (1980), 4, 12-13); R. Blobaum, et al., *A Potential for Applying Urban Wastes to Agricultural Land* (West Des Moines, Indiana: Robert Blobaum and Associates, 1979); C. Thomas, *Material Gains: Reclamation, Recycling and Reuse*, (London: Earth Resources Research Ltd., 1979); D. Hayes, *Repairs, Reuse, Recycling: First Steps Towards a Sustainable Society*, Worldwatch Paper #23 (Washington, D.C.: Worldwatch Institute, 1978); J. Vogler, *Muck and Brass: Domestic Waste Reclamation Strategy for Britain* (Oxford: Oxfam Public Affairs Unit, 1978); I. Pausacker, *Recycling: Is it the Solution for Australia?* (Harmondsworth: Penguin, 1978).

    27 A. Conacher and J. Conacher, *Organic Farming in Australia*, Geowest #18, Occasional Paper (Perth, Aust.: Department of Geography, University of Western Australia, 1982); C. Prendergast, *Introduction to Organic Gardening* (Los Angeles: Nash Publishing, 1971); R. Rodale, ed., *The Basic Book of Organic Gardening* (New York: Ballantine, 1971); H. Tyler, *Organic Gardening Without Poisons* (New York: Van Nostrand, 1971). Cf., Journals such as *Tilth* (Arlington, Washington).

    28 See: S. Hill and P. Ott, eds., *Basic Technics in Ecological Farming: The Maintenance of Soil Fertility*, Proceedings of the 2nd and 3rd International Conferences of the International Federation of Organic Agriculture Movement, held at Montreal (October 1978) and Brussels (September 1980) (Basel: Birkhäuser, 1982); B. Commoner, et al., *Report of a Comparative Study of Conventional and Organic Farms in the American Corn Belt*, Report #CBNS-AE-4 (Saint Louis, Mo.: Center for the Biology of Natural Systems, Washington University, 1975); G. M. Berardi, *Organic and Conventional Wheat Production: Examination of Energy and Economics* (Ithaca, N.Y.: Department of National Resources, Cornell University, 1977); United States Department of Agriculture, *Report and Recommendations on Organic Farming* (Washington, D.C.: United States Department of Agriculture, 1980); R. C. Oelhaf, *Organic Agriculture: Economic and Ecological Comparisons with Conventional Methods* (Montclair, N.J.: Allanheld, Osmun,

areas are also typical of the movement.[29] Many of the above activities require relatively sophisticated science or technology (i.e., sophisticated appropriate technology) if they are to be maintained in a viable manner.[30]

Radical Technology-practice

One stream of the movement in the North has been categorized under the rubric "radical technology", a phrase which has been adopted most vigorously by a group of protagonists associated with a British organization called *Undercurrents Limited*.[31] A book on the subject, published by this group, takes the form of a "grass-roots" encyclopaedic collage rather than a scholarly treatise, and begins with the statement:[32]

This is a book about technologies that could help create a less oppressive and more fulfilling society. It argues for the growth of small-scale techniques suitable for use by individuals and communities, in a wider

& Co., 1978); *Organic and Conventional Farming Compared* (Ames, Iowa: Council for Agricultural Science and Technology, Report #84, 1980).

[29] T. Fox, "Urban Agriculture as an Appropriate Technology", in *Fundamental Aspects of Appropriate Technology*, ed. by J. de Schutter and G. Bemer (Delft: Delft University Press and Sijthoff and Noordhoff International Publishers, 1978), pp. 39-51; Tout Va Bien, "The Gardens of Cockaigne", appended to *People's Technologies and People's Participation*, by P. de Pury (Geneva: World Council of Churches, 1983), pp. 157-162; W. Pierce, "Polyculture Farming in the Cities", in *Radical Agriculture*, ed. by R. Merrill (New York: Harper Colophon, 1976), pp. 224-256.

[30] See the following for examples: Office of Technology Assessment, *An Assessment of Technology for Local Development* (Washington D.C.: U. S. Government Printing Office, 1981)

[31] Boyle and Harper, *Radical Technology*. This group was set up in Gloucestershire in the early 1970s and has been the focus for "radical technology" activities in the London area; the group is also responsible for the journal *Undercurrents*.

[32] Boyle and Harper, *Radical Technology*, p. 5. The informal and vernacular style of the book (and of other groups associated with "radical technology") is illustrated by the following quotes from the end of the preface and the beginning of the introduction: "It took a lot more hard work than we expected to get this wretched book together, and we're sick of the sight of it, so now we're off for a bit of breakfast. Over to you." (p. 5); "Given that modern capitalist industrial societies are morally contemptible, ruthlessly exploitative, ecologically bankrupt, and a hell of a drag to live in, is there anything that we can do to change them?" (p. 6). While differing from much in orthodox Marxist socialism, *Undercurrents* is nevertheless explicitly committed to the ideals of the political left.

social context of humanized production under workers' and consumers' control.

Most streams and people in the Appropriate Technology movement would advocate the promotion of technologies compatible with this vision, but not all would adopt the same style of expression. The whole Appropriate Technology movement may be viewed as radical in the sense that fundamental re-examination of the role of technology in society is advocated (along with fundamental changes in the design and selection of technology); those who adopt the term "radical technology",however, claim to be distinctive.[33] The people associated with *Undercurrents* link themselves most closely with the anarcho-utopian stream of socialism, but with qualifications. They summarize their position as follows:[34]

> What emerges from these varied influences is a jumble of theories and elements: a theory of technology and society which insists that we can control technology, but if we don't it will control us; recognition of physical and biological constraints on human activity; social structure emphasizing group autonomy and control from the bottom up; a bias towards simplicity and frugality in life and technology whenever possible; preference for direct gratification in production rather than through the medium of commodities; an exploratory rather than dogmatic application of the theory (such as it is...); willingness to learn from unlikely sources such as "primitive" cultures and technologies, "mystical" experiences or abilities, and even liberal social theory. This may sound a strange chimera. Well, we have no monopoly on radical technology. Make up your own criteria if you don't like these.

In this section we attempt to draw attention to the chief concerns of the anarcho-utopian approach to Appropriate Technology, as typified by *Undercurrents*, but we will include other approaches of differing style but which are nevertheless radical in intention.[35]

---

[33] An attempt to outline the distinctive directions in which "radical technologists" should strive has been made by P. Harper, emphasizing organization rather than technological artefacts, social(ist) rather than ecological considerations and production rather than consumption (P. Harper, "What's Left of Alternative Technology?" *Undercurrents*, 6 (March/April 1974), 35-38.

[34] Boyle and Harper, *Radical Technology*, p. 8.

[35] Note: Boyle and Harper indicate (*ibid.*, p. 272) that they consider the approach of Schumacher and the I.T.D.G. not to be "radical" (i.e. anarcho-utopian); however Theodore Roszak ("Foreword", *Small is Beautiful* [1975]) explicitly places Schumacher

Advocates from within the broader spectrum of "radical" approaches tend to resist viewing technologies as discrete entities and tend to depict technology as an aspect of society.[36] Thus they avoid using "technology" with the narrow definition we raised in Chapter Two. Accordingly, technological change is not viewed as something which can occur exogenously vis-a-vis social and political change. We have adopted the phrase "radical technology-practice" here to cover this approach without violating our semantic conventions.

Advocacy of radical technology-practice is normally motivated by a negative assessment of the technological society and involves insistence upon the need for structural changes to the *whole* socio-technological system and its superstructure.[37] Changes to technology, independent of other action aimed at comprehensive changes to the *whole* social system, are viewed suspiciously. "Radical technology" is used as a shorthand term for what should more strictly be called "technology for radical social change". Thus, advocates of radical technology-practice support technology appropriate to an *alternative society*.[38] For this reason the term "alternative technology" is often used interchangeably with "radical technology".[39]

Advocates of the term "alternative technology" sometimes claim that it is more accurate and meaningful than "appropriate technology" because the latter is only valid if it is specified for whom and what purpose the technology should be "appropriate".[40] "Alternative", on

in the anarcho-utopian tradition; this type of enigma is not unusual in the "A.T." literature.

[36] An example of this broader approach, which contains a survey of its seminal influences, may be found in *Alternative Technology* by David Dickson; cf., D. Hart, "Concepts and Precepts in Alternative Technology" (Mimeo; Edinburgh: Science Studies Unit, Edinburgh University, 1975); R. Clarke, "The Pressing Need for Alternative Technology", *The Impact of Science on Society*, 23, 4 (1973), 257-271.

[37] Cf., Lodwick and Morrison, "Research Issues".

[38] This is brought out explicitly by R. Clarke in "Technology for an Alternative Society", published in a book edited by himself entitled *Notes for the Future: An Alternative History of the Past Decade* (London: Thames and Hudson, 1975), pp. 154-159. Concurrence with this view from the South may be found in the work of M. M. Hoda: "It must be understood clearly that appropriate technology is for the poor people and weak societies and should be developed with the help and participation of the people themselves. It is a technology of the people, for the people, and by the people. It is not only a technology but a total philosophy for an alternative society and different style of living and should be treated as such" ("Appropriate and Alternative Technology" [mimeo; Lucknow, India: Appropriate Technology Development Association, n.d.], pp. 5-6).

[39] See, Boyle and Harper, *Radical Technology*, pp. 268-269.

[40] E.g.: A. K. N. Reddy, interview in film by P. Kreig (*Tools of Change* [Freiburg: Teldok Film, 1978]). Reddy also suggests the term "inequality-reducing technologies" as

the other hand, is argued as having more substantive content because it directly emphasizes that the technology ought to be directed at changes to the status quo. It also implies that the "appropriate" technology would be different to dominant or currently existing technology. These criticisms are plausible. It should be added, however, that parallel criticisms may be directed at "alternative technology". Currently used technology could in fact be quite appropriate for the sort of alternative society envisaged by radical-cum-alternative technologists (on the assumption that the nexus between changes in technologies and broader sociopolitical changes is not rigid). Consequently, the adoption of an alternative technology to the present one might be counterproductive from the point of view of creating an alternative society.

"Alternative technology", in conclusion, needs to be subjected to the same semantic restriction placed on "appropriate technology" in Chapter Two: it should only be employed with a specific-characteristics meaning for specific situations for which the circumstances have been defined. "Appropriate technology", as defined in Chapter Two, is a more universally applicable rubric than "alternative technology"; it may include technologies appropriate to an alternative society (so long as this is clearly indicated) but may do so without the inconsistencies inherent in "alternative technology". In this study "alternative" will still be employed, however, but for the specific purpose of denoting options in technology other than the dominant or presently used kind.

The radical technology and alternative technology streams within the Appropriate Technology movement are unified in maintaining an overriding interest in the issue of *control*.[41] Control by "the people" over the forces of production which affect their lives appears to be the pivotal objective. Where alternative-cum-radical technologists appear to differ from mainstream Marxist socialism and other "peoples power" movements is on the insistence that technology requires special attention and plays a role in either enabling or restricting the attainment of broader political purposes. While accepting the traditional Marxist view that social control of production is desirable and requires a political struggle for its realization, alternative-cum-radical technologists place little faith in changes in the relations of production independently of appropriate changes in the means of production. They also have little faith in centralized and state-orchestrated political

an unambiguous expression of the ideas behind "alternative technology" (Reddy, "Alternative technology", p. 334).

[41] D. Elliot and R. Elliot, *The Control of Technology* (London: Wykeham 1976), esp. pp. 132-239.

"change" independent of popular political action at the local level. "Radical technologies" are viewed as providing a medium through which oppressed social classes may develop greater political control, although they are not viewed as a substitute for other political means.[42] Advocates of radical technology also tend to differ from mainstream Marxists by pointing to *capitalistic technological society* as a source of structural injustice rather than to capitalism *per se*.

Other notions, which reflect some of the concerns of radical technologists and alternative technologists but differ in style and emphasis, may be grouped under "radical technology-practice". Murray Bookchin is one thinker with anarcho-utopian commitments who endorses the approach of *Undercurrents* but who exhibits a different style.[43] Bookchin was a pioneer in environmentalism[44] and, reflecting themes raised elsewhere by others, has evolved a well articulated theoretical framework.[45] Central to Bookchin's perspective is the concept of *liberation* (or human freedom) and an analysis of the political-cum-technological conditions conducive to a libertarian society. He has coined the term "liberatory technology" to denote technology

---

[42] This perspective is reflected clearly in the following quote from a paper by D. Elliot entitled "Alternative technology: Production for Need" (presented to *A.T. in the'80s* Conference, London, 16th June 1984): "Technology - the hardware - by itself is not the issue - rather the question is which technology to develop, how and in whose interest. Technology simply provides a context for political struggle. To a degree the traditional A.T. shopping list, small scale windmills, solar collectors, etc., with their clear social and environmental implications, provide a symbol of the sort of society and associated values we feel is appropriate. ... The fight to get A.T. established ought to help create the political confidence and organization necessary for a genuinely democratic self managed society. In that sense A.T. is a 'radical technology' - a technology the struggle for which both pre-figures and brings about social change" (p. 5).

[43] M. Bookchin, "Self-Management and the New Technology", *Telos*, 41 (Fall 1979), 1-16 (esp. p. 15).

[44] See references above to works under his pseudonym, Lewis Herber.

[45] E.g.: M. Bookchin, *The Ecology of Freedom: The Emergence and Dissolution of Hierarchy* (Palo Alto, Cal.: Cheshire Books, 1980). The following sources of influence on Bookchin are listed here because they represent a broad intellectual influence on the "radical technology-practice" stream. See: M. Weber, *The Protestant Ethic and the Spirit of Capitalism*, trans. by T. Parsons (London: Unwin Books, 1930); M. Horkheimer: *Eclipse of Reason* (New York: Columbia University Press, 1947); *Critique of Instrumental Reason* (New York: Seabury, 1974); T. Adorno and M. Horkheimer, *Dialectic of Enlightenment* (New York: Seabury, 1972); H. Jonas, "Technology and Responsibility: Reflections on the New Tasks of Ethics", *Social Research*, 40 (1972), 31-54; P. Kropotkin, *Fields, Factories and Workshops*, ed. by C. Ward (London: Allen and Unwin, 1974; first published in 1899); M. Buber, *Paths in Utopia* (Boston: Beacon Press, 1958). Buber's book is an overview of the whole utopian traditon which has influenced the modern "anarcho-utopian" approach to technology.

which accords with this vision[46]; he stresses "self-management" by people in local communities as an attribute of a libertarian society, but insists that this concept is counterfeit if defined in economic or technical terms rather than ethical and social terms.[47] He also employs the term "people's technology".[48] Bookchin is strongly anti-hierarchical in his social philosophy and has analyzed the connections between liberatory technology and the embodiment of ecological principles in community structures. He takes the view that human appropriateness and environmental appropriateness vis-a-vis technology may not be achieved separately from each other. He employs the term "ecotechnologies" interchangeably with "liberatory technologies" to underline this point.[49]

The type of society and technology advocated by Ivan Illich and discussed in the previous chapter under the rubric of "convivial technology" could also be categorized as part of radical technology-practice.[50] "Convivial technology" (Illich) and "ecotechnology" (Bookchin) are very similar notions, although Bookchin appears to draw more heavily than Illich upon philosophical and ethical concerns about nature. The term "appropriable technology" has received attention as a semantically lucid term for denoting the radical technology-practice approach.[51] It has been taken-up mostly by the World Council of Churches as part of the Council's practical programs in development aid.[52] Appropriable technologies are ones that may be appropriated by communities themselves and by people themselves; the capacity for technologies to be appropriated by poor or oppressed people accords with the requirements of a participatory approach to social and economic development.[53]

---

[46] M. Bookchin, "Toward a Liberatory Technology", in *Post-Scarcity Anarchism* (San Francisco: Ramparts Press, 1971), pp. 83-139.

[47] Bookchin, "Self-Management".

[48] *Ibid.*, p. 15.

[49] M. Bookchin,"The Concept of Ecotechnologies and Ecocommunities", *Habitat*, 2, 1/2 (1977), 73-85.

[50] Illich, *Convivial Technology*.

[51] E.g.: G. Blanc, "Technologie 'Appropriée/Appropriable': Le Mot et la Chose", *Autrement*, 27 (October 1980), 13-23.

[52] Commission on the Churches' Participation in Development, *On Appropriate Technology #2; Directory of Appropriable Techniques* (1st ed.; Geneva: World Council of Churches, 1979), First Supplement (October 1980).

[53] CCPD Technical Services, *The Churches and People's Technologies*, report of the "Reference Group", Geneva, 3-5 July 1980 (Geneva: World Council of Churches, mimeo, 1980); cf., de Pury, *People's Technologies*.

Some of the deeper questions of social theory implied by the radical technology-practice approach will be addressed in later chapters. This section will be concluded now with the observation that it is only partially valid to portray the "alternative technology" and "radical technology" streams of the Appropriate Technology movement as more radical than other streams, such as have evolved under the rubric of "intermediate technology" for example. As demonstrated in Chapters Four and Five, Intermediate Technology is above all an approach for enabling disadvantaged people (in the South) to develop economic power through the capacity to control their own means of production. Behind the apparent differences the overall movement appears to be united by a concern for enhancing the *control* by communities and classes of people of their economic and socio-cultural life.

Energy Pathways

Concern about energy issues is almost universal throughout the Appropriate Technology movement. This is partly a reflection of the growing recognition in science and in public policy of the essential role of energy in physical processes, of the common link between different scientific disciplines which may be fulfilled by thermodynamics, and of the critical dependence of industrialized society on the ready availability of reliable energy supplies. The background to energy and Appropriate Technology discussed earlier vis-a-vis the South applies equally to the North. Concern with energy as a general theme has, however, been more pervasive and intense in the North (notwithstanding the importance of energy choices for the South).

The literature on energy and technology for the North is profuse. The sudden rise in OPEC oil prices in 1973 brought into sharp focus the connections between energy supply, economics and the vulnerability of societies which are dependent upon supplies of energy over which they have little control. Debates over the so-called energy crisis reinforced the need for diversification in the sources of energy supply for a country or region. The events of 1973 and following also served to demonstrate that energy supply may not be reduced to a merely technical or physical problem; social and political factors may lead to a real shortage in available supplies irrespective of physical constraints. The energy crisis led to a rapid expansion of investment into research on energy technologies for the purpose of averting a repetition of the events of 1973. Much of the literature on energy and technology bears little relation to the main concerns of the Appropriate Technology movement and does not warrant investigation here. It is worth noting, however, that advocates of Appropriate Technology have been leaders in addressing energy-futures issues.

Schumacher, for example, appears to have been the first person to emphasize the distinction between renewable and non-renewable resources and apply this to the field of energy (in a 1955 paper); in the same paper he predicts the certainty of an energy crisis.[54] He was a critic of the industrialized world's energy regime over a decade before such a stance became respectable. In a 1960 paper he argued that because the world's reserves of concentrated carbon deposits were of a finite quantity, which was not a very large multiple of the quantity being used each year, the essential non-renewability of fossil fuels possessed great significance.[55] In a 1961 feasibility study, based upon oil reserve estimates by the oil industry's leading geologists, Schumacher predicted an oil "crisis" for countries with high consumption levels and low indigenous supplies (e.g., Western Europe, Japan) by about 1980, at a point, not when world supplies were exhausted, but when they ceased to expand. In that study report he writes:[56]

> ...there can be no doubt that the oil industry will be able to sustain its established rate of growth for another ten years; there is considerable doubt whether it will be able to do so for twenty years; and there is almost a certainty that it will not be able to continue rapid growth beyond 1980.

Schumacher's warnings were treated with derision and contempt throughout the 1960s, until his predictions were vindicated in the 1970s.[57] Concern about energy was a central influence on the development by Schumacher of Appropriate Technology.[58]

The nexus between energy and Appropriate Technology in the North is best represented in the notions of *soft energy systems* and *hard*

---

[54] E. F. Schumacher, "Economics in a Buddhist Country", paper written for the Government of Burma, Rangoon, 1955 (published in his *Roots of Economic Growth* [Varanasi: Gandhian Institute of Studies, 1962]), pp. 6-7.

[55] E. F. Schumacher, "Coal - The Next Fifty Years", paper presented to the National Union of Mineworkers Study Conference, London, 25th, 26th March 1960 (published in *Britain's Coal*, Study Conference Report [London: National Union of Mineworkers, 1960]).

[56] E. F. Schumacher, *Prospect for Coal* (London: National Coal Board, 1961), p. 9.

[57] See "An Interview with E.F. Schumacher, Part I", *Manas*, 29, 20 (1976), 1-2, 7-8.

[58] See, Schumacher, *Small is Beautiful*, pp. 112-135.

energy systems espoused by Amory Lovins. The rest of this section will be restricted mostly to Lovins' work because: it is the most integrated, detailed and consistently argued (with some exceptions); it was the precursor for the work of many others; and, much of the subsequent literature adds little new substantive material.

Lovins, a physicist by training, adopts an interdisciplinary approach using energy as the integrating theme. As a central figure in Friends of the Earth he shares the general concerns of the environment movement and, in particular, severe misgivings about the prospects for a future based upon expanded use of nuclear energy and the proliferation of nuclear weapons.[59] Like Schumacher, he views decisions about energy futures as ethical decisions and considers the widespread adoption of nuclear energy technology to be objectionable on both ethical and technical-cum-economic grounds.[60]

Lovins rejects the view that options for energy supply and demand in the future are technologically determined and not open to human choice. He holds that there are alternative energy strategies available which require conscious choice.[61] In an influential article published in the United States during 1976 Lovins systematically articulates his view that urban-industrialized societies need to make a transition from the dominant ("hard") energy strategy to an alternative ("soft") energy strategy, and that the latter is eminently justifiable not only on ethical grounds but also on economic grounds. He argues that distortions in the market, such as from massive government subsidies for the "hard" strategy or from lack of adequate dissemination of knowl-

---

[59] For a global survey of the debate over nuclear power and of the burgeoning social movement associated with opposition to its proliferation, see J. Falk's book *Global Fission: The Battle Over Nuclear Power* (Melbourne: Oxford University Press, 1982). Falk writes: "Nuclear power is part of a more general trend in industrialized society, and opposition to nuclear power inevitably brings those who oppose it into opposition to other components of that trend" (p. 108).

[60] A. B. Lovins and J. H. Price, *Non-Nuclear Futures: The Case for an Ethical Energy Strategy* (San Francisco/Cambridge, Mass.: Friends of the Earth International/Ballinger, 1975). Schumacher presented a trenchant public criticism of nuclear power from both points of view as the Des Vœx Memorial Lecture (1967) in London (published with modifications as "Nuclear Energy - Salvation or Damnation?" in his book *Small is Beautiful: A Study of Economics as if People Mattered* (London: Blond and Briggs, 1973), pp. 124-135. Lovins indicates his sympathy with Schumacher in a paper entitled "Soft Energy Paths" published in *The Schumacher Lectures*, ed. by S. Kumar (London: Blond and Briggs, 1980), pp. 28-65. A philosophical analysis of the ethical issues associated with nuclear energy technology has been conducted by K. S. Shrader-Frechette (*Nuclear Power and Public Policy: The Social and Ethical Problems of Fusion Technology*; Pallas Paperbacks [Dordrecht: D. Reidel, 1983]).

[61] A. B. Lovins, *World Energy Strategies: Facts, Issues and Options* (San Francisco/Cambridge, Mass.: Friends of the Earth International/Ballinger, 1975).

edge, have prevented the "soft" strategy from being embraced.[62] With the publication of this article and the subsequent publication of a scholarly book which canvasses the ideas and evidence in more detail, considerable debate ensued, setting off a series of published criticisms of the ideas, accompanied by a series of rejoinders by Lovins.[63]

At the center of Lovins' propositions is the notion of the *energy path*, which refers not only to the direction of energy use but to the complex, mutually reinforcing set of factors which make up an energy system in a given society or community. Lovins distinguishes between the technologies which are essential components of the system and the overall system itself (which is *both* social and technological in nature). He claims that different energy paths have different social impacts and that energy policy issues are therefore also social policy issues. The appellations "soft" and "hard" refer primarily to the degree of disruption or impact inflicted upon a given society (and its environment) by an energy path. A hard energy path would thus have a disruptive or injurious impact while the impact of the soft energy path would be relatively benign.[64]

Lovins portrays the (dominant) hard energy path as involving systems which are capital-intensive, complex, large-scale, centralized, resource exogenous, resource intensive, resource depleting, resource degrading and which tend to alienate human beings and generate inequality. In contrast, the soft energy path he advocates involves systems which are light in capital, small in scale, decentralized, resource conserving, resource endogenous and which engender quality work and lifestyle opportunities and reinforce social equity. His image of the

---

[62] A. B. Lovins, "Energy Strategy: The Road Not Taken?" *Foreign Affairs*, 55, 1 (1976), 65-96.

[63] A. B. Lovins, *Soft Energy Paths: Toward a Durable Peace* (San Francisco/Cambridge, Mass.: Friends of the Earth International/Ballinger,1977). Cf., C.B. Yulish, ed., *Soft vs. Hard Energy Paths: 10 Critical Essays on Amory Lovins' "Energy Strategy: The Road Not Taken?"* (New York: Charles Yulish Assoc. Inc., 1977); H. Nash, ed., *The Energy Controversy: Soft Path Questions and Answers by Amory Lovins and his Critics* (San Francisco: Friends of the Earth, 1979). Note: by late 1977 two volumes of critiques of Lovins' work, with responses, totalling over 2200 pages had been published jointly by two U.S. Government Committees: United States Select Committee on Small Business and the Committee on Interior and Insular Affairs, *Alternative Long-Range Energy Strategies* (Washington, D.C.: U.S. Government Printing Office, 976); U.S.S.C. on S.B. and C.I.I.A., *Alternative Long-Range Energy Strategies: Additional Appendices* (Washington, D.C.: U.S. Government Publishing Office, 1977).

[64] While Lovins does not appear to make reference to earlier uses of "hard" and "soft" within the Appropriate Technology movement, he probably adopted the terms from their use by Clark ("Utopian Characteristics") and Dickson ("Alternative technology").

soft energy path is remarkably similar to that of the development strategy proposed by Schumacher for the South under the rubric of "Intermediate Technology" and later under the rubric of "Appropriate Technology".[65]

Lovins holds that each energy path is characterized by particular kinds of technologies. In 1978 he published a paper outlining the characteristics of what he terms "soft energy technologies", and which he later came to refer to simply as "soft technologies".[66] He defines these as technologies which use diverse renewable resources (direct solar, wind, biomass, waterpower, etc.), are relatively simple from a user's point of view (though often technically very sophisticated), and matched in both scale and quality to our range of end-use needs.[67] He argues that soft technologies are not necessarily incompatible with hard technologies, and that a conscious commitment is required at all levels of society to a *gradual transition* from a hard energy path to a soft energy path, during which a judicious combination of both hard technologies and soft technologies will need to be used.

The *soft technology* approach advocated by Lovins, while similar in many respects to the approach advocated under the same name by the radical technology-practice stream (discussed earlier), differs from the latter. Lovins expresses confidence in the possibility of a gradual and widespread transition to the adoption of soft technology through the mechanism of market forces, if current biases in government programs towards hard technologies were neutralized or redressed. The "radical" technologists, in contrast, tend to insist upon the need for socialization of the energy system and related forces of production (at either the state or community level) and upon a more consciously directed radical transformation of the economy.

The two main elements of the nexus between Appropriate Technology and energy are the objectives of increased *energy conservation* and increased use of *renewable energy*. Proponents of Appropriate Technology argue that advances in technology continually improve the

---

[65] See, Schumacher, *Good Work.*. Note: A detailed treatment of the contrasting impacts of the soft and hard paths has been achieved by Morrison ("Social Impacts", pp. 358-365; "Soft, Cutting Edge", pp. 290-291).

[66] A. B. Lovins, "Soft Energy Technologies", *Annual Review of Energy*, 3 (1978), 477-517; Lovins, "Soft Energy Paths", pp. 44-65.

[67] An international survey of technologies of this type which are either currently available or are coming on-line has been produced by the International Project for Soft Energy Paths, based in San Francisco (J. Harding, et al., *Tools for the Soft Path* [San Francisco: Friends of the Earth, 1982]).

competitiveness of these options.[68]   A number of empirically-based studies have been produced which indicate that a transition from a fossil-fuels/nuclear-fission based economy to an energy conservation/renewable-energy based economy is technically possible for most urban-industrialized countries within a generation and without a reduction in the standard of living.[69]   Market trends which affirm the conclusions of these studies will be examined later but it may be noted here, to illustrate this theme, that worldwide reliance on renewable energy sources has grown more than 10 percent per year since the late 1970s.[70]   In the United States, as a further example, renewable energy growth outpaced that of conventional sources from the late-1970s to the mid-1980s and reached about 10 percent of the national energy budget while nuclear power provides only 5 percent;   this trend has occurred despite the preferential government support for the technologies of conventional and nuclear power.[71]

Until recently it was almost universally held by economists and those within the energy industry that expansion of primary energy production is a prerequisite of economic development and employment generation.   A growing number of commentators, however, point to the fact that while there have been historical correlations between these factors, the correlations are not intrinsic;   the adoption of strategies based upon technologies for energy conservation and renewable energy usage make possible employment-cum-economic growth along with

---

[68] E.g.: W. Clark, *Energy for Survival: The Alternative to Extinction* (New York: Anchor, 1974);   D. Deudney and C. Flavin, *Renewable Energy* (New York: W. W. Norton, 1983);   E. P. Gyftopolous, L. J. Lazaridis and T. F. Widmar, *Potential Fuel Effectiveness in Industry*, report to the Energy Policy Project of the Ford Foundation (Cambridge, Mass.: Ballinger, 1974);   Office of Technology Assessment, *Application of Solar Energy to Today's Energy Needs* (2 vols.; Washington, D.C.: U.S. Government Printing Office, 1978).

[69] E.g.: Ford Foundation Energy Policy Project, *A Time to Choose* (Cambridge, Mass.: Ballinger, 1974);   D. Hayes, *Rays of Hope: The Transition to a Post-Petroleum World* (New York: W.W. Norton, 1977);   D. Hayes, *The Solar Energy Timetable*, Worldwatch Paper #19 (Washington, D.C.: Worldwatch Institute, 1978);   G. Leach, et al., *A Low Energy Strategy for the United Kingdom* (London: International Institute for Environment and Development, 1979);   M. Lönnroth, P. Steen and T. B. Johansson, *Energy in Transition* (Stockholm: Secretariat for Futures Studies, 1977);   M. E. Watt and K. R. Roby, "Survey of the Potential Prospects for Renewable Energy Resources in Western Australia", in *Solar Realities in Western Australia in the 1980's*, proceedings of a conference presented by the International Solar Energy Society (Western Australian Branch), ed. by J. E. D. Barker, et al. (Perth, Aust.: University of Western Australia, 1979), pp. 1-10.

[70] C. Flavin and C. Pollock, "Harnessing Renewable Energy" in *State of the World, 1985*, ed. by L. R. Brown, et al. (New York: W.W. Norton, 1985), p. 172.

[71] *Ibid.*, p. 198.

stagnation, or even decline, in energy use rates.[72] A recent study, for example, has shown that electricity use is growing half as fast in California as in Texas, despite nearly identical rates of building and economic growth.[73] The so-called soft energy pathways normally rely upon increased energy-related policy, planning, and action at the local level and by the community as a whole rather than by centralized energy utilities and other centralized bodies alone.[74]

In summary the Appropriate Technology movement in the North places a great deal of attention on energy, points to different pathways available for the development of energy technology, and claims that the "soft" pathway is superior on both normative grounds (especially vis-a-vis social impact) and on technical-cum-economic grounds.

Responses to Local Poverty and Unemployment

Concern about local and/or regional poverty and unemployment has been the main stimulus for Appropriate Technology in the South. A similar situation has not held for the North. This may be partly explained by the North's high wealth levels on an international scale and its dominance in the international economic arena, with the consequence that social welfare programs have been able to ameliorate much of the inequality-producing impacts of dominant economic growth strategies. The countries of the South, by contrast, are normally unable

---

[72] E.g.: D. Annandale, J. Wood and R. Greig, *Building Employment in New Energy Systems in Western Australia*, W.A.I.T. Discussion Paper (Perth, Aust.: Western Australian Institute of Technology, 1984); R. Grossman and G. Daneker, *Energy, Jobs and the Economy* (Boston: Alyson, 1979); S. Laitner, *The Impact of Solar Energy and Conservation Technologies on Employment* (Washington, D.C.: Critical Mass, 1976); Cf., H. E. Daly, et al., *Economics, Ecology, Ethics: Essays Toward a Steady-State Economy* (San Francisco: W.H. Freeman, 1970); Chandler, *Energy Productivity*.

[73] The study, conducted by A. Rosenfeld of the Lawrence Berkeley Laboratory, is reported by C. Flavin in "Reforming the Electric Power Industry" (Chapter 6 of *State of the World, 1986*, ed. by L. R. Brown, et al. [New York: W. W. Norton, 1986], p. 115).

[74] J. H. Alschuler, Jr., *Community Energy Strategies: A Preliminary View* (Hartford, Conn.: Hartford Policy Center, 1970); J. Ridgeway and C. S. Projansky, *Energy-Efficient Community Planning: A Guide to Saving Energy and Producing Power at the Local Level* (Emmaus, Pa.: The J. G. Press, n.d. [prob. 1980]); K. R. Roby, "Energy Prospects for Country Towns", in *Energy and Agriculture: The Impact of Changes in Energy Costs on the Rural Sector of the Australian Economy*, ed. by K. Howes and R. A. Rummery (Perth, Aust.: Commonwealth Scientific and Industrial Research Organization, 1980), pp. 185-199; *Energy Self-Reliance*, reprints from *Self-Reliance* (Washington, D.C.: Institute for Local Self-Reliance, 1979); R. D. Brunner and R. Sandenburgh, eds., *Community Energy Options: Getting Started in Ann Arbor* (Ann Arbor: The University of Michigan Press, 1982).

to generate sufficient publicly owned surplus to afford such redistributive welfare programs.

During the last decade, however, structural unemployment became a politically significant issue in the North; and economic recession simultaneously exacerbated unemployment problems and made it more difficult to fund redistributive welfare programs. Thus, the North appears to be faced with the paradox that anti-poverty welfare programs become less "affordable" the more they are "needed". It has also been recognized that the benefits and costs of orthodox economic growth (or recession) are distributed evenly neither throughout the population nor geographically. Despite their relative wealth, countries of the North have begun to exhibit certain problems of a structural nature which are similar to those which beset the South.[75]

The existence of a fundamentally new situation in the global economy has been recognized in recent studies by the Organization for Economic Cooperation and Development. The Delapalm Report, for example, puts it this way:[76]

> The conclusion we have reached after some two years of investigation is that the changes which constitute the new context, and the relations between them, are so vast and far-reaching that they can neither be reduced to their strictly economic aspects nor explained in purely conjunctual terms.

The report goes on to emphasize that the major relations and trends of the current climate "underline the specific and deeply structural nature of the present problems" as compared with those of the 1950s and 1960s.[77]

In view of these trends an open-ness to new approaches has begun to emerge in the North, although this preparedness to consider alternatives still appears to be a minority phenomenon. As McRobie points out, there have always been relatively underdeveloped regions within the wealthier countries, regions which have special economic needs which

---

[75] M. Castells, *The Economic Crisis and American Society* (Princeton, N.J.: Princeton Unversity Press, 1980); D. P. Moynihan, *Family and Nation* (San Diego: Harcourt Brace Jovanovich, 1986); M. Baldassare, *Trouble in Paradise* (New York: Columbia University press, 1986); P. Abrahamson, *et al., Poverty, Unemployment, Marginalization,* Research Report #1, University of Copenhagen, Department of Sociology, April 1988).

[76] B. Delapalme, et al., *Technical Change and Economic Policy: Science and Technology in the New Economic Context* (Paris: Organization for Economic Cooperation and Development, 1980), p. 14.

[77] *Ibid.,* p. 15 and following.

may differ from the aggregate needs of the national economy. McRobie indicates that these problems existed even during the high-growth period of the 1960s:[78]

> Not surprisingly, the first people in the rich countries to discern that they too needed something on the lines of Intermediate Technology were those living and working in the hinterlands of the large metropolitan economies. As early as 1968 I attended a meeting sponsored by the Memorial University of Newfoundland, and met people from western Scotland, north Norway, Iceland, north and west Canada and other territories - or parts of them - which are the poor areas of the rich countries. It is characteristic of such territories that they closely resemble colonies (which produce, as someone put it, what they do not consume, and consume what they do not produce); and the faster the metropolitan center that controls them grows, the more rapidly they deteriorate. If they are to do more than merely survive with the aid of welfare payments, such communities need technologies appropriate to their resources and lifestyles.

McRobie reveals that the small beginnings of an interest in the relationship between technology, local poverty and local economic development were apparent about two decades ago.

From the late 1970s onwards a small number of organizations in the North began to develop programs aimed at mobilizing technology especially suited to the needs of local communities and groups of people suffering from poverty and unemployment. A strategy of participatory and self-reliant economic improvement and employment generation is stressed. Within the United States the National Center for Appropriate Technology was established for the purpose.[79] Within Britain the Intermediate Technology development Group set up a unit to deal specifically with the U.K.[80] and the Greater London Council

---

[78] McRobie, *Small is Possible*, p. 76. Cf., M. R. Freedman, ed., *Intermediate Adaptation in Newfoundland and the Arctic* (St. John's, Newfoundland: Institute of Social and Economic Research, Memorial University of Newfoundland, 1969).

[79] National Center for Appropriate Technology, Information Brochure; F. Manley, ed., *Economic Development: Bibliography*, Publication #B004 (Butte, Montana: National Center for Appropriate Technology, 1978); D. Quammen, *Appropriate Jobs: Common Goals of Labour and Appropriate Technology*, NCAT Brief #3 (Butte, Montana: National Center for Appropriate Technology, 1980).

[80] Davis, *Changing World*; J. Davis, ed., *AT-UK Bulletin*, newsletter (London: Intermediate Technology Development Group); "Intermediate Technology News: U.K. Project Developments", *Appropriate Technology*, 7, 2 (1980), 33-35; "News from I.T.D.G. - U.K.", *Appropriate Technology*, 7, 4 (1981), 35; A. Bollard, *Industrial Employment*

established an enterprise board[81] which addresses this problem area. Collaboration occurs between these two bodies.[82] A further example from Britain is the Council for Small Industries in Rural Areas which has conducted a well organized service of technical assistance for the non-metropolitan areas of U.K. for many years prior to the growth of the current Appropriate Technology movement.[83] The Office of Technology Assessment of the U.S. Congress has conducted a serious assessment of programs in the U.S. concerned with technology for local development; the resultant report reveals the considerable potential for such technologies, but identifies the need for more concerted support from relevant agencies to enable a significantly wide range of benefits to accrue to needy communities.[84]

The programs in the North which take local *economic development* and *employment generation* as their *raison d'être* tend, at the level of actual projects, to dovetail closely with issues of concern to other streams of the Appropriate Technology movement. An emphasis on technologies for renewable energy usage, resource conservation and recycling, small-scale and low-capital ventures, and on technologies which are humanly attractive, is normal.

Self-sufficiency

Another stream within the Appropriate Technology movement comprises those who advocate self-sufficiency. Proponents of self-suffi-

through Appropriate Technology (London: Intermediate Technology Publications Ltd., 1983); A. Bollard, *Small Beginning: New Roles for British Businesses* (London: Intermediate Technology Publications Ltd., 1983); A. Bollard, *Just for Starters: A Handbook of Small-Scale Business Opportunities* (London: Intermediate Technology Publications Ltd., 1980).

[81] M. Ward and J. Smith, "Interview: The Role of the Greater London Enterprise Board", *The London Accountant* (April 1983), 16-18; Greater London Enterprise Board, *Technology Networks* (London: GLEB, 1983); Greater London Enterprise Board, *Saving Jobs ... Shaping the Future: An Introduction to Enterprise Planning* (London: GLEB, 1983). The Greater London council was eventually disbanded by the Thatcher Governemt, but the Greater London managed to remain in operation.

[82] Private communications: G. McRobie, November 1982; A. Bollard, January 1985.

[83] See: Council for Small Industries in Rural Areas, *Select List of Books and Information Sources on Trades, Crafts and Small Industries in Rural Areas*, (rev. ed.; London: CoSIRA, 1973; first published in 1968); The Development Commissioners, *Development Commission: Council for Small Industries in Rural Areas*, Forty-Second Report (London: Her Majesty's Stationery Office, 1984); private communication with Prof. P. Dunn (CoSIRA and University of Reading), August 1984.

[84] Office of Technology Assessment, *Local Development*.

ciency are sometimes referred to as "self-providers" or portrayed, with some justification, as attempting to "opt out". The self-sufficiency movement has links with environmentalism and various "back-to-the-land" movements, but ought to be differentiated in certain respects both from these and from the broader Appropriate Technology movement.

Self-sufficiency is typified by the rustic ideal of a totally (or nearly totally) self-sufficient small scale community, or family-sized group eking out its livelihood in a rural or semi-rural setting. Such ventures emphasize detachment from the economic system of industrial society and are often associated with the rejection of technology, or at least of modern technology. They also aim for the deployment of agricultural methods based upon low-impact "organic" principles, and involve the cultivation of a relatively simple lifestyle emphasizing "do-it-yourself" skills, as a conscious rejection of what is perceived as the privatized, materialistic culture of the technological society. In short, "self-sufficiency" normally denotes bucolic, communal autarchy. It is hailed as elitist by some and by others as the only viable alternative to the technological society.

There is a growing literature on self-sufficiency, much of which takes the form of popular manuals or magazines describing techniques or artefacts of potential use to self-providers. There are also models available as to how self-sufficiency might be practised in an urban setting. Under the rubric of "permaculture", for example, attempts have been made to identify techniques for self-sufficiency which enable environmental sustainability on small holdings without reliance upon heavy inputs of human labor; most experiments have nevertheless shown the need for a great deal of labor input. Speaking from a background of practical experience and academic learning, Allaby and Bunyard have identified a paradox in the self-sufficiency movement:[85]

> Thus, the self-sufficient person is out to shape his own destiny by getting back to basics, and he can only find his way through rejecting much of the technologies which he feels are inflicted willy-nilly on society. There are bound to be inconsistencies in that rejection. Yet, ironically, for man to become self-sufficient again he has to return to being *Homo faber*. It is only in doing for himself and in taking responsibility for his own actions that he can escape from the stranglehold of a society in which the prime movers are faceless, inanimate machines operated too often by mindless men.

---

[85] M. Allaby and P. Bunyard, *The Politics of Self-Sufficiency* (Oxford: Oxford University Press, 1980), p. 88.

On the one hand, rejection of dependence upon technology appears at the heart of the quest for autarchy in the self-sufficiency movement. On the other hand, the use of technology is essential for self-sufficiency to be even remotely practicable. The only demonstration projects which appear capable of replication on a broad scale, and of attaining "acceptable standards" of living, involve substantial inputs of science and technology beyond what may be generated in a situation of complete autarchy. For example, power supply through photovoltaic cells and battery-storage systems requires materials, components and skills which are only cost-effective, or perhaps only possible, in a larger more complex society than that of small-scale autarchy. Thus, it appears that complete self-sufficiency is probably impossible; and, in so far as it is possible, it is only possible for a minority who are able to draw upon the collective resources of a larger and sophisticated society.

Notwithstanding the proliferation of enthusiasts for self-sufficiency, almost all serious commentators and analysts advocate *self-reliance* as a goal in keeping with Appropriate Technology. Self-reliance is seen as a quite different concept to self-sufficiency. The former implies mutual cooperation in the sphere of technology-practice between mutually interdependent (and *relatively* independent) communities. The latter is neither replicable on a large scale nor affordable by the majority, and insofar as it implies *absolute* independence, is contradictory. On closer examination, most "self-providers" use the term "self-sufficiency" in a loose way and would probably in fact advocate a high degree of self-reliance rather than absolute self-sufficiency.

Democratization of Technology

The last two decades have witnessed an upsurge of interest in questions of public control of technology and of decision-making in technological development. With increasing awareness of the social costs of technological change has come calls for systematic efforts to assess technology before it is deployed *en masse*.[86] Observations that much technological decision-making appears to reside with technological

---

[86] F. Hetman, *Society and the Assessment of Technology: Premises, Concepts, Methodology, Experiments, Areas of Application* (Paris: Organization for Economic Cooperation and Development, 1973); V. T. Coates, *Technology and Public Policy: the Process of Technology Assessment in the Federal Government*, 3 vols., Program of Policy Studies in Science and Technology (Washington, D.C.: George Washington University Press, 1972); D. M. Freeman, *Technology and Society: Issues in Assessment, Conflict and Choice* (Chicago: Rand McNally College Publishing Co., 1974).

elites, rather than with people in the broader society who endure the consequences of technology choices, has raised questions as to whether current programs for the social assessment of technology tend to rein-force elite control of technology, despite their ostensible role of enhancing popular control of technological change.[87]

Criticism of "expert dominated' approaches to technological decision-making has brought forth frequent calls for more explicit public participation in technology-related decisions.[88] The concept of democracy has been re-examined in the context of modern technological society. It is now widely held that "ballot-box" democracy is not sufficient to ensure that the popular will is implemented, and that the principles of democracy ought to be applied in a concerted manner to technology choice.[89] Advocacy of technological democracy is related closely to movements in support of workplace democracy, industrial democracy and economic democracy.[90]

The notion of democratic technology has arisen amidst the broader push just mentioned, for participatory democracy in economic life.[91] The distinctive feature of the Appropriate Technology movement in this field has been the recognition that some technologies are more conducive to participatory democracy than others - because of such factors as cost, scale, complexity and the general capacity or incapacity for technologies to be comprehended and controlled by non-experts, workers or people at the local community level. Some technologies, such as nuclear fission power systems, are intrinsically difficult to subject to popu-

---

[87] D. Nelkin, *Technological Decisions and Democracy* (London: Sage, 1977); R. Sclove, "Decision-making in a Democracy", *Bulletin of the Atomic Scientists*, 38, 5 (1982), 44-49; L. Sklair, "Science, Technology and Democracy", in *The Politics of Technology*, ed. by G. Boyle, D. Elliot and R. Roy (London: Longman, 1977), pp. 172-185.

[88] K. G. Nichols, *Technology on Trial: Public Participation in Decision-Making Related to Science and Technology*, Report of the O.E.C.D. Committee for Scientific and Technological Policy (Paris: Organization for Economic Cooperation and Development, 1979).

[89] B. Wynne, "The Rhetoric of Consensus Politics: A Critical Review of Technology Assessment", *Research Policy*, 4 (1975), 108-158; M. Gibbons, "Technology Assessment: Information and Participation", in *Directing Technology*, ed. by R. Johnston and P. Gummett (London: Croom Helm, 1979), pp. 175-191.

[90] D. Zwerdling, *Workplace Democracy: A Guide to Workplace Ownership, Participation and Self-Management in the United States and Europe* (New York: Harper and Row, 1970); P. Blumberg, *Industrial Democracy* (New York: Schocken, 1973); M. Carnoy and D. Shearer, *Economic Democracy: The Challenge of the 1980's* (White Plains, N.Y.: M.E. Sharpe, Inc., 1980). Cf.: J. Vanck, *General Theory of Labour-Managed Market Economics* (Ithaca, N.Y.: Cornell University Press, 1970); J. Vanek, *The Participatory Economy* (Ithaca, N.Y.: Cornell University Press, 1971).

[91] See Chapter 5, "A Democratic Technology", in *Economic Democracy*, by Carnoy and Shearer (pp. 195-232)

lar control (in any tangible manner) because of their centralization and complex safety requirements, amongst other things.[92]

A seminal proponent of the democratic technology concept has been the historian and social critic Lewis Mumford. He foreshadowed the modern Appropriate Technology movement with his distinction between *megatechnics* and *polytechnics*.[93] Mumford regards technology as a formative part of human culture and not as a separate system, and he holds that technology and the direction of technological change may be deeply modified at each stage of history by the exercise of human imagination and will. He employs the term "megamachine" or "megatechnics" to represent one pathway in technology-practice based upon centralization, large scale operations, and the management of society itself along technical lines - as a machine. He argues that 20th century industrialized society is not alone in exhibiting megatechnic characteristics; previous civilizations, such as in Egypt developed this pathway to extremes. Modern technology, however, has been employed to resurrect authoritarian structures in Western society.[94] In contrast, Mumford also points to another tradition within civilization where technical modes of operation are not applied to the society as a whole and where technology-practice is based upon more organic, earth-centered, humane and multifarious principles. He labels this "polytechnics", emphasizing a pathway of decentralization, smallness, diversity and multiplicity in technology.

An important concept, for the purposes of this book, is the view contained in Mumford's work that it is virtually impossible to subject megatechnics to democratic control, no matter how committed a culture is to democracy *in principle*. Mumford does not deny the possibility of a resurgence of democracy within a megatechnics society; rather, he argues that this is only possible through a dissolution of megatechnics into a differently organized society based upon polytechnics. He also employs the terms "authoritarian technics" to denote megatechnics and "democratic technics" to denote polytechnics.[95]

---

[92] See: Lovins, *Soft Energy Paths*; Bookchin, *Ecology of Freedom*.

[93] L. Mumford: The *Myth of the Machine*, Vol I: *Technics and Human Development* (New York: Harcourt Brace Jovanovich, 1967), Vol II: *The Pentagon of Power* (New York: Harcourt Brace Jovanovich, 1970).

[94] Mumford, *Technics and Civilization*.

[95] Mumford: *Human Development*, pp. 234-242; cf., *Pentagon*.

Another influential advocate of the democratization of technology is the British technologist and industrial designer Mike Cooley. Cooley writes of technological decisions:[96]

> The choices are essentially political and ideological rather than technological. As we design technological systems, we are in fact designing sets of social relationships, and as we question those social relationships and attempt to design systems differently, we are then beginning to challenge in a political way, power structures in society.

He is a strong proponent of industrial democracy and the control of technology by those who are employed in its use, and by those who are affected by it.[97] Cooley does not limit his notion of industrial democracy to questions of who owns technology or who possesses formal power over decisions. He criticizes a trend in technology whereby workers become dehumanized by jobs which make them little more than extensions of machines. Cooley shows that it is possible to take a different approach to the design of industrial technology which enables the worker to cultivate his or her creative skills while at the same time using technology to reduce monotony and enhance efficiency. In effect he demonstrates that industrial and economic democracy are extremely limited notions unless combined with the design of human-centered technologies which enable democracy to operate in a tangible way in the workplace.

The activities of Cooley and colleagues represent the insight that the democratization of technology is only meaningful when real individual people have the capacity to be in command of machines and technical processes, not just as a formal right, but as a material reality. The form of technology is a limiting or enabling factor in the practice of participatory democracy.[98]

---

[96] M. Cooley, *Architect or Bee? The Human/Technology Relationship* (Sydney: Trans National Cooperative Ltd., 1980), p. 100.

[97] Cf., M. Cooley, "Contradictions of Science and Technology in the Productive Process", in *The Political Economy of Science*, ed. by H. Rose and S. Rose (London: MacMillan 1976), pp. 72-95.

[98] Cooley was part of a team of employees of Lucas Aerospace Ltd. which formed the Lucas Aerospace Combine Shop Stewards Committee. The Combine Committee has been a pathfinder in demonstrating how workers may mobilize their skills to develop economically viable plans for the production of socially useful products using human-centred technology. Cf., D. Elliot, *The Lucas Aerospace Workers Campaign,* Young Fabian Pamphlet #46 (London: Fabian Society, 1977).

Community Technology-practice

Given the diversity of viewpoints within the Appropriate Technology movement, and the fact that the movement exhibits concerns with more explicit content than the general principle that technology ought to be "appropriate", some people advocate a more specific rubric than "Appropriate Technology". "Community technology" is one which has received serious attention.[99] The term indicates three themes.

Firstly, the term reveals the normative commitment behind Appropriate Technology: technology ought to be consciously designed, developed and selected to address the true needs of the community at large, rather than those of an elite or rather than simply to meet some technology-based imperative. The term "community science" has been proposed by some to indicate that science may be (and ought to be) organized along similar lines.[100] While science is different to technology the two may be seen as potential partners in the bid to consciously serve community needs.

Secondly, "community technology" accords with the strong theme within the Appropriate Technology movement that individual local communities (i.e., *geographically* local communities) ought to receive special attention as socioeconomic units rather than be subsumed under national and international economic aggregates. "Community" here may refer to anything from an individual neighborhood to a city.[101]

---

[99] E.g.: Boyle, *Community Technology*; K. Hess, *Community Technology* (New York: Harper and Row, 1979); J. Simpson with K. Bossong, *Appropriate Community Technologies Sourcebook*, Vol. 1 (Washington, D.C.: Citizen's Energy Project, 1980).

[100] P. Harper, "In Search of Allies for the Soft Technologies", *Impact of Science on Society*, 23, 4 (1973), 287-305 (esp. pp. 293-294); K. R. Roby, "On the Aims of Education in Community Science", paper presented to Section 37 (History, Philosophy and Sociology of Science), A.N.Z.A.A.S. Jubilee Congress, Adelaide, May 1980. Note: following from the work of Dr. Keith Roby, Murdoch University in Western Australia has established a postgraduate diploma in *Community Science*.

[101] Three general reviews of this theme, from within the Appropriate Technology movement, are: L. Johnston, "Neighborhood Energy: Designing for Democracy in the 1980's", in *Stepping Stones: Appropriate Technology and Beyond*, ed. by L. de Moll and G. Coe (London: Marion Boyars, 1978), pp. 174-187; D. Morris, *Self-Reliant Cities: Energy and the Transformation of Urban America* (San Francisco: Sierra Club Books, 1982); D. Annandale, *Approaches and Initiatives in Regional Economic Development*, Community Employment Initiatives Unit Discussion Paper #9 (Perth: Department of Employment and Training, Government of Western Australia, 1985). The reviews by Johnston and Morris illustrate the tendency for Appropriate Technology to involve a synthesis of concerns

Advocates of Appropriate Technology generally take the view that particular communities have particular and often unique needs and problems, and therefore require correspondingly particular and unique strategies in technology-practice.[102] Many also adopt the view that particular communities often have immense resources for problem-solving and development which may not be utilized if standardized or exogenous types of technology-practice are adopted.

What we label here as "community technology-practice" is closely related to the upsurge in calls for neighborhood democracy or community governance which have arisen as part of the "new localism" of the last decade or so.[103] The Appropriate Technology movement in the North has raised questions as to the connections between politics and economics at the neighborhood level. Morris and Hess write:[104]

> Neighborhoods cannot survive by themselves, nor would they desire to do so even if they could. But it is just as clear that our communities can be a great deal more self-reliant than they have been. ... community control and local liberty can only be retained if they stem from a productive base. There must be partial economic independence or else the democracy becomes only an illusion.

---

from diverse fields (e.g. community development and energy). An integrated application of this theme to a particular city has been conducted by the staff of the U.S. Appropriate Technology organization, Rain (*Knowing Home: Studies for a Possible Portland*, special issue of *Rain*, 8, 3 [1981] [Portland, Oregon: Rain, 1981]); cf., Max-Neef, "Tiradentes Project".

[102] An awareness seems to be beginning to spread amongst mainstream industry planners and land-use planners that industrial innovation and technological innovation may be a key to regional economic development (See, J. Goddard, ed., *Industrial Innovation and Regional Economic Development*, special issue of *Regional Studies*, 14, 3 (1980) [Oxford: Pergamon, 1980]). However, it is mainly within the Appropriate Technology movement as such that a need for matching special technology to the special needs of individual communites is clearly recognized (see, E. F. Schumacher, "Using Intermediate Technologies", in *Strategies for Human Settlements: Habitat and Environment*, ed. by G. Bell [Honolulu: University Press of Hawaii, 1976], pp. 121-125). Occasional expressions of similar themes occurred prior to the growth of the movement (e.g., A. E. Morgan, *Industries for Small Communities* [Yellow Springs, Ohio: Community Service, Inc., 1953]).

[103] The most explicit discussion of this may be found in *Neighborhood Power: The New Localism* by D. Morris and K. Hess (Boston: Beacon Press, 1975). Cf.: D. Blunkett and G. Green, *Building from the Bottom: The Sheffield Experience*, Fabian Tract #491 (London: Fabian Society, 1983); N. Glazer, "Decentralization: A Case for Self-Help", *The Public Interest*, 70 (Winter 1983), 66-90.

[104] Morris and Hess, *Neighborhood Power*, p. 114.

We may view community technology-practice as the necessary means to provide the economic base from which neighborhoods and local communities may enhance their self-reliance and political power.

The third theme indicated by "community technology" is the rise of the *informal economy* alongside the formal economy of industrialized societies.[105] The informal economy refers to that sector of human activity which involves the conduct of work, the provision of services and the generation of real wealth, but which is neither institutionalized nor included as part of the national accounts. Examples of work from the informal economy include unpaid housework and child-rearing, back-yard food-raising, voluntary community work, bartering and the activities of self-providers. Some estimates indicate that about one third or even a half of the work conducted in urban-industrialized societies takes place in the informal sector.[106]

With the growth of structural unemployment many commentators argue for a redefinition of the nature of work, claiming that we are on the verge of *post-industrial* society, where wealth creation will decreasingly involve human work. While many laud this prospect, a number of Appropriate Technology proponents point to the inequitable aspects of the current trends toward post-industrial society, namely, that a disadvantaged sector of the population is excluded from both productive employment (while "employment" still remains the chief form of obtaining social status and income) and are deprived of the means of obtaining sufficient wealth for the maintenance of an acceptable standard of living.[107] Consequently, self-help activity within the informal economy may be seen as a necessary replacement of "employment" within the formal economy, for the purposes of both

[105] A spirited analysis of this is provided by I. Illich in his book *Shadow Work* (London: Marion Boyars, 1981).

[106] E.g.: G. Bäckstrand, "On the Informal Economy - or Work Without Employment" (mimeo; Stockholm: Secretariat for Futures Studies, 1982), p. 8; O. Hawrylyshyn, *Estimating the Value of Household Work in Canada, 1971* (Ottawa: Statistics Canada, June 1978); D. Ironmonger, "Economic Policies and People", in *The Nation is People*, Theme Lectures of the 55th Annual Summer School, University of Western Australia (Perth: University Extension, University of Western Australia, 1983), pp. 38-47. The informal nature of the informal economy makes obtaining reliable statistics extremely difficult and its actual scope is unclear; it includes both financial and non-financial transactions. A review of this theme and the statistical problems involved may be found in J. Gershuny's book, *Social Innovation and the Division of Labour* (Oxford: Oxford University Press, 1983), esp. pp. 32-49, 143-160.

[107] C. Hines and G. Searle, *Automatic Unemployment* (London: Earth Resources Research Ltd., 1979); K. W. Willoughby, *Technology for Employment Creation*, APACE Occasional Paper #1 (North Fremantle, Aust.: APACE Western Australia, 1984), esp. pp. 8-11; Schumacher, *Good Work*.

working (where work itself is seen as a worthwhile activity) and creating a reasonable livelihood.[108] Community technology-practice is the technological means by which people may generate wealth, reduce their cost of living, cultivate their labor creativity, and relate to other people within the informal economy.

Community technology-practice embraces all three of the above themes and reflects the insight that scope for the community to "appropriate" [verb] technology may be most realistic at the local level (where "community" exhibits the most tangible meaning) and in the informal sector.[109] The structural changes now occurring in the economies of the North may in fact leave no alternative but for disadvantaged people to resort to community technology-practice.

Response to Market Pressures

Much of the impetus for Appropriate Technology in the North has come from essentially political and normative concerns for people and the environment. The sheer pressure of market forces, however, has also played a significant role.

Market pressures have been particularly important in the field of energy technology: "appropriate technologies" or "soft energy technologies" are frequently cheaper than their alternatives.[110] In a paper which surveys applications for high or low temperature heat, portable liquid transport fuels, and electricity (and which includes analysis of public subsidies for "hard" energy technologies), Lovins writes:[111]

> ...although the soft technologies are *not cheap*, they are *cheaper than not having them*. They may or may not be cheaper than present oil or gas - some are and some are not - but what matters is that they are a lot cheaper than what it would cost otherwise to replace present oil and gas.

---

[108] G. Shankland, *Our Secret Economy: The Response of the Informal Economy to the Rise of Mans Unemployment* (London/Bonn: Anglo-German Foundation for the Study of Industrial Society, 1980); J. Robertson, *The Sane Alternative: Signposts to a Self-fulfilling Future* (London: James Robertson [7 St. Ann's Villas, London], 1978).

[109] This is illustrated by the work of the Massachusetts Self Reliance Project (*Hands-On: A Guidebook to Appropriate Technology in Massachusetts* [Amherst: University of Massachusetts, 1978]).

[110] This was indicated clearly during 1978 by Lovins ("Energy Technologies").

[111] Lovins, "Energy Paths", p. 50.

Lovins argues that it makes good business sense to compare all categories of investment opportunities with each other rather than (as is often the case with energy planners) only some with each other, and many with only the cheaper categories of oil and gas. He claims that when the former approach is adopted "by far the cheapest options are the efficiency improvements, then the soft and transitional technologies, then synthetic gas; and the dearest systems by far, even with heat pumps, are the central-electric systems".[112]

The general trend towards higher prices for fossil fuels (nothwithstanding fluctuations in this trend, and the current oil glut) combined with the general inability of nuclear-fission power to be cost effective (without massive subsidies)[113] has created a market climate for the uptake of more cost-effective alternatives. Chandler has shown that the adoption of technologies and integrated strategies for energy conservation is the cheapest energy option for boosting economic growth in O.E.C.D. countries.[114] Urban developments in the City of Davis, California, have proven that solar and conservation techniques in housing and community infrastructure may be market led and may save money for a community.[115] Energy utilities in California have provided financial rewards to municipalities which manage to *reduce* energy consumption, because the capital expenditure required for expansion of traditional sources of power leads to a higher marginal cost than for conservation options.[116] Utilities have also invested in photovoltaic and wind technologies for this reason.[117]

Market conditions have stimulated the spread of efficient small-scale power production technologies. Flavin writes:[118]

> This rush to small-scale power production is not being led by utilities.
> Leading the way instead are large industrial companies building their

---

[112] *Ibid.*, p. 51.

[113] C. Flavin, *Nuclear Power: The Market Test*, Worldwatch Paper #57 (Washington, D.C.: Worldwatch Institute, 1983).

[114] Chandler, *Energy Productivity*.

[115] See: M. N. Corbett, *A Better Place to Live* (Emmaus, Penn.: Rodale Press, 1981); Ridgeway and Projansky, *Energy-Efficient Community*, pp. 39-110.

[116] J. P. O'Sullivan, *The Impact of Community Scaled Energy Initiatives: A Comparative Case Study of the City of Davis, California and the City of Fremantle, Western Australia*, Honours Thesis (Murdoch University, Western Australia, 1983).

[117] *Ibid.*

[118] C. Flavin, *Electricity's Future: The Shift to Efficiency and Small-Scale Power*, Worldwatch Paper #61 (Washingon, D.C.: Worldwatch Institute, 1984), p. 7. Emphasis added.

own power systems and small firms created to tap new energy sources such as windpower and geothermal energy. Utilities buy power from the "small producers" and distribute it to customers. Behind much of this activity is legislation passed in the late seventies and court rulings in the early eighties that have ended utilities' monopoly control of power generation in the United States. ...*The resulting boom in small-scale power production is a good example of what can happen when rapid advances in technology are joined by entrepreneurial capitalism.* The cost of the new power sources is falling steadily. Some are already less expensive than recent coal and nuclear plants, and others soon will be.

In 1978 the U.S. Government enacted the Public Utility Regulatory Policies Act which directs utilities to link up with small scale independent power producers and to pay a fair market price for the electricity.[119] This legislation has helped break down some of the monopolistic control previously held over the energy industry in United States by large utilities. In California, where greater receptivity has emerged than in most other states to new alternatives for power supply and management, the greatest advances have been made in responding to the 1978 Act and opening up the energy industry to market forces. By mid-1985 sufficient generating capacity had been proposed by small independent power producers in California to meet 39 percent of the state's peak power needs and more than half of all U.S utilities obtained some of their power from such sources.[120] Many of the newer conservation or renewable technologies are economically viable at the smaller scale pursued by independent producers and are not feasible at the larger scale at which the utilities normally operate; the new legislation and strengthened market conditions in the United States have therefore helped strengthen the prospects for these technologies.

Another market pressure which has led to interest in Appropriate Technology has been the need for countries to innovate in technology to maintain healthy economies. Some comparative studies have shown that differences amongst firms and countries in the O.E.C.D. area, in their capacity for technical innovation, have a significant influence on international competitiveness, economic growth and living stan-

---

[119] See C. Flavin's work (e.g., "Reforming the Electric Power Industry", esp. pp. 108-118) for a review of this Act and its impact.

[120] *Ibid.*, p. 114 & 115.

dards.[121] The decline in Britain's economic performance for example, has been linked to the country's deficiencies in adapting industrial technology to changing circumstances.[122] Evidence is growing that innovative use of technology by small firms is a key to regional development and for decentralizing a country's economic base.[123] While "successful technical innovation", as widely understood, may not necessarily equate with Appropriate Technology, there are close links: Appropriate Technology requires ongoing innovation to ensure that technology is compatible with the real conditions prevailing in a given region. These links are confirmed by Bollard, who has shown that much of Britain's poor economic record (with particular reference to employment generation) relates to a bias in policies and actual business investment towards larger-scale, capital-intensive (and, by implication, often unresponsive) industries.[124]

Bollard has conducted a study of the wool textile industry[125] and has shown that many of the larger-scale, capital-intensive low-price woollen firms throughout Western Europe are in decline and are suffering through competition with the more efficient industry of the Prato region in Northern Italy. According to Bollard, the Pratese industry, employing about 46 000 people (1983), is completely decentralized and most production takes place in small workshops where from two to ten workers carry out the specialized production processes. There are about 10 000 of these tiny firms. They operate with very modern but small-scale equipment which, in turn, is the basis of a healthy local machinery manufacturing industry. There are about 1 200 Pratese firms, mostly small, making and repairing equipment. The capital cost of most of the plant is low enough to be funded by investments from locally produced financial surpluses.[126]

Examples may also be found in the field of agriculture, where economic advantages have recently been identified in moves away from

---

121 K. Pavitt, ed., *Technological Innovation and British Economic Performance* (London: MacMillan, 1980)

122 *Ibid.*.

123 R. Rothwell, "The Role of Technology in Industrial Change: Implications for Regional Policy", *Regional Studies*, 16, 5 (1982), 361-369.

124 Bollard, *Industrial Employment*.

125 "Picking Up the Threads: Small Firms in the Wool Textile Industry", Chapter 5 in Bollard, *Small Beginnnings*, pp. 139-174.

126 See *ibid.*, esp. pp. 157-159, for a description of the Prato industry. Cf., U. Colombo and D. Mazzonis, "Integration of Old and New Technolgies in the Italian (Prato) Textile Industry", in *Blending of New and Traditional Technologies: Case Studies*, ed., by A. Bhalla, D. James and Y. Stevens (Dublin: Tycooly, 1984), pp.107-116.

the trend towards ever-increasing farm sizes. Many farmers in Australia are now finding that their operations may be more competitive by limiting the size of their farms and, correspondingly, of their capital stock.[127]

The above examples illustrate cases where "appropriate" technology-practice has been adopted largely as a result of economic pressures rather than because of ideological commitments or political agendas.

Intelligent Engineering

A certain amount of support for Appropriate Technology has come directly from within the engineering profession itself and from a concern for high standards of engineering design and development work.[128] Technology which doesn't work very well when situated in a real community application may hardly be viewed as *good technology* no matter how efficiently it may perform a narrowly defined technical operation.[129]

Some of the criteria of appropriateness often advocated by proponents of Appropriate Technology on social or political grounds - such as simplicity - often turn out to require the very best of engineering skill. Schumacher retorts:[130] "Any third-rate engineer or researcher can increase complexity, but it takes a certain flair of real insight to make things simple again". Much of the renewable energy technology dubbed as "appropriate technology" requires high level technical research and development and much of it appears to be promoted because of its technical interest as much as for its social and environmental amenity.[131]

---

127 See: G. Ansley, "Get Big or Get Out: Why it Turned Sour", *National Farmer* (Australia), #17 (6-19 September 1984), 24-29; For American comparisons see, United States Department of Agriculture, *Economies of Size in Farming*, Agricultural Economic Report #107 (Washington, D.C.: U. S. Government Printing Office, 1969). Note: Greene has argued that limiting the size of farms tends to enhance the agricultural sector as a whole by maintaining genuine market conditions which in turn tend to assure a sufficient return on the farmers' investment and labour (S.L. Greene, "Corporate Accountability and the Family Farm", in *Radical Agriculture*, ed. by R. Merrill [New York: Harper Colophon, 1976], pp. 54-63).

128 See: Institution of Civil Engineers, *Appropriate Technology*; cf., J. D. Davis, *Appropriate Engineering*, monograph, reprinted from *Engineering*, December 1979 (distributed by Intermediate Technology Development. Group Ltd.).

129 Cf., V. Papanek, *Design for the Real World* (London: Thames and Hudson, 1971).

130 Schumacher, *Small is Beautiful*, p. 144.

131 This is evident in both the articles and advertisements which appear, for example, in the journal *Alternative Sources of Energy*.

Many of the practitioners with long track-records within Appropriate Technology exhibit outstanding engineering competence - a competence at *both* the technical level and at the level of comprehensive design within the contingencies of the "real world".[132] Within engineering circles it appears that many of the best aspects of appropriate technology are understood simply as "good, intelligent engineering".

---

[132] Good examples lie with the work of Peter Dunn (Professor of Engineering Science, University of Reading; see, Dunn, *Appropriate Technology*) or Meredith Thring (Professor of Engineering at Queen Mary College, University of London; see, M. W. Thring and E. R. Laithwaite, *How to Invent* [London: MacMillan, 1977]; M. W. Thring, *The Engineer's Conscience* [London: Northgate, 1980].

8

Emerging International Trends

Technology for One World

Appropriate Technology has evolved in two streams: in the North and in the South. The concerns of each stream differ, as do the resources for problem-solving and the relative positions in international political-economy of their respective groups of nations. Nevertheless significant parallels and areas of overlap exist between the two streams. This is also reflected in the writings of Schumacher, who applied his analysis with alacrity to both the North and the South. A trend now appears to be emerging where the two streams of Appropriate Technology are converging - or, at least, the areas of common interest and activity are becoming more apparent and well organized.

George McRobie, who has been part of the movement from its beginnings, and who has played an enormous role in its successful achievements to date, writes:[1]

> The idea that capital-intensive technology is appropriate for the 'developed' countries and intermediate technology for the developing countries must finally be recognized as false, and all technologies must be selected on the basis of whether they are truly appropriate for the particular situation for which they are intended. The qualities of smallness and simplicity, and of being capital- and resource-saving, and non-violent are as essential for the technologies of the First World as the Third, and both "Worlds" must move towards these technologies before a stable economy for the whole world can be created - an economy with a sustainable future and a human face.

---

[1] G. McRobie, "Technologies for 'One World'", *Appropriate Technology*, 11, 2 (1984), 4.

In the article from which this quote is drawn McRobie argues for the theme of "technologies for *one world*". His sentiments accord with those of the broader international movement concerned with global issues and the interdependence of nations - a perspective which has undergirded much of the work of the United Nations and which was encapsulated in the publication during 1972 of *Only One Earth* by Barbara Ward and René Dubos.[2] A global perspective on technology and development has now been taken up by a number of authorities in recognition that: wealthy countries cannot isolate themselves from the problems of the poor; the poor countries require the cooperation of the rich in the transition to Third World self-reliance; and, that truly *global* development is urgently required, not just on the grounds of justice, but as a factor affecting the survival of human society in a workable form.[3] A substantial number of international fora point to technological self-reliance or Appropriate Technology as necessary ingredients of the transition to a sustainable global pattern of development.[4]

The emergence of a tendency for Appropriate Technology in the North and the South to converge is partly due to an increasing awareness of the principles of ecology and of the fact that global dispersion of locally generated environmental problems is reaching potentially catastrophic levels. It also appears to stem from the fact that the North is increasingly experiencing problems previously associated mainly with the South, such as structural unemployment, severe urban decay and the growth of dual economy. Thirdly, a number of countries of the South (e.g., India) have evolved a substantial modern scientific and technical research capacity of their own, and some experience is

---

[2] B. Ward and R. Dubos, *Only One Earth: The Care and Maintenance of a Small Planet* (Harmondsworth: Penguin, 1972). Seven years later Ward published a study indicating hopeful prospects for changes in the direction of industrialized civilization and non-industrial communities - sufficient to avert global human and environmental catastrophes; her hope, however, was dependent upon the adoption of strategies based on the Appropriate Technology approach (*Progress for a Small Planet* [Harmondsworth: Penguin, 1979]).

[3] See: *Development Dialogue* (journal published by the Dag Hammarskjold Foundation [Uppsala, Sweden]); *Development Forum* (journal published by the United Nations Divison for Economic and Social Information and the United Nations University [New York and Tokyo]); A. Mattis, ed., *A Society for International Development: Prospectus 1984* (Durham, North Carolina: Duke University Press, 1983) (this is an overview and review of the work of the Society for International Development [based in Rome]).

[4] See: G. McRobie, "Why the World Will Shift to Intermediate Technology", an interview with G. McRobie by J. P. Drissell (*The Futurist* [April 1977], 83-89); M. Goldsmith, A. King and P. Lacoute, eds. *Science and Technology for Development: The Non-Governmental Approach*, proceedings of the CISTOD World Congress, Tunis, April 1983 (Dublin: Tycooly, 1984).

now accumulating as to how this may be mobilized to address local problems of underdevelopment.[5] Recent work by the International Labour Office and the United Nations Advisory Committee on Science and Technology for Development has revealed how the artful combination of emerging (often sophisticated) modern technologies with traditional techniques may enhance prospects for success in local economic development.[6] This theme accords with the goals of Appropriate Technology. There appears to be some convergence in Appropriate Technology at the technical level as well as at the level of the social movement.

The Second Generation

A number of commentators observe that the Appropriate Technology movement is maturing, gaining wider acceptance and is now poised for a new phase in its evolution.[7] Nicholas Jéquier has stressed that Appropriate Technology involves a cultural and political revolution (in the Kuhnian sense) and points out that support is now coming from "above", "below" and increasingly from many of the organizations and corporations which have been the target of the Appropriate Technology critique. He writes:[8]

These changes in attitudes are fundamental, and testify to the fact that the cultural revolution of A.T. is in the process of succeeding. This is in essence the key to what I would like to call the second generation in A.T.: the main proponents of A.T. are no longer the marginalists, the radicals and the misfits, but the ruling élite, be they industrial, political or technological, and it is likely that in the years to come the main innovators in A.T. will be national governments, industrial firms,

[5] This was illustrated in a paper presented by the late Prime Minister of India, Shrimati Indira Gandhi, to the Science Policy Foundation at the Royal Society, London, on 26 March 1982 (see: S. I. Gandhi, "Science for Social Change", *Science and Public Policy* [June 1982], 114-119).

[6] A. Bhalla, D. James and Y. Stevens, eds., *Blending of New and Traditional Technologies: Case Studies* (Dublin: Tycooly, 1984);   E. U. von Weizsäcker, M. S. Swaminathan and A. Lemma, eds., *New Frontiers in Technology Application: Integration of Emerging and Traditional Technologies* (Dublin: Tycooly, 1983).

[7] P. Baron, "Appropriate Technology Comes of Age: A Review of Some Recent Literature and Aid Policy Statements", *International Labour Review*, 117 (1978), 625-634.

[8] N. Jéquier, "Appropriate Technology: The Challenge of the Second Generation", *Proc. R. Soc. Lond. B*, 209 (1980), 9.

existing research institutions, foreign aid agencies and development banks.

It may be questioned how substantial the support really is which arises from these sources; we may concur with Jéquier, however, that the logic of Appropriate Technology is now widely acknowledged, wider support has been received, and the initial challenge of awareness raising has been reasonably successful. The challenge for the "second generation" appears to be one of ensuring large scale dissemination and effective implementation of Appropriate Technology. Jéquier questions whether the Appropriate Technology movement and its specialist organizations have sufficient resources and ability to succeed in this task. He is sanguine about future prospects for Appropriate Technology, but places his hopes in an enlightened response by the large mainstream and highly organized national and international institutions which have grown up as part of the industrial era. He views the Appropriate Technology groups as the harbingers of such a transformation.

Awareness of the challenges of the "second generation" is quite apparent within the movement itself. Professor Peter Dunn, a pioneer with the I.T.D.G., and based at the University of Reading, points to growing acceptance of the technical feasibility of Appropriate Technology, but also points to some apparent weaknesses of the movement. He claims that the following matters now need greater attention: better dissemination of information; better implementation of technologies (through training of personnel, extension services, and the building up of local sources for the manufacture and supply of equipment); improved collaboration between universities and other groups involved with implementation; and, serious programs to evaluate the failure and success of projects to date as a basis for more realistic future programs.[9]

It appears that the Appropriate Technology movement is now at a critical juncture in its evolution. It is not clear in which direction it will proceed.

---

[9] P. D. Dunn, "Appropriate Technology: Priorities Past, Present and Future", *Appropriate Technology*, 10, 3 (1983), 18-19.

Institutional Factors

Despite the eclecticism of the Appropriate Technology movement it does have a unifying thematic focus on issues directly concerned with technology. Much of the practical activity in the movement has been at the level of the design, development, testing and demonstration of actual technologies. The comments of Dunn, Jéquier and McRobie just considered stress that the institutional framework of Appropriate Technology deserves at least as much attention as the technologies themselves. Along with greater research and technical development work than at present, the chief characteristic of the "second generation" is likely to be a more concerted attempt to address the institutional requirements of Appropriate Technology.

There has always been a concern within the Appropriate Technology movement for institutional factors. The first major activity of Schumacher and colleagues was institutional - the establishment of an organization and a network of cooperative links with other organizations. This concern for institutional matters is readily apparent in much of the Appropriate Technology literature but is now being discussed more intensively.[10] The demand for advice on the institutional aspects of Appropriate Technology has burgeoned to such an extent that, in order to meet it, the I.T.D.G. has established an A.T. Institutions Program.[11]

In addition to the challenges of establishing and maintaining suitable organizations, the institutional aspect of Appropriate Technology also includes problems associated with the economic and political context in which Appropriate Technology goals are to be pursued. The role

---

[10] E.g.: A. S. Bhalla, ed., *Towards Global Action for Appropriate Technology* (Oxford: Pergamon, 1979); United Nations Industrial Development Organization, *Conceptual and Policy Framework for Appropriate Industrial Technology*, Monographs on Appropriate Industrial Technology #1 (New York: United Nations, 1979); G. McRobie, *Small is Possible* (London: Jonathan Cape, 1981); J. de Schutter and G. Bemer, *Fundamental Aspects of Appropriate Technology* (Delft: Delft University Press and Sijthoff and Noordhoff International Publishers, 1980), esp. pp. 53-143; F. Stewart, ed., *Macro-Policies for Appropriate Technology in Developing Countries* (Boulder: Westview Press, 1987); N. Jéquier and G. Blanc, *The World of Appropriate Technology* (Paris: Organization for Economic Cooperation and Development, 1983); W. Rohwedder, *Appropriate Technology in Transition: An Organizational Analysis*, doctoral dissertation, University of California at Berkeley, May 1987.

[11] A. Sinclair, *A Guide to Appropriate Technology Institutions* (London: Intermediate Technology Publications Limited, 1984).

of government in providing infrastructure to support the pursuit of particular social or environmental objectives, in compensating for market failures or in creating optimum market conditions, is an example of the type of institutional matter which will increasingly be addressed in debates over Appropriate Technology. This subject will be further addressed in the following chapters.

Integrated and Comprehensive Analysis

A fourth emerging trend in Appropriate Technology is for programs to be conceptualized from a comprehensive and integrated perspective. Technology is not a discrete entity and proponents of Appropriate Technology have been leaders in pointing to the social and environmental context which surrounds technologies. The movement has also stressed that technology is not "neutral" and that particular technologies embody in various ways the characteristics of the political setting from which they originate.

A growing number of studies are appearing which are interdisciplinary in nature, place technology at the center of their analyses, advocate a strategy in keeping with that of Appropriate Technology, and which attempt to adduce an integrated theoretical viewpoint.[12] Many of these aim not only at new additions to scholarship, but towards radical changes in ways of looking at the world.[13] They seek to integrate *inter alia* the perspectives of economics, technology, culture, metaphysics, environmental science, policy studies, sociology, physical science and the humanities.[14] Normative and empirical approaches

[12] E.g.: C. Norman, *The God that Limps: Science and Technology in the Eighties* (New York: W. W. Norton, 1981); A. Pacey, *The Culture of Technology* (Oxford: Basil Blackwell, 1983); D. Elliot and R. Elliot, *The Control of Technology* (London: Wykeham, 1976); M. Bookchin, *The Ecology of Freedom: The Emergence and Dissolution of Hierarchy* (Palo Alto, Calif.: Cheshire Books, 1982); C. A. Hooker, et al., *The Human Context for Science and Technology* (Ottawa: Canadian Social Sciences and Humanities Research Council, 1980).

[13] E.g.: H. Henderson, *Creating Alternative Futures: The End of Economics* (New York: Perigree, 1978); W. Leiss, *The Limits to Satisfaction: On Needs and Commodities* (London: Marion Boyars, 1976).

[14] E.g.: I. Sachs, *Strategies de l'Écodéveloppement: Economie et Humanisme* (Paris: Editions Ouvrières, 1979); L. R. Brown, *Building a Sustainable Society* (New York: W.W. Norton, 1981); L. R. Brown, et al., *State of the World*, annual publication of the Worldwatch Institute in Washington, D.C. (New York: W.W. Norton, 1984, 1985,1986); B. Glaeser, ed., *Ecodevelopment: Concepts, Projects, Strategies* (Oxford: Pergamon, 1984).

are often conducted concurrently, with convergent conclusions ensuing.[15] Appropriate Technology appears to be part of a trend in scholarship and polemics away from reliance upon reductionistic and single-discipline analyses towards multifaceted and holistic analyses.

Four Foci of Appropriate Technology

In addition to the basic distinction between the North and South streams, a distinction can be made between four foci around which various sub-streams of the Appropriate Technology movement are grouped. These sub-streams transcend the North-South distinction to some extent, and are not discrete. Many people and organizations within the movement in some way adopt all four of the foci. Most, however, tend to exhibit biases towards certain foci and allow one to predominate.

An *economics focus* is widespread, particularly with regard to the South. Intermediate Technology is essentially an economic concept and the economics focus of Appropriate Technology is not surprising in view of the concept's origin in Schumacher's work as a professional economist.

A *technology focus* is also widespread. It has grown from the fact that Schumacher's original economic analysis pointed to technology choice as one of the most important issues facing regions and countries. It also stems from the fact that the development of Appropriate Technology requires concerted effort in a very practical way with actual technologies - and requires considerable input from professional specialists in engineering and other technical disciplines.

An *environment focus* undergirds much of the support for Appropriate Technology - particularly, but by no means exclusively, in the North.

A *people focus* is the other main orientation in appropriate technology. Virtually all streams within the movement promote the need for technology to be consciously chosen to act as a servant of people. Some Appropriate Technoogy advocates place almost all of their

---

[15] See, H. E. Daly, et al., *Economics, Ecology, Ethics: Essays Toward a Steady-State Economy* (San Francisco: W. H. Freeman, 1980). Cf.: A. Sen, *Resources, Values and Development* (Oxford: Basil Blackwell, 1984); P. Abrecht, *et al., Faith, Science and the Future*, preparatory readings for the 1979 Conference at Massachusetts Institute of Technology on the Contribution of Faith, Science and Technology in the Struggle for a Just, Participatory and Sustainable Society (Geneva: World Council of Churches, 1978).

attention on the "people" aspects and very little on technology and technologies.

Some groups place balanced attention on all four foci. It remains to be seen whether the movement as a whole will evolve so as to achieve and maintain such a balance.

Appropriate Technology International

The influence of Schumacher and colleagues has been international in scope right from the beginning of their work. One example of this lies with the initiatives of the United States Federal Government in the mid-to-late 1970s. Not long before his death in 1977 Schumacher had a private meeting in the White House with President Carter and it appears that his views also held sway with a number of influential people in the Congress. The idea of Appropriate Technology was incorporated into the appropriations bills of no less than four federal agencies. The Agency for International Development received $20 million in its 1975 budget to establish an Intermediate Technology program in conjunction with the private sector; the Community Services Administration was granted $3 million to establish the National Center for Appropriate Technology; the House Committee on Science and Technology directed the National Science Foundation to give emphasis to intermediate technologies; and a similar emphasis was directed into the budget of the Energy Research and Development Administration.[16] Some of these initiatives have continued through changing administrations, in one form or another, and others have wilted. The initiatives which commenced under the auspices of the Agency for International Development (AID), and which have remained in operation up until the present despite stringent budget constraints, provide some particularly interesting material relevant to this study.

As a result of the mandate given to AID by Congress, a new body, Appropriate Technology International (ATI), was established in Washington, D.C. in 1976. ATI is a private, not-for-profit organization which was given the mission of experimenting with innovative approaches to technological development and the transfer of new tech-

---

[16] "Congress Buys Small is Beautiful", *Science*, **192** (June 1976), 1086.

nologies.[17] ATI works with a range of project partners, mostly local businesses and private voluntary organizations, in twenty countries throughout Latin America, the Caribbean, Asia and Africa. It aims to assist in the establishment of commercially viable enterprises by the poor in rural and peri-urban areas of the South. In addition to providing general financial, organizational, management and planning assistance, it is active in the assessment, adaptation, dissemination and transfer of technologies suitable for this purpose.[18] Funding for ATI's work comes primarily from AID. While it is rooted within the U.S. government's foreign assistance framework, ATI has always sought advice from those experienced in the broader international Appropriate Technology movement. George McRobie, for example, originally from the Intermediate Technology Development Group, is a member of ATI's Board of Trustees.

The work of ATI appears to have been complicated by the ambitiousness of the goals given to it by Congress, its declining core financial support, and the problems associated with operating under the oversight of a government agency. ATI has had to alter its objectives and activities over time, in response to evaluations by AID, but it basically acts with operational autonomy.[19]

ATI is required by AID to fund projects which have a core technology associated with them; and most of the group's work is high in risk because of the necessarily long time periods involved in the development of technology, exacerbated in communities which lack important infrastructure support for technology development. ATI places great stress on the importance of projects becoming commercially viable, and concentrates its work in three main technological areas: agricultural products processing and utilization of agricultural wastes; local mineral resources development; and, equipment and support for small farms. Notwithstanding the fact that many of ATI's projects are still at an early stage, and that a number are commercially in doubt, an in-

---

[17] See, U.S. Agency for International Development, "Proposal for a Program in Appropriate Technology" (Washington, D.C.: AID, 1977).

[18] Appropriate Technology International, *Appropriate Technology as a Strategy for Development*, 1984 Annual Report, ATI (Washington, D.C., 1985); Appropriate Technology International, *Technology for Small Scale Industry*, 1985 Annual Report, ATI (Washington, D.C., 1986); Appropriate Technology International, *Appropriate Technology for Small Enterprise Development*, 1986 Annual Report, ATI (Washington, D.C., 1987).

[19] See P. Delp, *et al.*, *Promoting Appropriate Technological Change in Small-Scale Enterprises: An Evaluation of Appropriate Technology International's Role*, AID Evaluation Special Study No. 45 (Washington, D.C.: U.S. Agency for International Development, November 1986).

dependent team of evaluators recently concluded, "Generally, the ATI projects are establishing productive activities with good prospects for commercial viability."[20] Seventy percent of the projects commenced by ATI during the four years to 1987 resulted in the establishment of profitable and viable enterprises.[21]

There are a number of important things to recognize here about Appropriate Technology International, besides that fact that it is a tangible expression of the international scope of the Appropriate Technology movement.

The first is that its performance provides confirmation that concerted and systematic efforts at applying the Appropriate Technology philosophy can actually work.[22]

The second is that the interpretation of Appropriate Technology followed by ATI is quite sophisticated, avoiding the simplistic "technology fix" approach which critics often accuse Appropriate Technology enthusiasts of following.   For example, the Executive Director of ATI, Ton de Wilde, describes his organization's programs using four levels of analysis, concerned with: (1) informal structural factors (the combination of individual, social, traditional and moral factors that influence the behavior of a small-scale entrepreneur); (2) technical/project factors (the combination of factors that influence the efficiency of an enterprise); (3) structural factors (the combination of formal and informal institutional and governmental arrangements that influence small enterprise development); and, (4) human value systems (factors influencing the value system of small entrepreneurs, their objectives in life and the environment affecting the motivation of entrepreneurs and workers).[23]

Third, ATI's work underlines the critical importance in the practice of Appropriate Technology of networks of special-purpose organizations.

Fourth, ATI has been forced to grapple with the many practical obstacles intrinsic to the serious application of technology choice as a policy principle, and has made significant progress by learning from its successes and failures.   Its work in identifying alternative detailed

---

[20] *Ibid.*, Appendix F, p. 5.

[21] T. de Wilde, "Small Enterprise Development: Changing Paradigms and Contexts", in *Appropriate Technology for Small Enterprise Development*, 1986 Annual Report, ATI (Washington, D.C., 1987), pp. 10-14.

[22] For a well researched set of case studies of projects funded by ATI, paying special attention to the policy environment as a determinant of the economic viability of Intermediate Technology enterprises, see F. Stewart, *Macro-Policies* (cited above).

[23] de Wilde, "Small Enterprise Development".

strategies for the practice of Appropriate Technology, with the larger institutional, political and commercial environment in mind, will be an important subject for study by governments and organizations in the Appropriate Technology movement in future years (especially given the substantial scale of ATI's efforts: about US$40 million expended in operations and grants from 1976 to 1988).

Finally, the work of ATI is an excellent example of an intellectually sound application of the "enlightened" interpretation of Appropriate Technology proposed by the author in Chapter Two. ATI is actually an Intermediate Technology organization which has arguably properly understood the Appropriate Technology rationale as first developed by Schumacher. It appears to be intelligently applying the "specific-characteristics" approach for carefully defined specific circumstances, while also fitting within the "general-principles" approach to Appropriate Technology defined in Chapter Two.

Review

Part Two has been lengthy and has involved an extensive literature survey. In the process it has been demonstrated that the Appropriate Technology movement is a substantial phenomenon with a very broad base of support with relevance to a broad range of fields. It is neither a minor school of thought nor just a passing fad. It is a significant constellation of socio-political forces, philosophical-cultural views, organizations, individuals, practical projects, artefacts and technical systems. In short, Appropriate Technology stems from a global network of initiatives which address fundamental problems of the 20th century and, as such, it calls for serious attention and analysis.

There are common and overlapping themes within the movement, themes which have been articulated by E. F. Schumacher and exemplified in Schumacher's Intermediate Technology. The description in the preceding four chapters of the various streams of Appropriate Technology indicates that the movement shows signs of convergence and of attaining a degree of orthodoxy and broad acceptance. It also pointed to available evidence for the technical feasibility of appropriate technologies.

Notwithstanding the Appropriate Technology movement's partial acceptance by the mainstream it still faces some significant difficulties. Despite the latent consensus within the movement there are also considerable differences between various streams and schools of thought. There is also no definitive, systematic and comprehensive theory of Appropriate Technology. It could be argued that, in his later years, Schumacher possessed a complete theoretical perspective on the subject; his writings do not, however, contain a readily accessible systematic statement of the unifying framework of his views. The diversity of the Appropriate Technology movement is complicated by the fact that much of its literature uses inconsistent terminology or unusual jargon, or consists of mixtures of partly incongruous views.

These difficulties do not invalidate the basic tenets of Appropriate Technology. They do, however, point to the need for a systematic and comprehensive theory as a basis for making distinctions between the tenable and untenable aspects of the Appropriate Technology movement. Given that Appropriate Technology draws heavily upon other sources, there is a need to identify that which distinguishes the notion from those sources, and that which is merely a restatement of other notions.

Before synthesizing an integrated framework for Appropriate Technology it is necessary to address a number of criticisms which have been directed at the concept.

PART THREE

Prospects for
Technology Choice

9

Criticisms of
Appropriate Technology

Introduction

In Parts One and Two it was demonstrated that Appropriate Technology is a serious idea with substantial support from a range of sources; and, that despite its diversity, the movement exhibits sufficient internal congruence to qualify as a movement. It has also been subject to extensive criticism.

Some of the criticisms concern minor weaknesses or contradictions in the movement and the unrealistic claims of some of its proponents.[1] Others raise fundamental doubts about the validity of Appropriate Technology *per se*.[2] These two types of criticisms will be addressed in

---

[1] E.g.: P. Bereano, "Alternative Technology: Is Less More?" *Science for the People*, 8, 5 (1976), 6-9, 34-35; R. J. Mitchell, ed., *Experiences in Appropriate Technology* (Ottawa: Canadian Hunger Foundation, 1980); C. Norman, *Soft Technologies, Hard Choices*, Worldwatch Paper #21 Washington, D. C.: Worldwatch Institute, 1978); D. E. Morrison, *Energy, Appropriate Technology and International Interdependence*, paper presented to the Society for the Study of Social Problems, San Francisco, September 1978; N. Jéquier, "The Major Policy Issues", in *Appropriate Technology: Problems and Promises* (Paris: Organization for Economic Cooperation and Development, 1976). A similar critique has been conducted by D. Zimmerman from an urban habitat perspective ("Small is Beautiful, But: An Appraisal of the Optimum City", *Humboldt Journal of Social Relations*, 9, 2 [1982], 120-142).

[2] An objection to the concept as a whole, on logico-philosophical grounds, has been articulated by H. Sachsse ("Comment: What is Alternative Technology? A Reply to Professor Stanley Carpenter", in *Philosophy and Technology*, ed. by P. T. Durbin and F. Rapp [Dordrecht: D. Reidel, 1983], pp. 137-139). A highly polemical and ardent attack on Appropriate Technology has been published by Arghiri Emmanuel of the University of Paris I (*Appropriate or Underdeveloped Technology?* trans. by T. E. A. Benjamin [Chichester: Wiley, 1982]); Emmanuel's book contains a number of criticisms similar to

this chapter and the latter criticisms will be addressed in greater length in subsequent chapters.

While many criticisms are directed at Appropriate Technology by antagonists from outside the movement, the majority appear to arise from the concerns of those who are either sympathetic to Appropriate Technology[3] or who are part of the movement.[4] There is some dogmatism within the movement, but in the main members of the Appropriate Technology movement exhibit considerable capacity for self-criticism. It is possible to define the scope of the movement to emphasize those themes which are put forward in a dogmatic and polemical way. To do so would be improper, however, given that Appropriate Technology literature of a serious and critical kind is at least if not more dominant than the more popular kind. Furthermore, much of the popular literature is popular in *style* without negating the contributions of rigorous scholarship towards consistency of *content*.[5]

This chapter will include analysis of the criticisms of Appropriate Technology but a fuller response to them will follow in the remaining chapters. The following categories overlap considerably but serve as useful approximations of the different types of criticism which have been raised.

---

those considered in this chapter, and is one of the few substantial publications which has as its main purpose the confutation of Appropriate Technology. Rejoinders to the claims of Emmanuel have been raised by C. Furtado and H. Elsenhans at a Conference of the European Association of Development Research and Training Institutes during October 1980 (see, *ibid.*, Part Two, pp. 117-186), and by F. Stewart (Book Review, *Appropriate Technology*, 9, 4 [1983], 23).

[3] E.g.: T. Kitwood, *Understanding Technical Change*, text of five lectures delivered at a workshop on Technology, Science and Development, Appropriate Technology Centre, Kenyatta University College, Nairobi, Kenya, 1981; B. R. Rao, "Two Faces of Appropriate Technology", *Mazingira*, 3, 3/4 (1981), 2-11.

[4] E.g.: M. Marien, "The Transformation as Sandbox Syndrome", *Rain* (November/December 1984), 4-9; W. Rybczynski, *Paper Heroes: A Review of Appropriate Technology* (Dorchester: Prism, 1980); P. D. Dunn, "Appropriate Technology: Priorities Past, Present and Future", *Appropriate Technology*, 10, 3 (1983), 18-19; M. M. Hoda, "Appropriate and Alternative Technology", mimeo (Lucknow, India: Appropriate Technology Development Association, n.d.); A. K. N. Reddy, *Technology, Development and the Environment: A Re-appraisal* (Nairobi: United Nations Environment Programme, 1979), esp., pp. 20-23.

[5] The writings of E. F. Schumacher illustrate this point; Schumacher was as much concerned with communicating his ideas to a wide audience as he was with ensuring that they were intellectually sound.

General Criticisms

Technical

It is commonly assumed that so-called "appropriate technologies" are by definition technically inefficient and by nature incapable of matching the supposedly superior productive capacity of so-called "high" or "modern" technologies.[6] This claim is especially directed at "intermediate technology", "village technology" or related notions, and appears to be based upon the assumption that there is a uni-directional linear progression in technological change from "low" (supposedly inefficient) technologies to "high" (supposedly most efficient) technologies.[7] Many critics equate high technical efficiency within a machine or process with the successful achievement of a specified human goal - and would therefore conclude that the most "appropriate" technology is that which is the most "efficient".[8] It would follow that the most "high" technology is automatically the most "appropriate". Qualifying "technology" with "appropriate" would therefore be meaningless, or else it would imply that something

---

[6] This is repeatedly claimed by Emmanuel (*Underdeveloped Technology?* passim). Cf.: R. Bayard, "No Growth Has to Mean Less and Less", *New York Times Magazine*, 2nd May (1976), 13, 72, 76, 80.

[7] E.g., R. Eckaus writes: "There is little evidence to suggest that major research efforts to find efficient 'intermediate' technologies ... would either be markedly successful or contribute substantially to development" (*Appropriate Technologies for Developing Countries* [Washington, D.C.: National Academy of Sciences, 1977], p. 2). Even those who acknowledge the importance Appropriate Technology appear to retain an assumption of "intermediate" meaning "less efficient" (see, P. Dasgupta, "On Appropriate Technology", in *Appropriate Technologies for Third World Development*, ed. by A. Robinson [London: MacMillan, 1979], p. 12). This assumption is stated explicitly by P. Mathias (in the Preface to *Technological Change: The United States and Britain in the 19th Century*, ed. with an introduction by S. B. Saul [London: Methuen, 1970], pp. vii-viii), and is apparent in the work of D. Ironmonger (see, "A Conceptual Framework for Modelling the Role of Technological Change", in *The Trouble with Technology*, ed. by S. MacDonald, D. McL. Lamberton and T. D. Mandeville [London: Frances Pinter, 1983], pp. 50-55, esp. p. 52).

[8] This attitude may also be found amongst people otherwise sympathetic to the Appropriate Technology approach (e.g., N. Hogbe-Nlend, "Technological Choices and Options for Development", in *Science and Technology for Development: The Non-Governmental Approach*, ed. by H. Goldsmith, A. King and P. Lacoute [Dublin: Tycooly, 1984], pp. 46-54, esp. pp. 49-50).

other was desired than the most effective means for achieving a specified human goal. From this perspective one could easily be tempted to think that "appropriate technologies" are by nature not "appropriate"! This apparent contradiction could lead many to reject Appropriate Technology before properly assessing the idea.

There are a number of weaknesses with the above technically oriented criticism of Appropriate Technology.

Firstly, there is no *a priori* reason why a technology tailored to fit a given context may not be highly efficient. There also appear to be no *a priori* reasons why any of the sub-notions of Appropriate Technology need imply less than adequate efficiency. In any case, as has already been demonstrated in Part Two, there is now ample empirical evidence of technically efficient appropriate technologies.

Secondly, the criticism does not make a distinction between the *internal* technical (or engineering) efficiency of a technology and the *external* efficiency of the same technology (i.e., its efficiency in achieving certain human or practical ends while operating in the complex "real" world). In other words the criticism fails to acknowledge that technologies operate as part of technology-practice (which in turn takes place in a larger human and natural environment) and that efficiency may vary between each level.

Thirdly, such criticisms are frequently couched in highly ambiguous language where terms like "high", "modern" or "advanced" are uncritically equated with "superior" and employed without being defined. It is often implied that appropriate technologies are somehow incompatible with whatever it is that makes "high" technologies "high". The vague, almost dogmatic use of "high technology" rhetoric, tends to obscure the possible contradictions and logical gaps in the criticisms of Appropriate Technology which charge it with intrinsic inefficiency.

Fourthly, the criticism ignores the fact that the seminal version of Appropriate Technology, Intermediate Technology, has as one of its most important functions that of increasing the efficiency of technology-practice employed within poor communities. It is true that many primitive small-scale, simple or low-cost technologies may be inefficient; but it does not follow that it is necessary to abandon these primary characteristics as a precondition of increasing efficiency. In the main Appropriate Technology is concerned with directing technological research, development and dissemination facilities towards raising the internal and external efficiency of technologies which exhibit these primary characteristics.

Another technically oriented criticism of Appropriate Technology is that significant *technology choice* is not really possible.[9] Appropriate Technology implies the need to carefully select from amongst alternative technological options, whereas some critics speak as if there is always only one viable way of efficiently conducting an activity.[10]

This criticism possesses a superficial air of validity in that at any one time in any one place there might only be one readily available technology for a task. It does not follow that this is necessarily the case, however, nor that there are binding technical reasons for such a lack of choice. The Appropriate Technology movement has as one of its main purposes the expansion of available technology choices in circumstances where presently they are in fact limited. Thus, the criticism is based upon a misunderstanding of the movement's *raison d'être*.

A further rejoinder to this criticism is that those who claim that choices in technology are not possible normally simply assert this without providing solid arguments or empirical evidence. The criticism tends to arise from personal bias rather than from a more objective basis. In contrast, there is considerable literature which provides serious arguments for the prospects for technology choice, drawing upon a steadily growing array of empirical evidence.[11]

In conclusion the claim that it is not possible to make choices between alternative efficient technologies is largely a crude assertion

---

[9] Decisions and plans produced by government policy-makers in urban-industrialized countries are, in the author's experience, based upon the assumption that technology choice is not really possible - i.e., that technology is "given". Interestingly, this assumption is rarely articulated formally or seriously argued in published literature, but is often voiced in face-to-face discussions. An exception is the criticism of Appropriate Technology by J. Ellul (*TheTechnological System* [New York: Continuum, 1980], pp. 191-195).

[10] See, J. Ellul, *The Technological Society*, translated by J. Wilkinson (New York: Knopf, 1964). S. R. Carpenter provides an interesting review of how most official "technology assessment" practitioners exclude from their analyses the insights of the Appropriate Technology movement vis-a-vis choice ("Technoaxiology: Appropriate Norms for Technology Assessment", in *Philosophy and Technology*, ed. by P. T. Durbin and F. Rapp [Dordrecht: D. Reidel, 1983], pp. 115-136).

[11] Since the pioneering work of A. K. Sen (*Choice of Techniques* [3rd ed.; Oxford: Basil Blackwell, 1968]) a number of publications have extended the basic evidence, e.g.: G. Jenkins, *Non-Agricultural Choice of Technique: An Annotated Bibliography of Empirical Studies* (Oxford: Institute of Commonwealth Studies, 1975); R. Mercier, "Some Reflections on the Choice of Appropriate Industrial Technology for Third World Development", in *Appropriate Technologies for Third World Development*, ed. by A. Robinson (London: MacMillan, 1979), pp. 203-218; B. G. Lucas and S. Freedman, eds., *Technology Choice and Change in Developing Countries: Internal and External Constraints* (Dublin: Tycooly, 1983).

with no foundation in serious evidence. It is also based upon a static view of technology which takes available technology as simply given, and ignores the dynamic context which may alter the available choices.[12] The criticism ignores the possibility that people may consciously influence the dynamic processes which determine the availability of technologies for the purpose of expanding technology choices.

Economic

There appears to be at least four types of criticisms of Appropriate Technology related to economics.

First Economic Criticism. The first type of criticism is that appropriate technologies are not economically competitive. As indicated earlier there is nothing in the basic Appropriate Technology concept which logically disposes it towards non-competitiveness. Intermediate Technology, as promulgated by Schumacher, is essentially an economic concept about how activities based upon relatively inefficient technologies may be made economically viable through technological innovations. This theme is not *negated* by the other streams of literature in the movement, even where relatively little interest is expressed in economic competitiveness.

It is mostly the emphasis on smallness, low capitalization levels and simplicity in Appropriate Technology which evokes a response by economists and quasi-economists that appropriate technologies are not "economic". This response is fuelled by the economic theory that increasing the scale of production and increasing the level of gross and per capita capitalization in a production process brings concomitant increases in productivity. Appropriate Technology appears to contradict such theory.

A first response to this apparent contradiction is that, insofar as economies of scale are obtainable for a given production process, these economies have limits: beyond an optimum level, marginal productivity often becomes negative. The notion that diseconomies of scale occur beyond a certain point is part of orthodox economic theory; but the actual optimum point may not be derived on the basis of theory alone (insofar as there is a single optimum point) and it varies with real world circumstances. It is possible, for reasons other than technical ne-

---

[12] A recent I.L.O. study has shown that even in industries dominated by large multinational enterprises the scope for technology choice is still considerable (N. Jéquier, et al., *Technology Choice and Employment Generation by Multinational Enterprises in Developing Countries* [Geneva: International Labour Office, 1984]).

cessity, for an enterprise to deploy technology which commits the enterprise to a production scale beyond the economic optimum (or beyond the range of optimums). The substantive claim of many Appropriate Technology advocates is that maximum economies may often be achieved independently of increases in scale. When economies of scale are possible, furthermore, the optimum may often be achieved at a scale lower than is commonly thought possible.

A second response is that there is a body of empirical evidence proving that the "economies of scale" doctrine frequently does not hold above a relatively small scale - or that in cases where economies from large-scale production are possible there are often alternative production processes available which enable comparable productivity increases without increases in the scale of production.[13]

A third response to the charge of "appropriate technology" not being competitive is that competitiveness itself is an ambiguous notion. The level of return required on capital investment, for example, is a prime determinant of an enterprise's potential economic viability. It may be possible for an enterprise to be *economically self-sustaining* without being *commercially viable*, where the latter is defined as the capacity of an enterprise to provide a return on capital at a rate at least as high as the "market rate".[14] If investors adopt the goal of maximizing the provision of useful and satisfying work opportunities, rather than financial profit *per se*, there is scope for an enterprise to be competitive which otherwise would not be competitive.[15] This is one reason why cooperatives and other forms of worker ownership in industry are receiving greater attention recently.[16] Within the framework of a worker-owned enterprise, in an economy faced with structural un-

---

[13] Cf.: B. Stein, *Size, Efficiency and Community Enterprise* (Cambridge, Mass.: Center for Community economic Development, 1974); A. Bollard, *Small Beginnings: New Roles for British Businesses* (London: Intermediate Technology.Publications Ltd., 1983).

[14] This distinction has been made by R. Stares in a paper entitled "Local Employment Intitatives in Theory and Practice", presented to the Seminar on Local Initiative for Employment Creation, Nice, 1983, organized by the European Centre for Work and Society (cited by G. McRobie in "Technologies for One World", *Appropriate Technology*, 11, 2 [1984], 4).

[15] See: P. Barnes, "Confessions of a Socialist Entrepreneur", *The Washington Monthly* (October 1983), 41-47; P. Barnes, "Wise Economics", *New Republic* (15 June 1974), 29-31.

[16] A review of both the Australian and international developments in this field has been conducted by P. Kenyon (*Co-operative, Community and Self-Employment Business Ventures*, report of the State Employment Task Force [Perth: Government of Western Australia, 1983]). Cf.: M. Carnoy and D. Shearer, *Economic Democracy: The Challenge of the 1980s* (White Plains, N.Y.: M. E. Sharpe, 1980); D. Wright, *Co-operatives and Community* (London: Bedford Square Press, 1979).

employment, it may be possible for an "employee" to obtain a larger net income by investing his or her available funds in a firm which will be able to provide that person with a job, rather than by investing the same funds elsewhere at a higher rate of return but in activities which will not provide him or her with employment. The same argument applies to a local community (e.g., a small rural town).[17] It is possible for the net wealth generated within a local community (which is eventually available for re-investment) to be maximized if local people invest in local enterprises rather than "foreign" enterprises.[18] This argument could still hold even when the interest rates from investments directed outside of the community may be higher. Thus, if economic sustainability is made possible without the necessity of generating "competitive" financial surpluses, the scope for the successful deployment of a range of alternative technologies may be enhanced.

In summary, there is no reason why technologies advocated under the rubric of Appropriate Technology may not be competitive within a capitalist-oriented market environment;   and others which are marginal may become competitive in a market environment if alternative ownership and investment patterns are adopted.

Second Economic Criticism. The second type of economics-related criticism is that Appropriate Technology is an anti-growth concept. There are two aspects to this criticism: the view that economic growth is incompatible with the use of appropriate technologies; and, the assumption that a healthy economy requires continual growth.

The Appropriate Technology movement exhibits a range of positions on these issues. Appropriate technologies in the South are normally seen as means of enabling economic growth whereas the move-

---

[17] A. H. Burnell, ed., *Today's Action Tomorrow's Profit: An Alternative Approach to Community Development* (Santa Barbara: Community Environmental Council, 1972); N. G. Kotler, *Neighborhood Economic Enterprise* (Washington, D.C.: National Association of Neighborhoods, 1978); W. Sarkissian, ed., *Employment Creation Through Community Economic Development* (Armidale, N.S.W.: Social Impacts Publications, 1983).

[18] A field of research concerned with "community economics" has now developed considerable sophistication, as evidenced by the following examples from the literature: E. Blakely, *Planning Local Economic Development* (Beverly Hills and London: Sage, 1989); B. Harrison, "Ghetto Economic Development: A Survey", *Journal of Economic Literature*, 12, 1 (1974), 1-37; Institute for Local Self-Reliance, "Methods for Measuring Community Cash Flows", *Self-Reliance*, 23 (1980), 1, 4-5, 11; R. L. Schaffer, *Income Flows in Urban Poverty Areas: A Comparison of the Community Income Accounts of Bedford-Stuyvesant and Borough Park* (Lexington: Lexington Books, 1973); C. Moore, G. Karaska and D. Bickford, *Impact of Banking on the Regional Income Multiplier* (Philadelphia: Regional Science Research Institute, 1979); A. H. Block, *Impact Analyses and Local Area Planning* (Cambridge, Mass.: Centre for Community Economic Development, 1977).

ment in the North tends to embody various critiques of economic growth. Most advocates of Appropriate Technology avoid the extremes of either pro-growth *per se* or anti-growth *per se* and stress the importance of qualitative distinctions vis-a-vis the content of growth.[19] Thus concern is directed at *what* it is that grows, *how* and *where* it grows, in what *direction* it occurs, *who* growth benefits, and at what *rate* growth takes place. Aggregate measures of growth often obscure the realities of economic life in particular communities within a nation.

The rate of growth of Gross National Product (the standard measure of economic growth) is an inadequate measure of economic wellbeing, as it excludes the informal economy, excludes many externalities (unaccounted social and environmental costs) and treats *any* addition to the aggregate flow of finance as a positive benefit. (e.g.: $1000 spent treating, or causing, iatrogenic disease is treated as of equal value to $1000 spent on obtaining a healthy diet; or, $1000 spent depleting non-renewable resources is treated as of equal value to $1000 spent in developing renewable-resource technology). The Appropriate Technology movement does not view the deployment of technology for the sole purpose of increasing G.N.P. as a very meaningful strategy. It tends to advocate *sustainable socio-economic development* rather than unqualified rapid growth.[20]

The notion of a *steady-state economy* is widely supported within the Appropriate Technology movement. The concept is different to the concept of *zero economic growth*. In a steady-state economy the flow of tangible throughputs and the size of the tangible stock remain at a minimum level consistent with meeting human needs; the growth of the economy takes place through so-called "intangibles" such as services, information exchange, art, philosophy and recreation and through low-energy human activities.

In summary there are some activities associated with supporters of Appropriate Technology which may be viewed, with some justification, as not useful for economic growth (narrowly defined). The main emphasis in the movement is on a *critical attitude* towards economic growth, however, rather than one of narrow-minded acceptance or re-

---

[19] E.g., Schumacher writes: "A small minority of economists is at present beginning to question how much further 'growth' will be possible, since infinite growth in a finite environment is an obvious impossibility; but even they cannot get away from the purely quantitative growth concept. Instead of insisting on *the primacy of qualitative distinctions*, they simply substitute non-growth for growth, that is to say, one emptiness for another" (*Small is Beautiful: A Study of Economics as if People Mattered* [London: Blond and Briggs, 1973], p. 44).

[20] See, A. K. Biswas, "Sustainable Development", *Mazingira*, 4, 1 (1980), 4-13.

jection. Notwithstanding the above, Appropriate Technology is compatible with economic growth (when qualified) and may often have economic growth as its main purpose.

Third Economic Criticism. The third type of criticism associated with economics is that Appropriate Technology is based upon inadequate or spurious economic theory. A full analysis of this objection is not possible here, in view of the high level of disagreement within the economics profession as to what constitutes sound economic theory. The following comments are most relevant.

Appropriate Technology, as exemplified by Intermediate Technology, has its roots in the field of development economics. The main point of departure from orthodox economics is that *technology choice* is viewed as a central variable in economic decision making. The tendency in economic theory, prior to the spread of Appropriate Technology, was to treat technology as an exogenous factor which could be taken as given. The work of Schumacher and followers, however, has stressed that technology is a dynamic factor within the economy and that it is *both* determined by other forces in the economy and in turn is a determinant of those forces; technology, in other words, ought not to be treated as exogenous. Thus, Appropriate Technology exhibits a focus on technology which has made it uncomfortable for mainstream economic theory. It would be wrong, however, to say that the notion is in conflict with mainstream theory.[21] More correctly, Appropriate Technology has introduced novel aspects to economic theory, making it a focus for debate in certain schools of thought.

Since the mid 1960s considerable debate and theoretical refinement has occurred within the profession of economics on the role of technology in economic development and employment generation.[22] It may be said that "hard" theory addressing technology choice and Appropriate Technology has begun to emerge as part of orthodoxy, although debate still proceeds. Within the neoclassical theoretical framework, in particular, a minor stream of interest in questions of technology choice has existed for some time. Within the last decade a general interest in the economics of technology has burgeoned. This trend has made the emphasis on technology by Appropriate Technology advocates part of the mainstream rather than a minority interest. There is not as yet a gen-

---

[21] This has been demonstrated cogently by F. Stewart (*Technology and Underdevelopment* [London: MacMillan, 1977]).

[22] E.g.: N. Rosenberg, *Perspectives on Technology* (Cambridge: Cambridge University Press, 1976); A. Heertje, *Economics and Technological Change* (London: Weidenfeld and Nicholson, 1977); B. Delapalm, et al., *Technical Change and Economic Policy* (Paris: Organization for Economic Cooperation and Development, 1980).

erally acknowledged theory of technology within economics; Appropriate Technology is part of the evolving research and discussion. The basic concepts of Appropriate Technology, vis-a-vis economics, are now rarely disputed on formal economic grounds. Rather, most debate revolves around empirical evidence for the *extent* to which Appropriate Technology has a viable role to play, and around the political constraints on its dissemination.

Fourth Economic Criticism. A fourth type of criticism related to economics concerns the relative importance of public planning versus the "hidden hand" of the market.

Some argue that public intervention in the economy is necessary to create the sort of environment where appropriate technologies would be able to flourish.[23] There is clearly some validity to this argument and for those who are opposed on principle to active or planned public participation in the economy, this amounts to a criticism of Appropriate Technology. It could be argued, in response, that public involvement in economic decision-making is unavoidable in the modern economic environment, and therefore that governments ought to include serious planning of technology strategies as part of their political agendas.

Notwithstanding the important role which government initiatives may play in ensuring the practicability of Appropriate Technology, it is clear that it does not depend completely upon such initiatives. Chapter Seven indicated how market forces have been the stimulus for the development of many appropriate technologies. In addition Appropriate Technology may be seen as essential to the maintenance of effective market mechanisms. Orthodox market theory points to the need for a large number of competing firms (each one small enough to prevent it being able to significantly alter market conditions by its actions) as a prerequisite to the effective operation of the market system - and hence to minimizing the need for government intervention. It is only possible to have a large number of competing small firms if the technology available for production purposes is capable of operating efficiently at a scale small enough (in physical and financial terms) to be affordable for a small firm. Thus, small-scale, low-capital-cost technologies (often considered especially pertinent within the Appropriate Technology approach) may be viewed as an essential ingredient of an authentic free enterprise economy.

---

[23] See, C. Cooper, "Choice of Techniques and Technological Change as Problems in Political Economy", *International Social Science Journal*, 25 (1973), 293-304, esp. pp. 293-295.

In summary, there are aspects of Appropriate Technology which appeal to people on either end of the competition/market - planning/intervention spectrum. Appropriate Technology does not provide a clear bias towards one type of economic decision-making system rather than another, although it does point to the need for government to create or maintain an economic environment conducive to relatively open technology choice.

In conclusion, the argument that Appropriate Technology does not make good economic sense may not be sustained. In such a generalized form the criticism is nonsensical, highly ambiguous and requires further qualification. There *are* examples within the Appropriate Technology literature of proposals which would not be economically competitive in most circumstances, but this in itself does not provide grounds for arguing that the whole concept is fallacious. There are many examples of modern technology, designed without adherence to the principles of Appropriate Technology, which would not be economically self-sustaining; yet, few people would wish to therefore argue against modern technology as such. Many criticisms of Appropriate Technology are based upon either ignorance of available empirical evidence, distortion of the claims of leading protagonists, or reliance upon examples from the literature which differ from the consensus of the movement but which suit the biases of the critic. Arghiri Emmanuel's mordant polemic is a noticeable example of this type of criticism.[24] He criticizes "appropriate" technologies as being only capable of providing employment at the cost of sacrificing output and income. Stewart responds in the following terms:[25]

> The attack is misconceived because the appropriate technologies advocated do not do this: they maximize employment and output. Lack of research in the area of A.T.s has undoubtedly meant that in many areas efficient appropriate techniques are non-existent. This is precisely why groups such as I.T.D.G. were established: their activities consist in developing and promoting efficient A.T.s which increase output as well as jobs. Many examples could be cited from cement production to sugar processing, egg tray manufacture to wind power. The argument for appropriate technology is not that jobs should be put before output, but that techniques can be developed which promote both.

---

[24] Emmanuel, *Appropriate or Underdeveloped.*
[25] Stewart, Book Review, p. 23.

While "appropriate" technologies may be vindicated on economic grounds, much of the appropriate technology literature incorporates critiques of mainstream economic dogma, implying that a broader range of criteria ought to be applied in technology choice than maximization of short-to-medium term financial surplus.

Physical

One possible criticism of Appropriate Technology is that physical resource constraints make it impracticable.[26] The argument goes that for poor countries, or countries with poor access to cheap sources of raw materials, it is necessary for available resources to be exploited with the highest possible efficiency. By assuming that modern technologies are superior for this purpose, and that "appropriate" technologies are somehow inconsistent with modernity, it follows that "appropriate" technologies would not be the most appropriate.

This criticism possesses a superficial attraction in that it appeals to the concern for resource conservation which characterizes the Appropriate Technology movement. It incorporates a logical fallacy, however, that appropriate technologies are neither modern nor efficient in their use of resources. It may be that in some circumstances the most resource-efficient technologies do not conform with criteria of Appropriate Technology related to scale or per-capita investment. In such cases a trade-off would be required between conflicting criteria. This problem, while probably frequently encountered, may be seen as an argument for more resources being invested in innovation aimed at resolving such conflict, rather than as a fundamental weakness of Appropriate Technology.

Cultural

The Appropriate Technology movement is sometimes criticized as possessing anti-technology sentiments.[27] As a rejoinder we should make

---

[26] The issues surrounding this criticism have been explored, for example, by M. R. Biswas and A.K. Biswas, in "Environment Implications of Development for the Third World", in *Technology Choice and Change in Developing Countries: Internal and External Constraints*, ed. by B. G. Lucas and S. Freedman (Dublin: Tycooly, 1983), pp. 117-127.

[27] This is reflected in the criticisms of H. Marcuse by J. Habermas (*Toward a Rational Society*, trans. by J. J. Shapiro [London: Heinemann, 1971] esp. pp. 85-87) and in Rybczynski's portrayal of "California Dreaming" (*Paper Heroes*, pp. 83-109).

the distinction between, on one hand, opposition to unbridled technological change and the invasion of technical modes of operation into every sphere of life and, on the other hand, opposition to technology *per se*. As indicated in Part Two, the movement does exhibit the former attitude. There is a significant anti-technology stream in Western culture and other cultures and some commentators group Appropriate Technology and its sub-streams, such as Alternative Technology, together with anti-technology cultural movements. While some people who employ "appropriate technology" rhetoric may in fact possess anti-technology sentiments, it does not follow that the movement as a whole is against technology. The survey of the Appropriate Technology movement in earlier chapters indicated that most of its protagonists affirm the value and necessity of technology. The survey also revealed the heterogeneity of the movement and the scope for contradictions between the movement's core and its periphery. It appears that the broad social movement, as opposed to the concept as defined in this book, exhibits some inconsistency at the level of cultural responses to technology.

A second cultural criticism is that Appropriate Technology is predominantly a passing fad rather than a serious or enduring phenomenon.[28] There is indeed some superficial "bandwagon" activity within the movement but, as indicated in Part Two, the "fringe" elements are not representative of the whole Appropriate Technology movement. A more legitimate criticism would be that greater work is required in distinguishing between the less serious elements of Appropriate Technology and what might now be identified as the orthodox elements.

A further criticism of Appropriate Technology is that it represents inferior technology to the modern technology of urban industrialized societies. Critics argue that Appropriate Technology is an attempt to provide mere palliatives in the form of "second rate" technologies capable only of maintaining communities in situations of dependency and economic stagnancy. On the basis of this view it is claimed that advocacy by policy makers of so-called "appropriate technology" for disad-

---

[28] Cf.: Rybczynski, *Paper Heroes*, pp. 28-39; W. Riedijk, "Appropriate Technology: Fashion or Need", in *Fundamental Aspects of Appropriate Technology*, ed. by J. de Schutter and G. Bemer (Delft: Delft University Press and Sijthoff and Noordhoff International, 1980), pp. 200-207.

vantaged communities, in either the South or the North, is thinly veiled paternalism.[29]

If "appropriate" technologies were in fact inferior the above sentiments would be well founded. The criticism is, however, misplaced. Appropriate Technology, as defined herein, implies that concerted efforts have been made to ensure that the choice of technology is the *best* for the particular circumstances in question. Virtually all major proponents of Appropriate Technology make this point. The criticism also has semantic problems: if one wished to have the best technology for given circumstances it would be absurd to advocate inferior technology and doubly absurd to then call it "appropriate", when, logically, it would not be the best available.

The only remaining grounds for declaring appropriate technologies to be inferior depend on the possibility that Appropriate Technology proponents are normally wrong in their analyses of real-world conditions or wrong in their perceptions of human needs and of the corresponding technological design criteria. This is an issue which has already been addressed and which will be further examined later. The charge that Appropriate Technology embodies invalid normative goals (a hidden assumption implied by the above criticism) is not open to testing on empirical grounds; and, in any case, is not a position which has been seriously argued. Part Three will provide a framework for considering such normative objections cohesively.

Some critics may argue that Appropriate Technology does not require the cultivation of significant scientific and technical skills and that it therefore defeats its own purpose by failing to act as a stimulus for local skill in technological innovation.[30] The charge that Appropriate Technology does not require significant scientific and technical skills is an assumption rather than a proven proposition and belies considerable practical evidence to the contrary. A common technology design criterion promulgated within the Appropriate Technology movement is that of simplicity. The experience of practitioners, however, has revealed that considerable sophistication is frequently required in the design, development, dissemination and initial deployment stages to produce a technology which exhibits simplicity

---

[29] This view is widely expressed by delegates from the South to international fora on science, technology and development, but has been articulated most vehemently by Emmanuel (*Appropriate or Underdeveloped*).

[30] Criticisms of this type have been raised and summarized by A. Eberhard (*Technological Change and Development: A Critical Review of the Literature*, Occasional Paper in Appropriate Technology, School of Engineering Science, University of Edinburgh, 1982).

in its form and operation. The charge also appears to reflect the earlier unfounded assumption that appropriate technologies are *ipso facto* inferior. Even in the case where an appropriate technology may not involve high levels of sophistication it does not follow that such a technology is not suitable for encouraging local innovation skills. In communities with low levels of scientific and technical capacity, a highly sophisticated and demanding type of technology-practice may do very little to build-up indigenous technological skill because of being beyond the apprehension of local people. A less sophisticated technology may invoke greater technological response from local people thereby initiating a greater degree of technical problem-solving ability and greater endogenous innovation.

Social

Some critics may protest that Appropriate Technology is really a social concept rather than a concept about technology. There is truth to this claim. Appropriate Technology is most visible as a social movement. In this book "appropriate technology" has been defined in such a way as to denote technologies rather than social institutions. Nevertheless, our definition connotes social factors and not just technical factors and our notion of technology-practice indicates that technology ought not to be viewed as independent of social influences, even though it may not be reduced to social phenomena. While it is true that Appropriate Technology embodies a social dimension it does not follow that this fact constitutes a meaningful criticism.

A more plausible reason for rejecting Appropriate Technology could be that the social dimension implies that social changes are required as a prerequisite to its effective dissemination and that such changes are impracticable. A variation on this criticism is that Appropriate Technology embodies a social vision which is neither attractive nor commendable. Both of these criticisms will be explored more fully in the following chapters. The social aspects of Appropriate Technology do make it problematical but that this does not by itself provide grounds for rejection of the concept. The parameters of a desirable and realistic social vision are widely debated and it is not possible to examine the debate comprehensively as part of this study. Nevertheless, given that technology is not socially neutral (a view now firmly entrenched in the literature) it may be seen as commendable that the Appropriate Technology movement addresses this fact openly.

Intellectual

A possible criticism from an intellectual perspective is that Appropriate Technology is a mixture of incomplete and sometimes incompatible ideas, combined with a collection of ambiguous symbols and poorly defined terms - and hence that it is not a coherent concept.[31] As with the above criticism this charge has some validity. Evidence has been provided earlier, nevertheless, that sufficient consensus exists within the movement (at least in an incipient form) for our stipulative definition to incorporate most of the viewpoints. The criticism points to the need for a more clearly articulated conceptual framework for Appropriate Technology than has hitherto been available. Constructing the outlines of such a framework will be the focus of the next chapter.

Another charge against Appropriate Technology is that there is nothing novel about it. Many appropriate technologies may be identified which have come about independently of the social movement and without evoking the label "appropriate technology". This criticism appears quite valid. Some may consequently charge that the concept is intellectually superfluous or perhaps even dishonest. As a rejoinder it could be argued that the existence of appropriate technologies independently of the social movement provides evidence for the practicability of the concept rather than for its redundancy. Furthermore, the need for a social movement and its concomitant methodologies and ideas may be seen as a response to the observation that appropriate technologies *do not always* come about automatically, and that, in the conditions which have prevailed in the twentieth century, conscious and concerted effort is required to encourage the development of appropriate technologies. Some Appropriate Technology protagonists do give the impression that they consider the phenomenon to be the exclusive preserve of the social movement. This view is fallacious but, at the same time, the fact that it appears to be held by *some* protagonists does not mean that Appropriate Technology as such is intellectually untenable.

---

[31] See: C. Baron, "Appropriate Technology Comes of Age: A Review of Some Recent Literature and Aid Policy Statements", *International Labour Review*, 117, 5 (1978), 626; M. Hollick, "The Appropriate Technology Movement and its Literature", *Technology in Society*, 4, 3 (1982), 213-229; P. Deising, *Science and Ideology in the Policy Sciences* (New York: Aldine, 1982), pp. 288-302; H. Brooks, "A Critique of the Concept of Appropriate Technology", in *Appropriate Technology and Social Values: A Critical Appraisal*, ed. by F. A. Long and A. Oleson (Cambridge, Mass.: Ballinger, 1980).

A further charge could be that Appropriate Technology is symptomatic of an inferior intellectual life because it is anti-science, anti-technology and anti-civilization. This is a complex charge but it has been largely addressed by the preceding discussion; the charge is based upon a distortion of Appropriate Technology. A general principle, appropriate technology requires participation by local people in its management or design; this fact, combined with the fact that its development often requires considerable scientific and technical knowledge together with incisive social awareness, makes Appropriate Technology conducive to general intellectual development.[32] An added feature of Appropriate Technology is that its scope for providing intellectual stimulus extends to the vernacular and "grass roots" sectors of society rather than to intellectual elites alone.[33]

Miscellaneous

The following are criticisms which do not fit easily into a general category but which have been raised in the literature and deserve some attention.

The theme of local or regional self-reliance runs throughout the Appropriate Technology movement and some people argue that the notion is either unrealistic or patently impossible.[34] A justification for this criticism is the fact many of the technologies held up as making self-reliance and self-sufficiency possible (e.g., photovoltaic/wind hybrid electricity generation systems) may not themselves be produced on a self-reliant or self-sufficient basis: an international network of research, development, investment and education programs is often necessary to produce a technology for local self-reliance. The assumption upon which the justification is based is valid; the view that self-reliance is therefore not practicable does not, however, logically follow. Firstly, experiments have shown that it is *possible* to develop not only economic activity but also some of the technology behind economic activity on a relatively self-reliant or self-sufficient basis (this is, of course, not always possible and it also depends upon the standard and

---

[32] For a review of this issue see A. Pacey, *The Culture of Technology* (Oxford: Basil Blackwell, 1983).

[33] This is demonstrated graphically through a comparison of the role of toys in black Africa and rich urban-industrialized societies in a film by P. Krieg (*Tools for Change* [Freiburg: Teldok Film, 1978]); cf., Pacey, *Culture of Technology*, pp.142-149.

[34] Rybczynski asserts: "Self-reliance at different levels simultaneously is a patent impossibility" (*Paper Heroes*, p. 154).

style of life which is considered to be acceptable).[35] Secondly, the basic criticism glosses over the fact that most advocates of Appropriate Technology make a distinction between absolute self-sufficiency, on one hand, and self-reliance as a general approach, on the other hand. The extreme criticism, that Appropriate Technology is not possible because it relies upon an impossible level of self-sufficiency, is somewhat irrelevant because it addresses a position which is not in the main taken seriously by proponents of Appropriate Technology. Appropriate Technology advocates are concerned about the *direction* of development that is facilitated by technology (i.e., whether the technology encourages a more locally-oriented economy or whether it favours an economy which is sustained only by a large dependency upon imports) rather than with the attainment of self-sufficiency.

Another criticism of Appropriate Technology is that the various criteria for deciding the appropriateness of technology are often in tension with one another.[36] For example, maximum use of renewable energy technology or maximum recycling of certain types of waste materials, may require the use of highly complicated or expensive technology. This could be a dilemma for those who advocate renewable energy, waste recycling *and* simplicity together with low cost. Two rejoinders are apt here. Firstly, it does not follow that conflict between assessment criteria provides grounds for rejection of the basic concept of Appropriate Technology. Such conflict points to the need for flexibility, compromise, artful judgement and decision-making on the basis of broad perspectives. Trade-offs between principles or forces in tension with each other is not uncommon in technology, social life or in natural settings. Secondly, the existence of contradictory criteria *for a given set of circumstances* does not mean that innovation may not take place to resolve such contradictions (e.g., improvement in photovoltaic technology to make it more cost-effective). The need for innovation of this type is part of the *raison d'être* of Appropriate Technology rather than the grounds for its abandonment.

A further assertion often intended as a rebuff against Appropriate Technology is the claim that both "high" technology and "low" technology is required, and that "low" technology is not adequate to address global and local problems (both economic and ecological). This criticism is a highly ambiguous assertion. The meaning of "high" and "low" is frequently not clear. Furthermore, insofar as the meaning of "high" and "low" is made clear, there is no compelling reason to associate ei-

---

[35] See material evinced in Part Two.
[36] Eckaus, *Appropriate Technologies*, pp. 2, 10-12.

ther of the two notions exclusively with "appropriate". It is not possible to encapsulate the Appropriate Technology concept on a simplistic, one-dimensional "high-low" spectrum.[37]

A criticism related to the previous one is that it is foolish to reject all the benefits which have been produced in modern society.[38] Such a comment stands to reason. It is, however, a pointless comment in this context because it is almost impossible to find a serious argument within the Appropriate Technology movement for the rejection of modernity *per se*; the movement's critique of technological society may not be reduced to such a simplistic "either/or" argument.

Political Criticisms

While a number of the above criticisms do address identifiable weaknesses in Appropriate Technology, as the movement currently stands, many of them are based upon flimsy evidence or blatant bias and are frequently lacking in substantive analytical support. There is, however, a further category of criticisms, labelled here as "political", which may not be easily dismissed and which demands more detailed consideration. The majority of scholarly or seriously argued criticisms of Appropriate Technology fall into the political category. Some of the political criticisms are little more than dogmatic assertions or tirades by commentators predisposed against the movement.[39] Others take the form of balanced or sympathetic reviews which bring out cogent political critiques for the purpose of enhancing the concept and movement.[40] Finally, some raise fundamental objections to

---

[37] Ambiguity as to the meaning of "appropriate technology", and as to the role of "heavy" or "high" technologies in socio-economic development, has led to much confusion and misunderstanding in debate over the value of nuclear-energy technology in the South. This is evidenced by the debate in the journal *Research Policy*: G. Bindon and S. Mukerji, "Canada-India Nuclear Cooperation", *Research Policy*, 7, 3 (1978), 220-238; R. W. Morrison and E. F. Wonder, "Canada-India Nuclear Cooperation: A Rebuttal", *Research Policy*, 8 (1979), 187-190; G. Bindon and S. Mukerji, "Canada-India Nuclear Cooperation: A Rejoinder to a Rebuttal", *Research Policy*, 8 (1979), 191-198.

[38] E.g.: Sachsse, "Comment".

[39] E.g.: D. Chidakel, "Small is Beautiful as a Book and as a Bum Steer", *Science for the People*, 7, 4 (1975), 17-19; Emmanuel, *Appropriate or Underdeveloped*.

[40] E.g.: L. Winner, "The Political Philosophy of Alternative Technology: Historical Roots and Present Prospects", *Technology in Society*, 1, 1 (1979), 75-86; D. Elliot, "Working on All Fronts", *Undercurrents*, 16 (1976), 12-14; Hollick, "A.T. Movement"; D. Burch, "Appropriate Technology for the Third World: Why the Will is Lacking", *The Ecologist*, 12, 2 (1982), 52-66.

Appropriate Technology on grounds which, while not necessarily compelling, exhibit a degree of tractability.[41] Most of the possible political criticisms have already been published in the literature, so this chapter will mainly summarize these rather than to raise new criticisms. A full response to the political criticisms will be left to subsequent chapters.

Problems of Dissemination

The majority of the political critiques acknowledge the evidence that there are no intrinsic technical reasons why appropriate technologies or intermediate technologies may not be developed. An increasing number also accept the view that there are, strictly speaking, no fundamental economic criticisms which are immutable. The debate has shifted from the issue of whether appropriate technologies are possible in principle to the issue of how widespread their adoption may be. It is widely claimed, by both antagonists and protagonists, that despite the technical and economic feasibility of appropriate technologies they have been disseminated to only a limited extent.[42] Many commentators relate this to political constraints which they claim proponents of Appropriate Technology fail to address. Other commentators argue that some proponents do address these political issues but fail to follow-up the policy implications in a systematic manner. Cooper writes:[43]

> The questions of choice of techniques and technical change are often discussed as economic problems in the underdeveloped countries, but economists have not systematically taken account of the systems of political economy to which their discussions refer. ... this failure to be systematic about questions of political economy may lead to simplistic recommendations about technology policies - and possibly to some rather ill-founded optimism about the [efficacy] of 'appropriate' technologies as a 'cure' for problems of unemployment and maldistribution of income.

---

[41] Ellul, *Technological System*, pp. 191-195; E. Entemann, et al., "Alternative Technology: Possibilities and Limitations", *Science for the People*, 8, 5 (1976), 10-13, 33.

[42] Dunn is an example of an Appropriate Technology protagonist who has argued this point (see,"Appropriate Technology: Priorities").

[43] C. Cooper, "Choice of Techniques and Technological Change as Problems in Political Economy", *International Social Science Journal*, 25 (1973), 243.

Most advocates of Appropriate Technology take it very seriously as a means for addressing globally significant problems. Political constraints against the widespread dissemination of appropriate technologies are therefore of central relevance to the movement.

We will now consider four aspects of the political critique.

Narrow Technicism

Appropriate Technology is often accused of being a narrowly technicist notion which ignores the social and political context of particular technologies.[44]

It is argued by critics that the fundamental problems which motivate the Appropriate Technology movement are social and political problems rather than problems of technology. This argument takes two forms. Firstly, some assume that technology is "neutral" and thus open to being deployed for either good or evil, violence or non-violence, etc., depending upon the intentions and interests of those people who use the technology.[45] Thus the changes required (to achieve a more humanly or environmentally sound socio-economic system) are seen as "human" problems which have no bearing on the professional activities of engineers, technologists and others whose work bears directly on the design and production of technology. Secondly, some claim that technology is not neutral and that it always embodies the political and social conditions of the society in which it was spawned or in which it is deployed. Thus, the required changes of direction *are* seen as bearing upon the professional activities of technologists, but in the final analysis are "human" problems - or at least socio-political problems - and not tech-

---

[44] Almost *all* of the "political" critiques of Appropriate Technology make this claim. In addition to sources already cited, examples include: A. Harris, "Appropriate Technology", *Chain Reaction*, 2, 4 (1977), 31-32; D. Dickson, *Alternative Technology and the Politics of Technical Change* (London: Fontana/Collins, 1974); R. Disney, "Steady Reddy!", *Undercurrents*, 16 (1976), 16; E. Korn, "Small is Small", *New Statesman* (7 October 1977), 481-482; B. O. Pettman, "Some Labour Aspects of the New International Economic Order", *International Journal of Social Economics*, 5, 2 (1978), 93-111; M. Bookchin, *The Ecology of Freedom* (Palo Alto: Cheshire Books, 1982), pp. 260-266; D. Elliot and R. Elliot, *The Control of Technology* (London: Wykeham, 1976), pp. 216-235; J. Hanlon, "Visiting Windmills in Wales", *New Scientist*, 76, 1075 (1977), 216-218.

[45] Note: this is actually stated explicitly by the Productivity Promotion Council of Australia (*People and Technology in the 80s* [Melbourne: Productivity Promotion Council of Australia, 1980], p. 103): "The technology itself is neutral. The crucial questions are what we do with it, how fast and how far do we go, and how do we distribute its benefits and help its casualties."

nological problems.[46] Criticisms from either of these two viewpoints are directed at Appropriate Technology implying that because the concept is vitally concerned with technology it is *ipso facto* flawed in terms of social and political theory.

The assumption that technology is neutral (i.e., apolitical) is almost universally rejected in the literature and in cases where it is not it normally receives no more substantiation than an *ipse dixit*.[47] Hence criticisms coming from the first viewpoint just outlined do not warrant serious attention at this juncture. The assumption behind the second viewpoint (that technology is not neutral) is shared by the Appropriate Technology movement; in fact, this assumption is pivotal to the whole concept. Reddy, a central proponent of Appropriate Technology, writes:[48]

> For Technology is like genetic material: it carries the code of the society in which it was produced and survived, and tries to replicate that society.

The criticism is therefore flawed. The second half of the criticism is not necessarily annulled by this factual error but it does nevertheless contain another flaw: it does not logically follow that, because Appropriate Technology is vitally concerned with technology-qua-technology, the concept conflicts with the supposition of socio-political factors being prime determinants of technology choice. Proponents of this criticism often appear to portray a serious respect for technology-qua-technology as being diametrically opposed to a serious respect for politics-qua-politics. In principle it appears unjustifiable to treat these two perspectives as mutually exclusive. While there are exceptions to the general rule, awareness of the mutual interaction between

---

[46] Dickson (*Alternative Technology*), Winner ("Political Philosophy"), Entemann, et al. ("Alternative Technology") or Elliot ("Alternative Technology: Production for Need", paper presented to the *A.T. in the 80s* Conference, London, 16th June 1984), appear to be representatives of this position.

[47] This is increasingly so for the general literature on "development" and not just for specialist literature on Technology. Cf.: J. Baranson, "The Cornucopian Politics of World Development", *Bulletin of the Atomic Scientists* (November 1978), 41-44.

[48] A. K. N. Reddy, a central proponent of Appropriate Technology, writes: "For Technology is like genetic material: it carries the code of the society in which it was produced and survived, and tries to replicate that society" ("Alternative Technology: A Viewpoint from India", *Social Studies of Science*, 5 [1975], 332).

technology and socio-political forces is readily apparent throughout the Appropriate Technology literature.[49]

Notwithstanding the above rejoinders to the criticism of Appropriate Technology being a narrowly technical notion, we should acknowledge that the criticism is in fact not uncommon. It would appear that the thematic and semantic heterogeneity of the movement's literature provides *some* justification for the criticism. This provides further arguments for the articulation of a definitive and systematic theory of Appropriate Technology.

Finally, it is not difficult to find criticisms of Appropriate Technology which charge it with lacking a serious interest in technology.[50] The existence of contradictory criticisms confirms comments in earlier chapters about the heterogeneity of the movement. It also points to the wisdom of stipulating Appropriate Technology to be a mode of technology-practice rather than a collection of technologies. Appropriate Technology is not a narrowly technicist notion.

Technological Determinism

A related criticism to the one just considered is the charge that Appropriate Technology involves a commitment to the notion of *technological determinism*. The term "technological determinism" is frequently used with imprecision but generally refers to the view that socio-political structures are determined by technology and that desired socio-political changes may be brought about by technological means.[51]

Eberhard, who claims that technological determinism is pervasive throughout the Appropriate Technology movement, writes:[52]

---

[49] Cf., E. F. Schumacher, "Technology and Political Change", *Resurgence*, 7 (Nov/Dec 1976), 20-22: "If our technology has been created mainly by the capitalist system, is it not probable that it bears the marks of its origin, a technology for the few at the expense of the masses, a technology of exploitation, a technology that is class-orientated, undemocratic ..." (p. 20).

[50] E.g.: Rybczynski, *Paper Heroes*, esp. pp. 83-109; R. Vacca, *Modest Technologies for a Complicated World* (Oxford: Pergamon, 1980), pp 24-37. This criticism is often expressed verbally or informally by those who do not take Appropriate Technology seriously.

[51] The most comprehensive treatment of the subject has been conducted by L. Winner (*Autonomous Technology: Technics-Out-of-Control as a Theme in Political Thought* [Cambridge, Mass.: MIT Press, 1977]).

[52] Eberhard, *Technological Change*, p. 70.

Technological determinists have seen technology or technological change as the prime source of values and goals, as well as of material change, within a society. But in so doing they have ignored the historical roots of technological change and have obscured the relations of production which fundamentally structure the direction and nature of choice and innovation of technology.

The prime difficulty with technological determinism, according to its opponents, is that it lacks empirical support: technological determinism is propounded by those who ignore historical evidence to the contrary.[53] The second difficulty is that absolute technological determinism rules out the role of political action as a source of desired social change.[54] On the basis of this observation it would follow that the adoption of a technological determinist stance by an Appropriate Technology advocate is inherently contradictory: if technology absolutely determines social and human conditions then it would be absurd to expect that independent human initiative to make technology "fit" autonomously constituted social goals could be possible.

On the surface these criticisms appear compelling: it is not difficult to find historical evidence to confute the technological determinist doctrine; the notion does not sit easily with common sense; and, the technological determinism which Appropriate Technology is accused of endorsing would make the concept (of Appropriate Technology) untenable. On closer examination, however, the criticisms appear misdirected. The basic rejoinder to the technological-determinism criticisms directed against Appropriate Technology is that they are based upon a largely fictitious portrayal of Appropriate Technology and an inconsistent, or at least highly ambiguous, use of "technological determinism".

An article by the social anthropologist, M. Howes, as typical of critiques which exhibit these faults.[55] Howes takes Schumacher as the exemplar of Appropriate Technology and asserts that Schumacher's arguments assume "technology may be regarded as a wholly indepen-

---

[53] H. Braverman (*Labour and Monopoly Capital: The Degradation of Work in the Twentieth Century* [New York: Monthly Review Press, 1974]) is widely quoted as an authoritative source for claims to this effect.

[54] Cf.: S. Hill, "Technology and Society", in *Future Tense? Technology in Australia*, ed. by S. Hill and R. Johnston (St. Lucia: University of queensland Press, 1983), pp. 27-46; R. Johnston and P. Gummet, eds., *Directing Technology* (London: Croom Helm, 1979), esp. "Introduction", pp. 9-17.

[55] M. Howes, "Appropriate Technology: A Critical Evaluation of the Concept and the Movement", *Development and Change*, 10 (1979), 115-124.

dent variable in the development process, determining social economic and political relations, but in no sense being determined by them."[56] The striking feature of this assertion is that it is substantiated neither by actual quotes from Schumacher's writings nor by detailed analysis. Howes simply accuses Schumacher of technological determinism and uses this unsubstantiated assertion as the basis for further arguments that Schumacher's ideas are inconsistent. We should also note that Howes accuses Schumacher as being guilty not just of technological determinism but of *absolute* technological determinism; this is indicated by Howes' choice of the words "*wholly* independent" and "*in no sense* being determined". Such an *absolute* technological determinist position is relatively easy to debunk - but this is a relatively pointless exercise because it is extremely difficult to find a scholar who would espouse such a position and it is certainly not possible to mount solid evidence for the claim that Schumacher does so (this fact is no doubt the reason why Howes does not provide any such evidence).

There is no scholarly study which documents substantive evidence that the Appropriate Technology movement supports technological determinism in the above sense. David Dickson's book, *Alternative Technology and the Politics of Technological Change*, is, however, repeatedly quoted as an authoritative source for such a criticism. Dickson does intimate such a criticism when referring to the analysis of underdevelopment embodied in Intermediate Technology and Appropriate Technology. He states:[57]

> ... there is often an implication of technological determinism, by which the changes will emerge almost automatically once the new production techniques associated with Intermediate Technology have been introduced.
>
> This analysis corresponds to the general approach to social development that implies a functional interdependence between the processes of technological and economic development. In other words, social development is claimed to result from developing a technology considered appropriate to a particular set of social and economic conditions. Technology is seen as a neutral element in this process divorced from any direct social or political considerations.

The main problem with Dickson's comments is that while he provides some cogent arguments against technological determinism and

---

[56] *Ibid.*, pp. 158-159.

[57] Dickson, *Alternative Technology,* p. 159.

some reasoned cautions against turning Intermediate Technology into a mere "technological fix" strategy, his attempts to link Schumacher's views with technological determinism are little more than insinuations. As indicated by a qualifying remark in the same passage from which the above quote is taken, Dickson is well aware of this problem, viz.:[58]

> This is not to say that those concerned with implementing intermediate technology are unaware of the need for accompanying changes in social organization to enable intermediate technology to work effectively.

The paucity of support for the insinuation of technological determinism in Appropriate Technology is further illustrated by the logical lacunae in the above quote: Dickson leaps from comments about *implied functional interdependence* between technological and economic processes to the conclusion that technology is seen by Appropriate Technology advocates as neutral. While Dickson's book is a notable study of the political economy of technological change, citation of the book as a supposed proof of Appropriate Technology being a technological determinist doctrine is, at best, a poor substitute for good scholarship, and, at worst, dishonest. If there are grounds to substantiate Dickson's guarded comments it is reasonable to expect commentators such as Howes to document them.

The misleading nature of the "technological determinism" criticism is partly obscured by the ambiguity with which the term is used. A distinction should be drawn between a commitment to *absolute* technological determinism, on one hand, and the view that technology may exert *some* determining influences upon the economic, social and political environment in which it operates, on the other hand. Howes, Dickson and many others fail either to acknowledge or to adequately address this distinction. Appropriate Technology does embrace the latter view, but not the former. It is not uncommon for critics to cite evidence of the latter view being part of the Appropriate Technology concept and to then proceed to criticize the concept on the grounds of it incorporating the former view.

An interesting illustration of this type of approach is Brian Martin's censure of Amory Lovins for supposedly possessing a technolog-

---

[58] *Ibid.*, pp. 158-159.

ical determinist stance.[59] Martin makes it explicit that he considers
the forces influencing the development, choice and promotion of a particular form of technology to be *both* political *and* technological-cum-
economic.[60] He states that "while particular technologies lend
themselves to particular social and political structures, the connection
is not automatic".[61] This view is in keeping with that which prevails
throughout the Appropriate Technology literature, and in keeping
with the views of Amory Lovins. Oddly, Martin does not *explicitly*
claim otherwise. He points out that "soft technologies" may be employed within a political framework more in keeping with a "hard energy pathway" than a "soft energy pathway" and he stresses that
"politics" ought not to be subsumed under "technology". Martin resorts
to innuendo, implying that because Lovins focuses on technological matters this amounts to ignoring politics-qua-politics. Closer analysis reveals that the substantive views of Martin and Lovins are very similar.[62] It appears that in order to bolster the apparent distinctiveness of
his own discourse Martin resorts to the device of repudiating a theoretical position which he links by association, rather than fact or logic,
to the views of another person. Lovins simply does not view the introduction of soft technologies as guaranteeing a solution to the broader political matters raised by Martin.[63]

Published attempts to denounce Appropriate Technology advocates
for being technological determinists are, in the main, just poor scholarship. They also tend to artificially separate technological factors from
political factors, thereby generating a false dilemma between political
action and technology choice. While Appropriate Technology clearly
concentrates on technology and technologies, most of its protagonists ac-

---

[59] B. Martin, "Is Alternative Technology Enough? Amory Lovins: The Line Not
Taken?", *Chain Reaction*, 3, 2 (1977), 17-21. Martin's article is part of a series of three
articles on the politics of "alternative energy" produced by Friends of the Earth,
Canberra (see: "Queanbeyan Soft Drink Factory", *Chain Reaction*, 3, 2 [1977], 22-23;
"What Sort of Society is Possible?" *Chain Reaction*, 3, 2 [1977], 24-25). A later version of
Martin's paper was published as "Soft Energy, Hard Politics", *Undercurrents*, 27 (1978),
10-13.

[60] Note: Martin quotes David Dickson as his source of authority for this view
("Line Not Taken", pp. 17, 21).

[61] *Ibid.*, p. 18.

[62] This is apparent in a response by Lovins to Martin's claims (A. Lovins, "Lovins
Replies", *Chain Reaction*, 3, 2 [1977], 21)

[63] Lovins does not believe (apparently in contrast to Martin) that market forces are
intrinsic obstacles to a transition to a soft energy pathway; this does not mean that
Lovins disregards the political obstacles to such a transition (see pp. 60-62 of A. Lovins,
"Soft Energy Paths", in *The Schumacher Lectures*, ed. by S. Kumar [London: Blond and
Briggs, 1980]).

knowledge that, while technology may play a role in shaping society, the reverse process is *at least* as important. This is illustrated by the following quote from an address by Schumacher to an international conference:[64]

> All I am asking is not to extend the perfectly legitimate political and sociological argument in a way that invalidates the technological approach. Both are necessary. A political change, unless supported by an appropriate technology, will be just a paper change, will be just another gang of people trying to do different things... There is no contradiction between saying "an appropriate technology must be created" and saying "the basic problems are political and sociological" because there's an inherent relationship between the two.

Four comments may be offered in summary. Firstly, Appropriate Technology is not a disguised doctrine of technological determinism.[65] Secondly, the movement nevertheless views technology as a formative influence on society, but at the same time sees the need for conscious social and political action as a formative influence on technology. Thirdly, the concept of Appropriate Technology avoids creating a dichotomy between the category "politics" and the category "technology". Fourthly, most of the technological-determinism criticisms directed at Appropriate Technology are based less on substantive grounds than on the need for "ideological culprits" against which political and sociological theorists may contrast their views.

The issue of technological determinism and Appropriate Technology will be explored in more detail in Chapter Eleven.

Dependency, Inequality and Vested Interests

A major argument against the practicability of Appropriate Technology revolves around the claim that certain social-cum-political institutions, which have accompanied the development of dominant

---

[64] E. F. Schumacher, "Using Intermediate Technologies", in *Strategies for Human Settlements: Habitat and Environment*, ed. by G. Bell (Honolulu: University Press of Hawaii, 1976), pp. 124-125.

[65] P. Deising's study of the attitudes to science and technology of various schools of thought in economics, political science and sociology, concurs with this conclusion. In a comparison of New Left Marxism and Appropriate Technology he writes: "... both reject the technological determinism that treats technological growth as an autonomous, self-generated force in history" (*Science and Ideology in the Policy Sciences* [New York: Aldine, 1982], p. 301).

technology-practice in modern industrialism, are now firmly in place, and effectively preclude the development of potentially viable technological alternatives.

Two main variations of this argument may be identified. The first relies upon the supposition that Appropriate Technology is narrowly technical in orientation and guilty of technological determinism, with the implication that the movement ignores the need to address the institutional resistance it is likely (or bound) to encounter.[66] This supposition was examined above and was shown not to be valid. The second variation does not rely upon such a supposition but nevertheless remains skeptical about the movement's capacity to surmount the institutional problems. In other words, even given the possibility that the Appropriate Technology movement was fully cognizant of the political dimensions of its objectives and had organized its activities accordingly, some critics argue that the institutional inertia of the status quo is so great that anything more than marginal changes would be unrealizable. This latter variation on the main argument is, in the present author's opinion, the most serious objection which has been raised in the literature and as such will now be examined. It has been frequently used by critics and its salient points will now be paraphrased.[67]

The argument is based upon the observation that the growth of international and intranational capitalism, combined with the existence of elitism and hierarchical organization in traditional societies, has led to major imbalances in the distribution of economic power. Minority groups of wealthy and powerful people have gained positions of dominance in economic decision-making and have developed vested interests in the maintenance of the status quo. The inequitable distribution of

---

[66] The majority of sources cited with reference to "narrow technicism" and "technological determinism" also raise this criticism. A strong example is C. German and K. Alper, "Alternative Technology: Not a revolutionary Strategy", *Science for the People*, 8, 5 (1976), 14-17. Cf.: A. Cockburn and J. Ridgeway, "The Myth of Appropriate Technology", *Politicks and Other Human Interests*, 1, 3 (1977), 28.

[67] E.g.: L. de Sebastián, "Appropriate Technology in Developing Countries: Some Political and Economic Considerations", in *Mobilizing Technology for World Development*, ed. by J. Ramesh and C. Weiss, Jr. (New York: Praeger, 1979), pp. 66-73; J. Hanlon, "Does A.T. Walk on Plastic Sandles?" *New Scientist*, 74, 1053 (1977), 467-469; J. Matthews, "Marxism, Energy and Technological Change", in *Politics and Power, One*, ed. by D. Adlum, et al., (London: Routledge and Kegan Paul, 1980), pp. 19-37; P. Harper, "What's Left of Alternative Technology?", *Undercurrents*, 6 (March/April 1974), 35-38; Dickson, *Alternative Technology*; Burch, "Appropriate Technology"; Cooper, "Choice of Techniques"; Chidakel, "Small is Beautiful"; Entemann, et al., "Alternative Technology"; Eberhard, *Technological Change*; Emmanuel, *Appropriate or Underdeveloped*; A. Schnaiberg, "Did you Ever Meet a Payroll? Contradictions in the Structure of the Appropriate Technology Movement", *Humboldt Journal of Social Relations*, 9, 2 (1982), 38-62.

economic power is apparent in the relations between the North and the South, but is also noticeable within nations; the gap between the rich minority and the poor majority is normally at its greatest in the countries of the South. Thus, political and social institutions which favour the interests of dominant elites have become entrenched.

Given the growing importance which technology plays in economic activity, it follows that decisions about the direction of technological development may not be divorced from the political forces inherent in conditions of institutionalized maldistribution of economic power. It is further argued that the current dominant patterns of technological innovation, allocation of resources for research and development, and allocation of investment in new technology, not only reflect the aforementioned institutional environment but are derived from it. In other words the technology in use within a particular society is determined by and embodies the political, social and economic institutions of that society. Accordingly the dominant technology-practice in most countries is bound by the inertia of the status quo: powerful elites which benefit from the maintenance of established economic practices have vested interests in the dominant types of technology-practice being maintained. It follows that, insofar as Appropriate Technology represents a departure from dominant technology-practice, it will be opposed by powerful elites.[68]

Tactics for the suppression of appropriate technologies could involve, for example, manipulation of the market by monopolies and oligopolies, large companies putting small ones out of business through undercutting the prices of appropriate technologies (products and process) by temporarily running at a loss, the buying-up and subsequent closing down of small "appropriate technology" businesses, or by vested interests unfairly influencing the availability of credit. Large vested

---

[68] Such elites may not necessarily be representative of *private* interests. The Tanzanian Government, for example, has had an explicit policy commitment to the Appropriate Technology approach since the mid 1960s. F. C. Perkins has conducted a fascinating empirical study of technologies selected by private and public enterprises in ten major Tanzanian industries. He found that enterprises using technically and economically efficient labour-intensive, simple and often small-scale technologies, accounted for a large proportion of the output and employment in most of those industries. Despite the availability of such technologies, however, the publicly-owned enterprises had in most cases chosen large-scale, capital-intensive and usually technically and economically inferior technologies. Biases in the planning and management practices of government personnel led the bureaucracy to support less "appropriate" and less efficient technologies - in opposition to the Government's stated objectives and in contrast to private industry. See: F. C. Perkins, "Technology Choice, Industrialization and Development Experiences in Tanzania", *The Journal of Development Studies*, 19, 2 (1983), 213-243.

interests may also be able to damage the prospects for Appropriate Technology by influencing the allocation of funding for research and education away from Appropriate Technology and towards the enhancement of dominant technologies.

These political constraints to the dissemination of Appropriate Technology are exacerbated by the connections between technologies (in their capacity as means of production) and the end products of the production processes. It is not uncommon for the manufacture of certain products to be feasible only with highly standardized, "dominant" technologies; viz., process innovations and product innovations are often interdependent. Dominant business interests are able to employ a range of methods by which their products (usually popular "brand name" products) come to dominate consumer preferences. Consequently, "appropriate" production technologies may fail to survive in the market; this may be due, not to inferior technical efficiency, but to the fact that "appropriate" products do not adequately match dominant advertising-reinforced consumer preferences. Furthermore, especially in the South, the market for consumer goods is often biased heavily in favour of the preferences of elites; this bias is reflected in the choices of technology and investment decisions in the manufacturing industries of a given country. With growing pressure for domestic manufacturing enterprises to be internationally competitive comes a further pressure for them to adopt similar production technologies to foreign competitors, thereby adding to the bias in favour of dominant technologies.

As a consequence of the inequitable distribution of economic power, discussed above, relations of dependency by the poor upon the wealthy and powerful become entrenched. This economic dependency is reflected in technological dependency. Many political-economists argue that technological dependency is now a self-reinforcing feature of the relationship between non-dominant and dominant communities and that this situation may not be altered without a radical and comprehensive transformation of the whole system of political-economy (i.e., an anti-capitalist revolution). This viewpoint is based upon the supposition that technology is a dependent variable in systems of political economy rather than a dynamic variable open to manipulation in the absence of prior action to alter the whole system or other non-technological variables.

Many commentators therefore point to the need for "political" action to overcome the power of vested interests which maintain patterns of technology-practice conducive to inequality and dependency by non-dominant communities. On the surface this would appear to be a straightforward argument, given the inherently political nature of technology. The call for "political " action is frequently portrayed, however, as something separate from "technological" action.[69] Critics who separate politics from technology in this way tend to be either antagonistic towards Appropriate Technology or highly pessimistic about its prospects for success.

Such antagonism and pessimism is associated with the view that so called "appropriate technologies" tend to reinforce the status quo rather than lead to its transformation. The comments of L. de Sebastián, from El Salvador, are representative of those which accord with this view:[70]

> "Appropriate technology" is the latest, or one of the latest, strategies devised to soothe the social conscience of the affluent countries, which do not cease to protest against the enormous human suffering that economic underdevelopment perpetuates - suffering that cohabits in this world with exaggerated and, so to speak, insulting luxury.
>
> What has occurred for decades can be likened to new wine being poured into old wineskins: new technological strategies being poured into the same old, dependent, autocratic social structures of most underdeveloped countries.

Four propositions may be identified in support of the view typified by de Sebastian's comments.

Proposition #1. The first is that the technological emphasis in Appropriate Technology deflects attention from more fundamental "political" changes.

Proposition #2. The second is that, by helping to improve the economic circumstances of the poor and dependent, appropriate technologies placate the dissatisfaction of such people, leading to a reduced desire to bring about radical transformation of the whole structure of their society.

Proposition #3. The third proposition is that appropriate technologies will simply lead to a new type of technological dependency relationship. The argument goes that, because most scientific and technological resources are concentrated in the North, and to some extent in

---

[69] Disagreement on this point is reflected in British debate over the technology policies of the Greater London Enterprise Board. See: L. Levidow, "We Won't Be Fooled Again? Economic Planning and Left Strategies", *Radical Science Journal*, 13 (1983), 28-38; P. Clark, "Working With People", *Radical Science Journal*, 13 (1983), 101-104; M. Ince, "Economic Planning and Left Strategies: A Reply", *Radical Science Journal*, 14 (1984), 158-159; L. Levidow, "Reply to Critics", *Radical Science Journal*, 14 (1984), 160-161.

[70] de Sebastián, "Appropriate Technology", p. 72.

the hands of the South's elite, the widespread development of appropriate technology will itself depend upon a mobilization of "dominant" technical problem-solving resources in a different direction. This would in turn mean that "appropriate technologies" would generally need to be *transferred* from regions of technological dominance, thereby indirectly strengthening the knowledge and experience of the dominant regions and exacerbating the dependency of the non-dominant regions.

Proposition #4. The fourth proposition is that appropriate technologies may only be widely disseminated insofar as they form part of dominant technology-practice and, as such, reinforce the status quo; accordingly, some critics would suggest that the more successful so-called "appropriate technologies" become, ironically, the more they cease to serve the basic reformist goals of the Appropriate Technology movement.

These propositions may not receive a full response independently of articulating an integrated framework for Appropriate Technology. A brief response only will therefore be provided at this point.

Although there may be cases where the first proposition may valid, it is not generally applicable. Within the Appropriate Technology movement, technology is seen as a *medium* for addressing political problems, not as a channel for bypassing these problems

In response to the second proposition it should be asked whether real improvements to the condition of poor people, or dependent communities, may not be viewed as part of the process of transforming a society. The second proposition amounts to a cogent refutation of Appropriate Technology only given the assumption that institutional change needs to be total and abrupt in order to be authentic. While some people hold to such an assumption, it is not self-evident and is by no means uniformly accepted by social and political theorists. Furthermore, it depends upon the normative judgement that improvement in the welfare of people currently living ought to be sacrificed for the sake of greater improvements for future generations; this judgement is also open to question.

The third proposition reveals the possibility that, in the short term at least, compromises might need to be accepted between different objectives of the Appropriate Technology movement, e.g., between self-reliance in the production of technology and economic self-reliance. The Appropriate Technology movement appears to be aware of this problem, however, and it is not clear that the problem is immutable. Activities within the movement are generally designed to assist underdeveloped communities to develop greater endogenous technological capacity. Despite these rejoinders it appears that the Appropriate

Technology movement does indeed need to be cautious about the risk of exacerbating technological dependency by shifting the dependency from one type of technology-practice to another.

The fourth proposition stems to a large extent from ambiguity in the semantics of Appropriate Technology identified in Chapter Two. It also reveals ambiguity in the literature about whether Appropriate Technology amounts to a totally different form of technology-practice to that of the mainstream, or whether it may be continuous in some way with the mainstream. If *all* the literature associated with Appropriate Technology is included, then it would appear that the movement is divided over this point. The "mainstream" of the Appropriate Technology movement nevertheless clearly sees the need to draw upon the resources of mainstream science and technology for the provision of appropriate technologies.[71]

The fourth proposition also implies that Appropriate Technology is *by definition* destined to remain a minority movement. In one sense this is a completely false representation of the movement: its literature reveals an ambition that the rationale and methodology of Appropriate Technology become normal practice in the mainstream of the North and the South. Despite this fact the proposition partly reflects a self-negating aspect of Appropriate Technology, an aspect which is most apparent with Intermediate Technology. Singer has portrayed this matter in the following terms:[72]

> As developing countries succeed in achieving development, their factor proportions will become more similar to those of industrialized countries, and the differences in the technologies appropriate for the two groups of countries will diminish and perhaps finally disappear. One can say that the purpose of those who emphasize that the technology appropriate for developing countries is different from the technology appropriate for rich industrialized countries is precisely to make their own statement redundant. As development is successfully achieved by means of appropriate technology, the need for a different appropriate technology for developing countries will gradually disappear.

---

[71] See, e.g., E. F. Schumacher, "Intermediate Technology: Its Meaning and Purpose", mimeo, (London: Intermediate Technology Development Group Ltd., 1973).

[72] H. Singer, *Technologies for Basic Needs* (Geneva: International Labour Office, 1978), p. 9.

Appropriate Technology is generally intended by its protagonists to be interpreted as a progressive notion which forms part of a general movement for social reform or improvement. This implies that Appropriate Technology is intended to both act as a counterpoint to dominant technology-practice *and* become part of dominant technology-practice. Accordingly, it is not legitimate to bifurcate Appropriate Technology and mainstream technology-practice, although it is also not legitimate to assume that there is no tension between the two. Admittedly, there is some quasi-romantic rhetoric associated with the Appropriate Technology movement which does embody the mistake of bifurcating the two two types of technology-practice. This rhetoric draws upon the anti-technology sentiments which have become part of urban-industrialized cultures. It also draws upon the belief that mainstream technology has become a monolithic socio-technical system, and that it is impossible to participate in mainstream technology without becoming part of that system, thereby negating the principles of choice, assessment, propensity for local appropriation and human control of technology inherent in the Appropriate Technology movement. This belief is complex and contentious and will be further examined in Chapter Eleven.

A general conclusion to the immediately foregoing discussion, however, is that technologies promulgated under the rubric of "appropriate technology" may in fact be used to reinforce the technological status quo. Such an outcome is not intrinsic to Appropriate Technology, however, and is at odds with readily identifiable progressive and reformist themes within the movement. The reformist emphasis of the movement is also qualified by a recognition that not all aspects of the status quo in all communities necessarily warrant radical change. Furthermore, the criticism that Appropriate Technology *per se* is obstructive to needed social and political reforms is confounded by the inconsistent terminology employed by protagonists and antagonists of Appropriate Technology.

The political critiques of Appropriate Technology point to some widely documented institutional obstacles to the full benefits of technology accruing to communities and classes of people whose needs appear to conflict with the interests of certain powerful elites. There is no need here to dispute the merits or otherwise of the schools of thought from which such critiques are derived. The present author contends, however, that many political criticisms of Appropriate Technology are based upon misrepresentation of the movement's activities and literature and upon a failure to comprehend the particular political framework of the movement.

Overview of Criticisms

Misdirected Polemic

A considerable number of objections to Appropriate Technology have been raised. These come from a variety of sources and deal with a variety of themes. The extent of the criticisms reflects the extent and impact of the Appropriate Technology movement itself; it also confirms that the movement is a substantial phenomenon which deserves serious attention.

Most of the objections are readily matched by fairly straightforward rejoinders. While they may serve to highlight weaknesses in both the theory and practice of Appropriate Technology, the objections do not perforce invalidate the concept.

Many of the criticisms are superficial, ambiguous, based upon misinformation or prejudice, sometimes dishonest and sometimes incoherent. This is partly due to poor scholarship by critics and partly due to ideological bias. It is also due to the heterogeneity of the Appropriate Technology movement and to the fact that it is an evolving, dynamic social movement. This is compounded by inconsistency in terminology within the movement and by the fact that its literature combines popular rhetoric together with scholarly studies and policy discussions. The lack of a readily identifiable and widely accepted formal theory of Appropriate Technology creates further scope for the proliferation of confusing or misplaced criticisms.

There are, nevertheless, a number of plausible criticisms of Appropriate Technology. Most of these are concerned with political constraints to the effective dissemination of appropriate technologies or to the hoped-for achievement of the movement's goals. Notwithstanding the serious nature (in the main) of the political criticisms of Appropriate Technology, many are highly dogmatic or uncritical and do not withstand counter-criticism. Of those political criticisms which are relatively cogent, a number embody assumptions about the nature of technology and politics which are open to question; this matter will be pursued in more detail in Chapter Eleven.

The argument that there are political institutions and forces which may provide real obstacles to Appropriate Technology appears well founded, in the main, and there is no reason to take issue in this study with the general empirical and theoretical justification for the argument. There are no grounds, however, for viewing the argument as absolutely and universally applicable. In discussing the experiences of the I.T.D.G. McRobie writes:[73]

> We have had cases where people in big industry tried to oppose A.T. But they were half-hearted. We haven't met great opposition - but that may be a function of the fact that the movement is still very small. Even so, our experience is that a number of governments have done fantastic work without great upheavals. ... Where you have particular types of structure, as used to be the case in Ethiopia and Iran, where the structure itself was violent toward the people, there is no chance for non-violent change. But where you've got a relatively open political structure, there are lots of things that can be done within that to make it easier for the poor to help themselves.

Arguments against the likelihood of the ubiquitous implementation of Appropriate Technology do not provide sufficient grounds for holding that obstacles are universally insurmountable.[74] While accepting the basic thrust of the political objection to Appropriate Technology, the present author considers that it requires further investigation and refinement; and that it needs to be extended beyond a simple criticism to form a practicable strategy for technology-practice.

In addition to these qualifying remarks there is a further weakness with most of the political criticisms. They tend to rely upon distortions or misrepresentations of actual writings and activities of Appropriate Technology advocates. For the critics this has the advantage of making their own arguments appear more distinctive, more important and, on first appearance, more cogent than would otherwise be the case. An attempt has been made here to demonstrate that the concept of Appropriate Technology is founded upon an appreciation of the political-cum-institutional structures identified by the critics and is viewed by its proponents as part of a broader strategy for addressing those

---

[73] G. McRobie, "Not Revolution ... But Changing the Rules" in the *A.T. Reader: Theory and Practice in Appropriate Technology*, ed. by M. Carr (London: Intermediate Technology Publications, Ltd., 1985), p. 31.

[74] E.g., J. Wong has shown that industrial experiments in China under the slogan of "walking on two legs" - and the adoption of what was virtually an Appropriate Technology strategy - were *not* unmitigated successes and that they relied upon the political and ideological milieu created under Mao; however he shows that, nevertheless, they were *relatively successful* (*Appropriate Technology and Development: A Review of the Chinese Experience*, ERC Occasional Paper Series #3 [Singapore: Chopmen Publishers, 1980]).

structures. Furthermore, most proponents of Appropriate Technology do not disregard the real limitations of the Appropriate Technology approach and the need for political action with a broader scope than an exclusively technological orientation. The comments of F. Stewart, a leading scholar in the field of Appropriate Technology, are illustrative:[75]

> ... designing policies independently of actual political realities may often lead to ineffective recommendations. The area of political economy impinges on the timing and sequencing of particular strategies. Certain developments (e.g. technical choices) establish a nexus of interests which then become powerful in influencing future decisions - from both a technical and a political point of view.

Most political criticisms of Appropriate Technology conveniently bypass passages such as this from the literature, probably for reasons of sophistry.[76] Such criticisms are misdirected and counterproductive for the purpose of evolving more politically practicable strategies vis-a-vis Appropriate Technology.

While claiming that there is no intrinsic and fundamental conflict between the frameworks of the (cogently argued) political critiques, on one hand, and Appropriate Technology, on the other hand, the present author does not wish to imply that there are no differences in emphasis or approach. The technologically-oriented perspective on political problems adopted by the Appropriate Technology movement may be viewed in itself as a distinctive type of politics; and, serious examination of this possibility could lead to some useful advances in political theory. Unfortunately the misdirected polemic which obfuscates much of the ongoing criticism (both external and internal) of Appropriate Technology has hindered such advances. The lack of a widely acknowledged formal theory of Appropriate Technology has made it easier for misdirected polemic to proliferate.

---

[75] F. Stewart, "Macro-policies for Appropriate Technology: An Introductory Classification", *International Labour Review*, 122, 3 (1983), 286.

[76] A further example is the otherwise admirable paper by J. Clayson ("Local Innovation: A Neglected Source of Economic Self-Sufficiency", *Impact of Science on Society*, 28, 4 [1978] 349-358). Clayson outlines a strategy for technological innovation which involves activities at the micro-economic level in developing countries leading to multiplier effects in the macro-economy. He describes an approach to indigenous economic growth almost identical to that advocated by Schumacher, but accuses Schumacher and colleagues of failing to comprehend such an approach and of completely neglecting the "human" and "management" aspects of indigenous technological/economic development (cf., pp. 349, 354).

There is a need for a systematic-theoretical statement of Appropriate Technology as a complement to the more descriptive-narrative statement of Part two. In the next chapter a synthesis will therefore be constructed of the material thus far discussed, thereby forming an integrated framework for Appropriate Technology.

Appropriate Technology and Hope

Those criticisms of Appropriate Technology which qualify as both serious and cogent do not *ipso facto* invalidate the concept. Their main value revolves around the fact that they raise pertinent questions about the *future prospects* for the movement and for the uptake of its ideas as part of the mainstream. Much of the remainder of this study will grapple with this issue. An attempt to identify the potential power of Appropriate Technology to overcome the obstacles to its effective development and dissemination. We will be concerned, therefore, not just with whether Appropriate Technology is a "good idea", but with whether it may provide a viable framework for policy and practical action.

The foregoing analysis evokes consideration of the subject of *hope* - especially as it relates to future options for technology-practice. Subsequent chapters will form a response to the question: what are the grounds for hope that the goals of Appropriate Technology may be substantially achieved?

The difficulty with adopting a stance of either simple optimism, on one hand, or simple pessimism, on the other hand, is illustrated by a paper written by C. Cooper on the political-economy of technology choice - one of the more cogent of the published criticisms of Appropriate Technology. Cooper writes:[77]

> The main assumption which is implicit in a great number of the policy proposals that are made about technology policies, and here I am speaking generally, is that income distribution objectives are easier to achieve by using appropriate technologies and by increasing employment, than by fiscal or other methods. Obviously fiscal methods have been a rather weak tool for redistribution. All I want to suggest here is that technology policy may not be much more effective when we look at political realities, and that we have tended to overlook the

77 Cooper, "Choice of Techniques", p. 301.

problems of political and social organization in our proposals about 'appropriate technology' and the like.

Cooper expresses pessimism here about the prospects for Appropriate Technology. He also expresses pessimism, however, about the prospects for means other than Appropriate Technology being able to achieve the socioeconomic goals in question. In other words his political-economy criticisms lead to a comprehensive pessimism. Most of the political or political-economy critiques of Appropriate Technology exhibit a similar difficulty. Identifying limitations or weaknesses of Appropriate Technology does not constitute a satisfactory response to the issues addressed by Appropriate Technology.

The prospects for Appropriate Technology appear to be contentious and are not conducive to straightforward analysis. Hope for Appropriate Technology ought to be examined as part of the broader issue of the grounds of hope for the fulfillment of human goals in circumstances which provide a prima facie case for pessimism. This is a philosophical problem which requires more attention than is feasible in this study. Nevertheless, the conceptual synthesis in the next chapter will be extended later on to examine the nexus between optimism, pessimism and the choice of technology.

The enigmatic nature of the issue of hope and technology is illustrated by the response of George McRobie to a question of whether he was optimistic or pessimistic about the future. He replied:[78] "I'm neither. I think I can see the kind of things that need to be done, but they may not be done quickly enough." Is the position adopted by McRobie merely evasive, or is it a logical outworking of the nature of Appropriate Technology? The rest of the book will seek to answer this question.

---

[78] "Why the World Will Shift to Intermediate Technology" (an interview with G. McRobie by J. P. Drissell), *The Futurist* (April 1977), 89.

10

An Integrated Framework

There is sufficient unity within the Appropriate Technology movement for it to be recognized as a movement and for its members to communicate with each other or organize common activities. As demonstrated in Part Two, however, there is also great diversity in the movement which has led to much debate and misunderstanding. The weaknesses of the movement have been exacerbated by the lack of an authoritative and comprehensive theoretical formulation of Appropriate Technology. The nomenclature proposed in Part One made it possible to discuss the whole Appropriate Technology movement coherently, thus confirming the possibility of a comprehensive theory of Appropriate Technology. The consensus of ideas upon which such a theory could be based is, nevertheless, it must be admitted, only latent and imperfect. The comprehensive definition of Appropriate Technology put forward in Chapter Two must therefore be viewed as stipulative rather than descriptive.

This chapter aims to enrich the ongoing debate in the field by articulating an integrated framework for Appropriate Technology, based upon a synthesis of the ideas described in Part Two, refined by consideration of the criticisms of Appropriate Technology covered in the previous chapter. Most of the following ideas, at least in a nascent form, are contained in the material already outlined. In this chapter they will be made explicit where previously they might have been largely implicit, and will be articulated systematically rather than in the kaleidoscopic way in which they appear in the literature.

Three Dimensional Mode of Technology-practice

The first major element of the integrated framework is that Appropriate Technology be viewed as involving three dimensions: a technical-empirical dimension, a socio-political dimension and an ethical-personal dimension. It could also be argued that all technology-practice, whether conducted under the rubric of Appropriate Technology or not, involves these three dimensions in varying degrees. The necessary and distinctive feature of Appropriate Technology is that the three dimensions be *harmoniously* integrated. Some proponents of Appropriate Technology focus on one dimension rather than the others, but there is an emerging tendency in the movement for all three dimensions to be taken seriously and to be understood as being interdependent. The purpose of this chapter is to develop this concept systematically and to illustrate how it may clarify the debate on Appropriate Technology and provide a basis for adequately addressing criticisms considered in the previous chapter.

The political criticisms tend to selectively focus on the technical-empirical dimension of Appropriate Technology and argue that socio-political factors are decisive - as if Appropriate Technology does not incorporate a socio-political dimension. Some critics with a technical-empirical orientation, in contrast, point to apparent weaknesses in the Appropriate Technology movement because of its apparent neglect of technical-empirical factors in favour of ethical-personal or socio-political polemic. In his later writings Schumacher places a great deal of emphasis on ethical-personal factors but few commentators effectively draw the links between this aspect of his work and his other work which deals more directly with technology and social-cum-political factors. It is suggested here that the critical discussion of Appropriate Technology would be more fruitful if it recognized the inherent tripartite nature of Appropriate Technology. A framework which incorporates all three interdependent dimensions of technology-practice in Appropriate Technology is illustrated in Figure 10.1.

"Technical-empirical dimension" refers to that aspect of Appropriate Technology concerned with: technologies as artefacts; the tangible components of technical design processes; and the empirically derived knowledge which forms an objective part of technology-practice. "Socio-political dimension" refers to that aspect of Appropriate Technology concerned with: the strategic action of classes of people;

social institutions; organizations; and structures which form the main fabric of human corporate life. "Ethical-personal dimension" refers to that aspect of Appropriate Technology concerned with: normative factors; metaphysics; and matters related to the subjective experience of people or the inner experience of persons, particularly as it relates to their capacity for volition and to act as autonomous centers of power. Each dimension corresponds (very approximately) to the subject matter of technological science (*technologie*), the social sciences and the humanities, respectively.[1]

Figure 10.1 *Three Dimensions of Appropriate Technology*

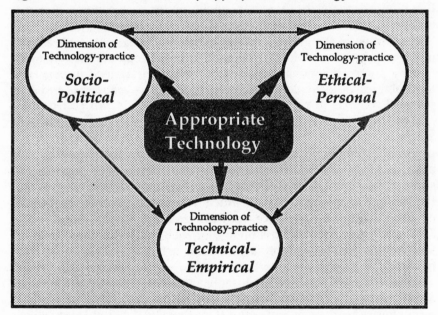

The chief claim of this chapter is that a cogent theory of Appropriate Technology becomes feasible if Appropriate Technology is understood as consisting of three mutually interdependent dimensions of technology practice. Theories which exclude any of the dimensions or

---

[1] The three dimensions correspond loosely to the "technical", "organizational" and "cultural" aspects of technology-practice identified by A. Pacey (*The Culture of Technology* [Oxford: Basil Blackwell, 1983]).

ignore the mutual interdependency of the dimensions will not be satisfactory, even if such theories exhibit a degree of internal consistency.

The criticisms considered in the previous chapter reinforce the importance of the distinction made in Chapter Two between the specific-characteristics and the general-principles approaches to defining appropriate technology. The specific-characteristics type of definition, it was argued, ought only to be used for specific contexts in which the circumstances have been clearly defined. The inappropriate use of the specific-characteristics approach by Appropriate Technology advocates adds some plausibility to the criticisms which have been raised. For example, speaking of "appropriate technologies" in such a way that they must by definition always be small, leads to the possibility that one may thereby promote inequality-inducing technologies which hinder both political reform and economic development.[2]

Adoption of the general-principles approach leads towards a view of Appropriate Technology as a mode of technology-practice exemplified by a special innovation strategy. We have defined appropriate technologies as technologies tailored to fit the psychosocial and biophysical contexts in particular locations and periods, where technologies are defined as artefacts intended to function as relatively efficient means. Because we have adopted an artefacts-based definition of appropriate technology, and because there is an incalculable number of possible contexts in which artefacts may be deployed, it is important to stress the distinction between particular appropriate technologies and the strategy or methodology by which they arise. In other words, it is of fundamental importance to distinguish between *Appropriate Technology* (the innovation strategy, or mode of technology-practice) and *appropriate technology* (a particular artefact or system of artefacts).

The approach adopted here implies that the rationale and method for developing, implementing, maintaining and modifying appropriate technologies is more fundamental than the actual technologies which typify Appropriate Technology at a given time and place. It was demonstrated in Chapters Four and Five that the concept of Intermediate Technology was in fact derived by Schumacher from the

---

2 The work of Prof. A.K.N. Reddy and colleagues from the Indian Institute of Science in Bangalore illustrates this point with reference to the use of Gobar biogas digestors. They have discovered that in situations where traditional feudal-style land ownership patterns prevail the smaller-scale digestors exacerbate inequality between landowners and poor peasants. Maximum benefits to the poor would accrue from the use of larger-scale digestors owned and operated on a cooperative basis by groups of peasants. See, P. Krieg, *Tools of Change*, film, (Freiburg: Teldok Film, 1978).

dynamic principles of Appropriate Technology (understood as an innovation strategy) rather than from a prior commitment to a given scale of technology viewed as "appropriate" independently of circumstances. Much of the subsequent literature of the Appropriate Technology movement has obscured this essentially dynamic aspect of Appropriate Technology.

A static conception of Appropriate Technology is precluded by our stipulated definition of appropriate technology. Unless the process of tailoring technologies to fit their contexts has been pursued, technologies should not be deemed to be "appropriate". Because the biophysical and psychosocial contexts of technology are themselves not static the process of ensuring that technologies are appropriate is necessarily dynamic. Furthermore, the complexity implied by "psychosocial context" indicates that a technology which has been deployed as a consequence of the Appropriate Technology innovation strategy (as described herein), has been selected amidst consideration of the social and political factors which impinge upon that choice. Consideration of social and political factors is not only compatible with the Appropriate Technology concept, but is an essential ingredient of the whole approach.

The *sine qua non* of the integrated framework is that Appropriate Technology be viewed as an innovation strategy aimed at ensuring that technological means are compatible with their context, where "context" is taken to include social and political factors and associated normative goals.

Technological Fit

The second major element of the integrated framework proposed here is the notion of the *technological fit*. This involves the two subnotions of *technological means* and *technological niche*.[3]

---

[3] The importance of the notion of "fit" for Appropriate Technology has been acknowledged, for example, by G. McRobie (*Small is Possible* [London: Jonathan Cape, 1981], p. 28). A. R. Drengson has featured the notion prominently in his philosophical outline of Appropriate Technology (see, "Toward a Philosophy of Appropriate Technology", *Humboldt Journal of Social Relations*, 9, 2 [1982], 161-176).

Technological Means

Qualifying the word "technology" with the adjective "appropriate" implies that technology cannot be properly assessed or evaluated without reference to something other than itself. To be appropriate, technology must be appropriate *to* something or appropriate *for* some purpose. The notion of Appropriate Technology stresses that technology does not exist in a material and social vacuum; it stresses technology's function as an instrument or a means.

Means are defined in the *Oxford* dictionary as that by which results are brought about. The etymology of the singular term "mean" refers to that which in some way mediates or occupies a middle position among things or between two extremes; in its plural form, "means", it denotes that through which an end is attained.[4] Hence, when used to denote a process or instrument (and when used as a statistical concept - in the singular) the term "means" always invokes the quantities or factors between which the means mediates. Defining technology *inter alia* as means reinforces the emphasis on the context of technology which is contained in the concept of Appropriate Technology. These semantic considerations are not included here for reasons of pedantry, but because of their implications for technology-practice: the very concept of technology points to the need for skills in assessing meta-technical factors. Thus, by definition, Appropriate Technology may not be practiced solely by technical specialists in their capacity as technical specialists.

The notion of the technological fit derives from the view that technology ought to be understood in relation to something other than itself. The word "fit" implies that one factor *corresponds* to another. The technological fit is therefore the degree to which a technology corresponds with its context.

Technological Niche

The notion of the technological fit may be explained by borrowing terminology from the science of ecology. In ecology particular species are understood by how well they are adapted to particular ecological niches. A species' survival prospects are enhanced by it being well

---

[4] D. D. Runes, et al., *Dictionary of Philosophy* (Totowa, N.J.: Littlefield Adams, 1962), pp. 192-193.

adapted to its niche. The niche is understood as either a subdivision of a habitat or as the "role" of a species in the environment. Some schools of thought argue that only one species may occupy each niche, and others argue that inter-species competition may exist within a niche.[5]

The technological equivalent of the species is a particular technological artefact (understood as a means). Similarly, the circumstances in which a particular technology is to operate may be viewed as a kind of niche - a *technological niche*. Thus, a technology would correspond to a particular technological niche. The particular circumstances which constitute a technological niche may be referred to as the *psychosocial* and *biophysical* context of the technology in question. Psychosocial includes the ethical goals, political framework, economic structures, social institutions, philosophical perspectives, ideological commitments, aesthetic sensibilities, personal aspirations or psychic needs of people. Biophysical includes the physical-cum-biological needs of people and other species, geographical parameters, the availability of physical resource endowments, thermodynamic principles, environmental limits, physical constraints and the overall ecological profile of a region. The term "technological niche" may be used to denote a psychosocial and biophysical context which prevails at a particular location and time.

In the same way that a species may be described in terms of how well it occupies an ecological niche, the appropriateness of a technology may be described in terms of its compatibility with a given psychosocial and biophysical context - that is, in terms of how well it occupies a given technological niche. A good technological fit is achieved when there is a high degree of compatibility and a poor technological fit is achieved when there is a low degree of compatibility.

The idea that a poor technological fit is possible implies that an "inappropriate technology" may still operate or be practicable - but only up to a point. It also implies that the costs of an inappropriate technology choice may not be immediately apparent or may not necessarily surface at the point of application. The possible hiatus between the deployment of a "poorly fitting" technology and its harmful or undesired impacts is a reason why the appellation "appropriate" is not superfluous. A good technological fit does not occur automatically, partly because the feedback to those who choose a technology is gener-

---

[5] For a detailed discussion of the concepts of *species* and *niches*, see: C. J. Krebbs, *Ecology: The Experimental Analysis of Distribution and Abundance* (New York: Harper and Row, 1972), pp. 211-242; M. Allaby, *A Dictionary of the Environment* (London: MacMillan, 1977), pp. 110, 337 and 452.

ally neither immediate nor adequate. By the time the investment of resources to a technological project is completed the technology may exhibit a degree of infrangibility, making it incapable of modification to achieve a better technological fit.

The arguments just applied to a particular technological artefact would also apply to a technological system, so long as that system could be contained within the environment which constituted its context. Our notions of the technological niche and of the technological fit point to the inherently problem-filled nature of technology choice. The fact that technology is part of technology-practice and that a particular technology is always situated in a particular context implies that decision-making criteria may not be reduced to a simple technical format.

Dynamism of the Technological Fit Concept

Some commentators may charge that the artefacts-based definition of technology adopted herein leads to a static conception of Appropriate Technology. It could be suggested that the focus on niches neglects the complex dynamics of the real world. This criticism is misdirected. Insofar as the psychosocial and biophysical conditions of a given place alter over time, then the technological niches associated with that place also alter over time. The notion of the technological fit which underpins our concept of Appropriate Technology requires that assessments of a technology's appropriateness be time-specific and location-specific. Changes in time and place will alter the design parameters of technology. The notion of technological fit does not allow for specific technology design parameters to be applied statically. Appropriate Technology points to the need for ongoing technological innovation in a given region and ongoing revision of design parameters rather than reliance upon a standard once-off "technical fix".

The dynamism of the technological fit notion may be illustrated further by the allegory contained in the phrase "tailored to fit" in our definition of appropriate technology. The association with the *tailoring* of clothes is intentional. Tailoring involves the use of formal techniques and rules which, while essential, are not sufficient to ensure a high quality suit of clothes. Quality tailoring requires interaction between the tailor and the client to determine his or her shape, size, tastes and requirements. This is made even more important by the fact that peoples' dimensions and preferences change with age. A durable suit that looks splendid on the hanger but which does not fit is of little value to the client. Tailoring is a dynamic activity which requires the exercise of professional judgement by the tailor - not just the relatively routine tasks of following pre-set patterns and assembling components. The phrase "tailored to fit" stresses the human factor in technology-practice.

Appropriate Technology, as we have defined it, points to the need for dynamic interaction between technologists, the users of technology, the environment of the technology and the technology itself.

Corollaries of Appropriate Technology

The features of Appropriate Technology which form the foundation of the integrated framework are that it is a mode of technology-practice characterized by an innovation strategy aimed at achieving a good technological fit.

When portrayed in the foregoing terms Appropriate Technology may appear to be self-evident and little more than common sense. It does not follow, however, that because ideas may appeal to common sense they are *ipso facto* properly understood and thoroughly implemented in society at large. One of the main contributions of the Appropriate Technology movement has been to point out that, despite the immense growth in the deployment of technology throughout the world, most basic human problems - such as poverty, maldistribution of wealth, political malaise or environmental degradation - persist. In other words, while the basic principles of Appropriate Technology may be no more than common sense they do not appear to have been *apprehended* in normal technology-practice on a widespread basis. In addition to the socio-political constraints considered earlier, a partial explanation for this may lie with a failure to appreciate certain corollaries of the concept. Some of these corollaries will now be considered before explicating additional features of the Appropriate Technology innovation strategy.

Non-neutrality of Technology

Appropriate Technology is incompatible with the view that technology is "neutral" in socio-political and normative terms.

It is often suggested that technologies are equally open to use for either good or bad, or that the social and political status of technology is dependent entirely upon the manner in which it is used. In other words, it is suggested that particular technologies do not possess *intrinsic bias* vis-a-vis human interests and environmental impacts. Similar comments are made with regard to the environment; viz. that technologies do not in themselves lead to certain types of environmental impacts. In contrast, the concept of Appropriate Technology implies that technologies have an *intrinsic propensity* towards certain types of social and physical impacts rather than others.

Technological science (i.e., *technologie*), on the other hand, appears much more open to being employed for a variety of different human purposes or environmental regimes than do actual technologies. The extent to which this is so is open to debate because technological science takes place within a social context and inevitably reflects something of that context. Technologies, however, in their capacity as artefacts are not, in the main, malleable and are therefore not neutral. Technological science could be said to possess a limited degree of neutrality. Although its neutrality is not unequivocal, the degree of flexibility which technological science does possess at the level of application is a key to the viability of the Appropriate Technology innovation strategy.

An example may illustrate these points. A typical modern nuclear-fission electrical power plant (i.e., a technology) may not readily be used for anything other than producing electricity (or radioactive materials).[6] It also embodies characteristics of the social setting in which it was spawned: e.g., a high degree of organization, centralization of control over energy production and use, extensive grid-based electricity distribution, sophisticated technical back-up facilities, necessity for strict safety procedures and discipline, high levels of capital accumulation, and necessity of relatively stable access to specialized sources of high-grade materials. Such a plant may not be readily operated in a setting which does not exhibit similar characteristics; hence, its possible benefits may not be enjoyed freely by all communities alike. Furthermore, when introduced into a region which does not already possess the appropriate conditions, the technology will create pressures for such conditions to be generated. Technological science - incorporating inputs from the science of thermodynamics, mechanical, civil and electrical engineering, or from cybernetics - is capable of producing a range of different energy technologies. These may involve: production of liquid fuels from biomass; a variety of fossil fuel systems; hy-

---

6 For an exposition of the inflexibility of nuclear power technology (especially breeder technology) and of the consequent difficulty it presents for political control when adopted, see D. Collingridge's book, *Technology in the Policy Process: The Control of Nuclear Power* (London: Frances Pinter, 1983)

dropower; energy efficiency devices; photovoltaic or solar-thermal systems; aerogenerators; and human or animal powered machines. Each of these products from technological science may reflect a different range of corresponding conditions.

The foregoing theme may be stated in a different way. To speak of a technology as being "neutral" is to imply that any range of ends may be attained by the deployment of given means. That is to say, that there is no functional and intrinsic connection between ends and means. Appropriate Technology, in contrast, implies that an intrinsic and functional relationship exists between ends and means: the two are not independent. In other words, ends are inherent in means; given means evoke certain ends.

Technology as a Determining Factor

A second corollary of Appropriate Technology is that technology-practice, incorporating particular technologies, may act as a determining factor in society. That is, the introduction of a new technology-practice may exert a dynamic influence on the structure of society. Adherents of Appropriate Technology tend to believe that technology is a key ingredient in the achievement of social goals.

Describing technology-practice as a determining factor in society does not require adopting a commitment to the doctrine of technological determinism. It is possible for technology-practice, or even technology, to exert a determining influence upon other factors in society while at the same time incorporating reflections of those factors in its structure. In other words Appropriate Technology implies *mutual interdependence* between technology and other factors in society. This picture of technology in society is not only promulgated by those operating under the rubric of "Appropriate Technology". Hill is an example of a scholar critical of technological determinism who also shares this perspective:[7]

> Technologies and technical change are ... not autonomous forces, but are produced *within* a social, economic and cultural context; thus their meaning to society depends upon the world views that society holds at that time. Both the type and pervasiveness of technical

---

[7] S. Hill, "Technology and Society", in *Future Tense? Technology in Australia*, ed. by S. Hill and R. Johnston (St. Lucia: University of Queensland Press, 1983), pp. 28-29; Hill's use of "techniques" and "technical" corresponds to our use of "technologies" and "technological".

change are limited by the wider social and economic conditions. Equally, however, change in the basic techniques of a society is likely to create pervasive change throughout the society's economic and social structure, culture and world views.

Appropriate Technology implies that the role of technology in society is dynamic and not simply passive. The movement's rhetoric, nevertheless, does portray technology an increasingly dominant factor in the dynamics of society.

Heterogeneity of Technology

The designation of one technology as "appropriate " and another as "inappropriate" implies that technology is not a monolith, incapable of differentiation.   Thus, the third corollary of Appropriate Technology is that technology is heterogeneous.

The popular usage of generic phrases like "*the* new technology" or "technology's impact on society" reveals a tacit assumption that technology is some kind of single entity, exogenous to the processes of human society. In colloquial terms, technology is often portrayed as "a thing out there". It is frequently treated as a homogeneous phenomenon to be either accepted or rejected *in toto*. Technical specialists often speak of having discovered or engineered *the* technically correct solution to a problem, as if technology could somehow be isolated from the complexity of its cultural, political and material context. This tendency to reify technology into a monolith reinforces the sense of powerlessness of the individual which may increasingly be found in urban-industrialized cultures. It is also reflected in political rhetoric, where debate is often polarized between those who "favour" technological development and those who are "against" it.

Appropriate Technology transcends the pro-technology versus anti-technology debate by rejecting the view that technology is homogeneous. The concept assumes that technology is a heterogeneous collection of phenomena and that it is possible for technologists and others to develop a diversity of technologies to match the diversity of psychosocial and biophysical contexts (i.e., technological niches) in a region. This principle is embodied in the adage, "there is more than one way to skin a cat"!

The emphasis in Appropriate Technology on the heterogeneity of technology does not require that there be no unity to the diversity of technology nor that technology does not exhibit systemic tendencies.[8] Rather, it implies diversity within unity. It also implies that the increasing tendency for modern technologies to form part of a *system* of technologies, while a dominant feature of technological society, is not absolute. Not all technologies require substantial systems for their successful operation; and, in cases where appropriate technologies do necessarily form part of modern systems, it does not follow that such systems negate the principles of the technological fit. There appears to be no reason why a diversity of *technological systems* is not possible - in addition to a diversity of *technologies*.[9]

Appropriate Technology may be seen as a response to the convergent and unifying tendencies in modern technology; but, it is a response which recognizes the historical heterogeneity of technology and which aims at cultivating this aspect of technology.

Technological Context

The fourth corollary of Appropriate Technology is that technology always operates in a context of some kind and that this context ought to be uppermost amongst factors affecting the design or choice of particular technologies. This supposition was implied by the earlier outline of the technological niche concept and does not require much elaboration. Two further points, however, should be raised here.

Firstly, it is facile to speak of "advanced" or "sophisticated" technology, for example, without reference to the context of the technology. Given that a technology may be viewed as a means towards an end, it follows that assessments of its sophistication should involve examination not only of the complexity of the artefact as a discrete entity, but

---

[8] The subject of the systemic nature of technology deserves considerably more attention than is possible in this study. J. Ellul (*The Technological System* [New York: Continuum, 1980]) has written extensively about the threats to human autonomy of the "technological system" and his arguments do raise serious questions about the meaning of the "technological fit" concept. For example, he writes, "It is absolutely useless to regard one technology or one technological effect separately; it makes no sense at all. Anybody doing that has simply no understanding of what technology is all about, and he will find lots of cheap consolations" (p. 107). We should simply note, at this juncture, that this synthesis of Appropriate Technology does not depend upon a non-systemic, atomistic approach to technology.

[9] Even Ellul, the most ardent exponent of the systemic interpretation of technology, does not rule out the role of people in affecting technological systems; e.g., "Thus, in describing the system, I do not exclude the initiatives and choices of individuals, but only the possibility that everything boils down to them" (*ibid.*, p. 87).

examination of how well it achieves the end which it is ostensibly meant to serve.

It was argued earlier that technological ends are, in one sense, immanent in the means. That discussion may be extended by making a distinction between the end immanent in the means (labelled "intrinsic end") and the end which is held as the human purpose for which the means are employed (labelled "extrinsic end"). In principle it is possible for the intrinsic and extrinsic ends of a technology to differ, in which case a poor technological fit is attained. The concept of Appropriate Technology implies that true technological sophistication is attained when the extrinsic and intrinsic ends of a technology are identical or very similar. Thus, it could be said that a technology "works" when a match is achieved between its intrinsic and extrinsic ends.

It could be suggested that there is little value in what appears to be a pedantic distinction between intrinsic and extrinsic ends, because all good technologists would automatically ensure that a match between the two is achieved; viz., it may be taken for granted that technologists will provide technologies which work properly. This suggestion pinpoints one of the main substantive claims of the Appropriate Technology movement: it *may not* be taken for granted that technological means will necessarily effectively serve the extrinsic ends for which they are intended. A technology may "work" properly in the sense that it functions efficiently to attain its intrinsic end (which may be equivalent to an extrinsic end adopted by the technologist), but without "working" properly from the point of view of the extrinsic end of the technology users.

Appropriate Technology embodies a broader concept of efficiency than may be held by a technologist concerned mostly with "intrinsic" efficiency. A truly sophisticated technology consists of means which efficiently serve both extrinsic ends and intrinsic ends. In other words, the *content* of technology ought to bear some relation to its *context*.

The second main point about the context of technology concerns the "psychosocial context". The psychosocial context of technology is not a static set of conditions; it includes the political and normative goals of a particular community. The definition of Appropriate Technology we have adopted therefore requires that unless a technology has been selected to serve the political and normative goals of the community or group of people in question it should not rightly be deemed "appropriate". Our concept of Appropriate Technology precludes, by definition, a narrow technicist approach, indifferent to political factors.

Technology Choice

A fifth corollary is that once the context of technology has been identified and the objectives for technological innovation articulated, there still remains a choice amongst alternative technological means to achieve those objectives. *Technology choice* is a cardinal feature of the Appropriate Technology innovation strategy.

Evidence for the possibility of technology choices has been surveyed in earlier chapters. Two examples here may help illustrate the character of such choices. Firstly, given an objective of preparing to adequately meet a region's energy-supply requirements by early in the twenty-first century, a choice may still remain between, for example, a fossil-fuels, a nuclear or a solar/conservation oriented strategy - with the concomitant technology mixes.[10] Choices between alternative energy policies and associated mixes of energy technologies are not just about choices of technology; they are also choices about such matters as the quality of life and the style or structure of society.[11] Another example of the possibility of choice lies with the problem of the disposal of human waste. The technology exists for: centralized sewage systems, which often dump partly treated wastes at sea or in rivers; local-community wastewater recycling depots; or, domestic grey-water recycling systems and compost toilets.[12] Each of these options is feasible from a strictly technical point of view. The choices themselves, however, are not reducible to a technical format. Each community must make a choice (or have the choice made on its behalf) about which option or combination of options it will adopt. Such choices are not straightforward, however, because they require comprehensive consideration of the parameters of the psychosocial and biophysical context of the community.

The distinction made in Chapter Two between *technical* and *technological* is important. The scope for real choice between alternative

---

[10] Cf.: K. R. Roby, "Towards a Sustainable Energy Society", in *Prospect 2000: A Conference on the Future*, ed. by S. T. Waddell (Perth, Aust.: Australian and New Zealand Association for the Advancement of Science, 1979); M. Diesendorf, ed., *Energy and People: Social Implications of Different Energy Futures* (Canberra: Society for Social Responsibility in Science, 1979).

[11] C. A. Hooker, *Energy and the Quality of Life: Understanding Energy Policy* (Toronto: University of Toronto Press, 1981).

[12] Office of Technology Assessment, *An Assessment of Technology for Local Development* (Washington, D.C.: U.S. Government printing Office, 1981), esp. pp. 147-194.

options may differ in each case. Technical phenomena are *dedicated* to efficient, rational, instrumental, specific, precise and goal-oriented operations; viz., technical phenomena exhibit a high degree of technicity. Technicity is a necessary feature of technology. The degree to which technology is dominated by technicity, however, may vary considerably between different technologies. When, in the course of history, the *available* technology for a particular field of technology-practice becomes dominated by technicity, very few significant choices may actually be *available* - even if there are no physical reasons, *in principle*, why such choices may not exist. For example, technology-practice in the field of banking appears to be growing increasingly technological - as expressed in the use of electronic means for funds transfer, credit provision, accounting and customer services. The need for compatibility between the technological systems of different financial organizations appears to be forcing international convergence in banking technology-practice and higher levels of technicity in that technology-practice. If these trends continue they could reduce the available choice to organizations and individuals of means for conducting financial transactions; e.g., it could become difficult to make purchases in some communities without access to electronic funds-transfer-terminals and the appropriate magnetic credit/identification card.

Even in fields where technology-practice is highly technical, e.g., electronic computing, it does not appear that technology choice is completely excluded. It is possible for an organization to choose between the use of a main-frame computer with a series of terminals or a collection of microcomputers (operated either separately or connected in a network). Furthermore, even for a given piece of computer hardware there may be a wide choice of software packages available for a given activity (such as word-processing) - each of which is highly technical but nevertheless different.

If "technology" refers to an individual technical process (e.g., direct conversion of solar radiation to electricity by amorphous silicon cells) there will probably only be limited room for choice within that process; although, even with a technical process such as this it is not obvious that *one* option is the superior option. If, however, "technology" refers to a general field of technology-practice (e.g., conversion of solar energy into a form of energy which is useful to human beings) the range of possible choices is broadened.

Ellul's somewhat prodigious writings on technology are marred by an apparent failure to apply the distinction between technology and technicity consistently.[13] For example, he avers:[14]

> There is no real choice, strictly speaking, about size; between three and four, four is bigger than three. This is not contingent on anybody, no one can change it or say the opposite or personally escape it. Any decision about technology is now of the same order. *There is no choice between two technological methods*: One foists itself inevitably because its results are counted, are measured, are obvious and indisputable. ... there we have a decisive aspect of technological automatism: it is now technology that makes the choice *ipso facto*, with no remission, no possible discussion, among the means to be used. Man is absolutely not the agent of choice.

His views appear to contradict one of the main tenets of Appropriate Technology - that people may make genuine technology choices. It would appear, however, if we allow him certain literary licence, that Ellul's analysis applies primarily to technological phenomena and technological environments which exhibit a high degree of technicity. Technology choice, vis-a-vis Appropriate Technology, applies most readily within general fields of technology-practice rather than within specific highly technical processes. Within a given field of technology-practice the dominance of technicity may vary considerably from case to case, leading to varying scope for technology choice from case to case. As the empirical evidence surveyed in earlier chapters indicates, Ellul's repudiation of technology choice is largely rhetorical and may not be sustained at the level of real technology-practice.

In conclusion, our synthesis of Appropriate Technology, strengthened by consistent use of the semantic conventions adopted in Chapter Two points to the importance of technology choice for sophisticated technological innovation.

---

[13] See, e.g.: *The Technological Society* (New York: Knopf, 1964) and *The Technological System* (as cited above). This apparent failure may perhaps be explained partly by Ellul's highly idiomatic style, but even allowing for this it is often not clear whether he is referring to technique, technology, technicity, technology-practice or his ubiquitous *La Technique.*

[14] *Technological System*, pp. 238-239 (emphasis added).

Control of Technology

A sixth corollary of Appropriate Technology is that it is actually possible for people to control technology. The control of technology is closely related to the heterogeneity of technology and the possibility of technology choice. Unless technology may be controlled by people the notion of tailoring technology to fit its psychosocial and biophysical context makes very little sense.

Control of technology, in this context, involves more than the possibility of people being able to manipulate technologies according to technical rules within the framework of a technological system. It involves people being able to master and direct technology in accordance with principles which are derived independently of the imperatives of the technology or technical system in question. The extent to which this is possible and the conditions under which it might be possible are contentious and will be considered more fully in the next chapter.[15] It should be noted here that the capacity of people to control technology is not automatic; it is highly contingent upon the exercise of certain human potentialities such as volition, political acumen, critical reflection, technological prowess, and organizational imagination. These potentialities require cultivation, and therefore provide no *guarantee* of the success of efforts aimed at the social control of technology. The grounds of hope for the future of Appropriate Technology, however, remained linked to the capacity of human beings to control technology.

Technology Assessment

The seventh corollary of Appropriate Technology is that the attainment of a good technological fit will normally require the use of technology assessment procedures.

Tailoring technology to fit the context in which it is to operate requires assessment of the range of technologies available, the nature of

---

[15] A notable debate on this matter is emerging in the literature. See, e.g.: D. A. MacKenzie and J. Wajcman, eds., *The Social Shaping of Technology* (Milton Keynes: Open University Press, 1985); D. Collingridge, *The Social Control of Technology* (London: Frances Pinter, 1980); R. Johnston, "Controlling Technology: An Issue for the Social Studies of Science", *Social Studies of Science*, 14 (1984), 97-113; D. Collingridge, "Controlling Technology (Response to Johnston)", *Social Studies of Science*, 15 (1985), 373-380; R. Johnston, "The Social Character of Technology (Reply to Collingridge)", *Social Studies of Science*, 15 (1985), 381-383.

the technological niches in question, and the likely impacts of the introduction of alternative technologies. The complexity of modern urban-industrialized societies, and of their impact on traditional societies, means that technology assessment cannot be conducted adequately on an ad hoc basis.

Acknowledgement of the need for technology assessment procedures is not unique to the Appropriate Technology movement. It has received serious attention at an international level.[16] Considerable debate has ensued over the effectiveness of technology assessment as a tool for addressing complex social and environmental problems, and some doubts have been raised as to whether it may be cost-effective (should a comprehensive assessment be desired) and whether the scope of available methodologies is great enough to make them generally applicable. A full review of the technology assessment literature is not possible here.[17]

There are at least two important implications of Appropriate Technology for technology assessment. The first concerns methodology and the second concerns assessment criteria.

Some reviews of dominant technology assessment practices suggest that the methodologies employed *a priori* effectively preclude consideration of policy options which accord with Appropriate Technology.[18] This is partly due to a tendency to exclude information from assessments which cannot be readily quantified. Many of the factors which constitute the psychosocial and biophysical context of technology may not be reduced to a numerical form. Thus, Appropriate Technology

---

[16] See: F. Hetman, *Society and the Assessment of Technology: Premises, Concepts, Methodology, Experiments, Areas of Application* (Paris: Organization for Economic Cooperation and Development, 1973); Organization for Economic Cooperation and Development, *Methodological Guidelines for Social Assessment of Technology* (Paris: Organization for Economic Cooperation and Development, 1975); Coates, *Federal Government*; P. Behr, "Office of Technology Assessment", *Environment*, 20, 10 (1978), 36-38; R. Ishida and H. Eto, "Integrating Assessment in National Technological Policy", *Impact of Science on Society*, 28, 2 (1978), 139-146.

[17] The following are useful sources: H. Brooks, "Technology Assessment as a Process", *International Social Science Journal*, 25, 3 (1973), 247-256; W. F. Hederman, *Assessing Technology Assessment* (Santa Monica, Cal.: Rand Corporation, 1975); J. I. Gershuny, "Technology Assessment: Oversold and Under-achieving: Second International Conference on Technology Assessment", *Futures*, 9, 1 (1977), 74-76; A. Porter, et al. (*A Guidebook for Technology Assessment and Impact Analysis* [New York: North Holland, 1980]); R. Kasper, ed., *Technology Assessment* (New York: Praeger, 1972); F. T. Ayers, "The Management of Technological Risk", *Research Management*, 20, 6 (1977), 24-28.

[18] See: Carpenter, "Technoaxiology". Cf.: S. R. Carpenter, "Philosophical Issues in Technology Assessment", *Philosophy of Science*, 44, 4 (1977), 574-593; K. S. Schrader-Frechette, "Technology Assessment as Applied Philosophy of Science", *Science, Technology and Human Values*, 33 (Fall 1980), 33-50.

points to the need for more comprehensive approaches which incorporate qualitative assessments as well as quantitative assessments. The importance of people and of local communities in the Appropriate Technology innovation strategy also suggests that technology assessment procedures conducted by technocratic elites, independently of participation by people from the communities where proposed technologies are deployed, may be inadequate and probably counterproductive.

Appropriate Technology also points to the need for careful reflection upon the range of criteria used to guide assessments; viz., all procedures are based upon some criteria which, in turn, reflect certain socio-political interests and normative biases. It is important for assessment criteria to be examined to prevent inappropriate *de facto* criteria being adopted tacitly.

From the perspective of Appropriate Technology, unifactorial approaches to technology assessment which depend upon a single criterion, such as internal technical efficiency or short term financial profitability, ought to be avoided. Appropriate Technology requires a multifactorial approach which takes into account a diverse range of assessment criteria. Criteria which might be included in such an approach could address the following issues: technical efficiency; economic status; socio-economic bias; cultural compatibility; environmental impact; resource requirements; ownership potential; scientific input; aesthetics; durability; social value; capital cost; political bias; origins; employment impact; technical sophistication; development pattern; or, scale.

Local Focus in Technology-practice

Finally, the basic concept of Appropriate Technology, combined with the aforementioned corollaries, points irrevocably to the importance of a local or regional focus in technology-practice.

A focus on local conditions and the parameters of actual local communities was demonstrated earlier to be a central aspect of Intermediate Technology. The theme recurred in our survey of the broader Appropriate Technology movement. Hence, by emphasizing the local focus of the movement, we are reiterating an observation of historical fact. The local focus is also *logically* implied by the very concept of Appropriate Technology.

The notion of the technological fit is not very meaningful if technological niches are not understood as geographically based niches.

Appropriate Technology would be quite prosaic if it were not for the fact that it embodies the presumption that technological niches vary between geographical locations. The fact that biophysical and psychosocial conditions vary immensely between locations - both urban and non-urban - may be taken as generally accepted and as requiring no further justification here. It is not so widely comprehended, however, that this has major ramifications for technology-practice. The diversity of localities, both between and within countries, means that a technology which exhibits a good "fit" in one location will not necessarily achieve the same in another. The parameters of technological niches cannot be deduced from abstract principles - they need to be based upon observation of real conditions in *particular* places. This is the prime implication of the general-principles approach to Appropriate Technology.

The tendency for a number of Appropriate Technology advocates to adopt the specific-characteristics approach to defining Appropriate Technology, while nevertheless inadequate in certain respects, reflects the local focus inherent in Appropriate Technology. In the final analysis, the general-principles approach is vacuous unless translated into tractable and specific terms at the level of particular localities and regions.

Endogenous Technological Development

In this chapter it has been proposed that Appropriate Technology be viewed above all as a mode of technology-practice aimed at achieving a good technological fit. Having articulated some important corollaries of this view we are now in a position to outline another essential theme which forms part of the integrated framework for Appropriate Technology: endogenous technological development. At least four important aspects of endogenous technological development may be identified: endogenous innovation; self-reliance; community development; and, the technological mix.

Endogenous Innovation

"Endogenous technological development" refers to social and economic development in which technology plays a significant role and which is generated and sustained primarily by dynamics which emanate from within the country or community in question.

Innovation is an important part of technological development because, amongst other reasons, the environment in which most economies operate is not static. Continual innovation is required to ensure that technology and the industry with which it is associated are adapted to their changing environment. This is not to imply that the environment in which technology is deployed is not itself determined by that technology, but rather that it is also subject to many other determinants (e.g.: foreign competition, resource depletion, changes in consumer preferences or other social pressures). The technology in a community may become inappropriate to its environment - even in cases where it may have originally been selected according to the principles of Appropriate Technology. Thus, endogenous innovation is an essential part of endogenous technology development.

Endogenous innovation is the process whereby the impetus and resources for the transformation of a community's technology stem from within that community rather than from an exogenous source. A policy commitment to a local focus in economic development leads directly to the notion of endogenous innovation. This is partly because technology is much more likely to be appropriate to a locality if it has been developed with the involvement of local people and with accurate knowledge of local conditions. Local people may be more likely than others to be concerned with the longer term impact of technology on their community; the "externalities" of an enterprise will most likely be "internalized" in the community in due course and hence will not be excluded from cost-benefit assessments as readily as might otherwise be the case.

Another reason why the above policy commitments lead directly to the notion of endogenous innovation relates to the importance of technological skill. Innovation, especially when aimed at achieving a good technological fit, requires a great deal of human skill. If the impetus for technological innovation within a region normally stems from exogenous sources, the accumulation of relevant skills will tend to accrue outside of the region. Consequently, it would appear that neglect of endogenous innovation may institute a self-reinforcing decline in the availability of skills to adequately address local technological problems.

Economic Self-Reliance

Economic self-reliance is a theme which runs throughout the Appropriate Technology movement and which forms an important aspect of endogenous technological development.

The theme stems from an analysis of the global problematique upon which many Appropriate Technology programmes are based. Appropriate Technology has been promulgated by people who acknowledge the persistence of serious problems throughout the world - problems, for example, of social anomie, economic decay, ecological destruction and resource depletion - and who do not see these problems as likely to be resolved without conscious and concerted human action towards this end. Appropriate Technology is a response to the existence of *structural* problems in modern industrial civilization which do not appear likely to disappear if current dominant trends continue. The growing trend towards structural interdependence between the economies of different countries, and the growing prominence of what may be termed the "world economy", creates special difficulties for local communities, local regions and their respective economies. Local communities are increasingly dependent for the health of their economies upon forces in the world economy and the economies of other regions over which they may exert very little control. The persistence of unemployment, economic stagnation, and the lack of resources to adequately confront local environmental problems in many poorer communities, has led many to question the wisdom of reliance upon exogenous economic growth and exogenous technological vitalization for the solution of local economic problems. McRobie points to this emerging attitude in the following way:[19]

> The insistence that economics and technology must spring from local culture and not dominate it runs counter to the centralist trend in all societies. Fortunately, people still object to being made the objects of nationalized production, especially if their lives are controlled by some remote and authoritarian body. This is the real pressure underlying demands for economic and political self-determination that have emerged in Scotland, Wales, Brittany, the Basque country and elsewhere - and also why, incidentally, Tasmania was the first relatively poor area of a rich country to set up an appropriate technology

---

[19] McRobie, *Small is Possible*, p. 76.

organization. There are, of course, many more of these 'mini-economies' than meet the eye: only a few have the political and cultural cohesion sufficient to demand more self-determination and, in more extreme cases, political separation.

For local communities and regions, economic self-reliance is an alternative to dependence upon the capricious forces of external economies. Economic self-reliance is no instant remedy for underdevelopment, but it may be more effective in the long term than passivity or complacency. Economic self-reliance may be thought of as an endogenous rather than exogenous mode of economic development.

Economic development may not be sustained without adequate access to resources. Economic self-reliance requires that, whenever practicable, greater prominence be given to the employment of local resources than exotic resources. The strategy works from the assumption that there are often substantial reserves of underutilized resources within local regions which are bypassed by dominant (normally exogenous) economic strategies; and, that these resources may frequently be mobilized for local economic development. Such resources may include: the wasted talent and labour of unemployed and underemployed people; underutilized land; waste materials; public-cum-municipal infrastructure (including buildings, capital equipment and organizations); and, local financial surpluses which may be recirculated locally with the assistance of suitable regionally-oriented banking and investment mechanisms. From the perspective of economic self-reliance, the limiting factor to local economic development may often lie more with a failure to generate the commitment and institutions to *mobilize* local resources than with a shortage of resources *per se*.

The Appropriate Technology innovation strategy is based upon the supposition that some types of technology-practice may be more suitable than others for economic self-reliance. Some technologies may be totally inappropriate for a particular community if that community is unable to obtain or cannot afford to obtain the resources necessary for the operation of those technologies.

An excellent example of this principle may be found in the joint development of a micro-hydroelectricity project by the Australian Appropriate Technology organization (APACE) and the people of a village in the Solomon Islands (Iriri, on the island of Kolombangera). Iriri, like many island villages in the Pacific region, had been suffering from social, environmental and economic decay in response to the impact of "western" cultural and economic pressures in the region. In an attempt to obtain money to, amongst other things, pay for fuel to power engines for transport and electricity generation, many Pacific islanders have allowed massive deforestation to occur on their land. A common byproduct of this strategy is the destruction of the traditional habitat and the resources it provided for the local economy; this often leads to tragic results when the viability of "selling off" timber ceases (because of reduced stock) and villagers become dependent upon imported fuel, the purchase of which they can no longer cover from their income. Iriri opted for a self-reliant approach to the development of their economy - in contrast to the practice of other Pacific island communities which turn to exogenous business ventures (often operated and owned by multinational companies).[20]

Iriri, with technical assistance from APACE, developed and installed a small-scale hydro-electricity system, in contrast to the usual diesel-driven systems. This provides a low-cost supply of electricity to the village (approximately 120 people and 30 houses) for lighting, coolroom and freezer facilities, recreational activities and machinery for small industries (e.g.: sawmilling, woodworking and copra drying). Iriri's economy has subsequently thrived and diversified relative to other villages in the region, and the traditional habitat has been largely maintained. Iriri has also adopted a self-reliant approach to the provision of its food supply. The village introduced an "organic" market garden based upon the use of locally available organic materials, and rejected methods which depend upon imported fertilizers and pesticides. The maintenance of the villagers' traditional habitat, made possible with the micro hydro-electricity scheme, has been an important factor enabling the organic methods to work properly. The village succeeds in meeting its own food needs and now exports a sizeable surplus in exchange for goods it cannot manufacture locally.

The economic success experienced in Iriri may not be explained exclusively by the selection of a small hydro-electricity system. It does illustrate, however, that economic self-reliance may be a workable option for local communities and that careful selection of technology which is tailored to fit the local context is a key to success. Appropriate technologies may be viewed as a means for mobilizing local resources for local economic development.

---

20 This review, covering the decade up to 1985, is based upon the following sources: private communication with Dr. R. Waddell, President, APACE, Sydney (December 1982, November 1984); private communication, J. Tutua, Co-ordinator, Western Solomons community projects (December 1982); K. Offord and P. Bryce, "Microhydroelectric Design Program for Solomon Islands Village", paper presented to *E.F.Schumacher Memorial Conference on Appropriate Technology*, Macquarie University Sydney, December 1-5, 1982.

Appropriate Technology calls for self-reliance at many levels, ranging from the sub-global, at the grandest level, down through the nation, state/province, region, city and local community, to the local organization, at the local level - and even to the level of the individual person. A theme within the Appropriate Technology movement innovation strategy is that self-reliance at any one level may help reinforce self-reliance at the others. Thus, self-reliant activity by individuals within a community may help that community to become more self-reliant as a whole, and a nation comprised of self-reliant communities may in turn be more self-reliant than otherwise. Likewise, it may be argued that self-reliance amongst local communities may be an important condition for effective self-reliance by individuals, and that national self-reliance may be an important condition for effective self-reliance by communities within nations.[21]

As stressed in an earlier chapter, self-reliance is different to absolute self-sufficiency, the latter being a goal which very few serious writers within the Appropriate Technology movement advocate or believe practicable. For example, self-reliance in the manufacture of one product (e.g., electrical consumer goods) may require the use of production technologies which have been manufactured elsewhere. This illustrates how absolute self-sufficiency may not be a practicable policy, but it does not undermine the importance of self-reliance as a direction of striving, and does not rule out the validity of endogenous technological development as a strategy.

Community Development

A theme very closely related to economic self-reliance is that of community development. The main features of community development, as a policy objective and as an orientation for action, are: an emphasis on actual communities, particularly local communities, as the focus for development; and, an integrated approach which includes not only economics, but also cultural, social and other human factors.

---

[21] W. Rybczynski (*Paper Heroes: A Review of Appropriate Technology* [Dorchester: Prism, 1980] p. 154) disputes this point; he writes: "Self-reliance at different levels simultaneously is a patent impossibility". He does not, however, substantiate his claim. Research in support of the multi-layered approach to self-reliance has been published in: J. Galtung, *Self-Reliance* [Oslo: University of Oslo, Chair in Conflict and Peace Research, 1976]; J. Galtung, P. O'Brien and R. Preiswerk, eds., *Self-Reliance: A Strategy for Development* [Geneva: Institute of Development Studies, 1980]; K. R. Hope, "Self-Reliance and Participation of the Poor in the Development Process in the Third World", *Futures*, 15, 6 (1983), 455-462.

Appropriate Technology requires a balanced approach to community development.

Three approaches to development may be distinguished, according to the relative importance placed upon either the "bottom" or the "top" of the economy.

An approach, which may be labelled "mainstream economics", is the most widely followed of the three and is mainly concerned with the aggregate level of the economy and with such aggregate notions as Gross National Product, national inflation rate, national balance of trade, or national economic growth rates. It generally fails to account for qualitative and quantitative distinctions between local communities and assumes that what benefits the aggregate will also benefit the particular and the local. Economic policies which concentrate on the aggregate level in this way are normally oriented towards the centralized concentrations of economic power which reside with governments, large corporations or major population centers. Development, defined in narrow economic terms, is thought of as best encouraged by providing stimuli at the "top" of the economy. It is assumed that increased activity at the top or "center" of the economy will eventually "trickle down" to the "bottom" or "periphery" of the economy. This approach aims at "development-from-above" and is characteristic of the methods adopted under dominant aid programmes from the North to the South in the decades since the Second World War. Most of the literature emanating under the rubrics of the "North/South Dialogue" or the "New International Economic Order" points to the general failure of this approach. In countries of the South development-from-above has been successful mainly only for those classes or interest groups which operate at the top of the economy. Development-from-above has also tended to be the dominant approach of policy makers in the North.

An alternative approach, labelled here as the "community initiatives approach", is diametrically opposed to the one just outlined and takes as its starting point real people, organizations and communities at the local level of the economy. It rejects development-from-above in favour of "development-from-below" and an emphasis on the initiative and entrepreneurship of people at the periphery or bottom of the economy. It is based upon the view that multiplier effects, of both a narrow economic and broader socio-cultural kind, may emanate from the bottom of the economy (or society) as well as from the top. One important feature of the community initiatives approach is that it relies upon the mobilization of important biophysical and psychosocial resources from the local community level - a task for which the mainstream economics approach is not well suited. In this sense, it has value not only because of its direct benefit to classes of people normally bypassed by development-from-above, but because it enables more *efficient* use of certain underutilized resources.

The third approach may be labelled "balanced development" and is the one which most accords with the Appropriate Technology innovation strategy. Although Appropriate Technology leans towards development-from-below, its successful implementation requires the balancing of inputs from both the top and the bottom of the economy. It appears that some products and some industries of importance to a country's economic self-reliance may not be established without significant inputs from the top of the economy (e.g., alumina refining) and that some of the technologies suitable for development-from-below require the use of materials or components which may only be available from industries at the top of the economy (e.g., high grade materials for solar energy absorption devices). Furthermore, many opportunities for the development of small industries at the bottom of the economy may rely upon a market for their products being generated by the activities of enterprises emanating from the top of the economy; this is particularly so for economies already dominated by the development-from-above approach or which lack industrial diversity.

Technological Mix

A fourth major aspect of endogenous technological development is the attainment of a good technological mix - a mix which is both suited to the circumstances of the community or region in question and which enables diversity in economic and social life.

The importance of attaining a good technological mix follows from the basic concept of the technological fit and its corollaries. The context in which technologies operate is a diverse mixture of psychosocial and biophysical factors. The notion of the technological fit may apply not only to individual technologies but also to the whole blend of technologies employed within a community: this blend should reflect the complexity of its context.

The ecological metaphors used to explain the technological fit concept may be extended here. The health and sustainability of ecosystems are viewed by ecologists as being dependent upon the maintenance of ecosystem diversity. If the diversity of an ecosystem is markedly reduced, its capacity for homeostasis is also reduced. There is a growing tendency in the literature for economics and technology-practice to be viewed in systems terms, subject to the principles of general systems.[22] To the extent that the systems of technology-practice are similar to ecological systems, it follows that cultivation and maintenance of technological diversity is a key to their stability and sustainability. Appropriate Technology is conjoint with the view that the longer-term technological and economic capacity of a community is related to the diversity of its technological base.

A good technological mix is essential to the self-reliance of communities; the capacity to draw upon a *range* of technological skills and economic activities is a key to effective local innovation. If the technological base of a community is very narrow, the ability of that community to adapt to changing circumstances may be severely limited; this is particularly true in cases where a community's economy is based primarily on a small number of primary industries and where the technology for those industries (e.g., minerals extraction industries) is not generated endogenously. If the international market for the products of those industries alters significantly, for example, the repercussions throughout the economy of that community could be devastating. A broad technological mix would provide greater opportunity for a community to absorb the loss of one of its economic activities, through either expansion of one of its other industries or the pursuit of new opportunities.

Appropriate Technology, if adopted by a community as an innovation strategy, would require that gaps in that community's technology-practice be identified and that efforts be directed towards filling those gaps with either new, locally designed technology, or by transfer of technology suited to filling those gaps from elsewhere. In either case a multifactorial technology assessment procedure would be important as a tool for optimum technology choice.

Practical Holism

Another major aspect of the integrated framework for Appropriate Technology is what may called "practical holism". By this is meant a mode of praxis based upon a holistic approach to society, the environment, technology and other factors. Such an approach avoids considering individual phenomena apart from their relationships to other phe-

---

[22] See, e.g., K. E. Boulding, *The World as a Total System* (London: Sage, 1985).

nomena and to the total environment in which they are situated. Four aspects of this theme will now be considered.

Radical Critique

One aspect of practical holism in Appropriate Technology is the radical critique which characterizes the concept and the movement. There are several ways in which the critique may be thought of as radical.

Firstly, the Appropriate Technology critique may be thought of as radical in the sense that it seeks to address the *roots* of problems rather than just symptoms. The mode of technology-practice represented by Appropriate Technology does not take the situation in question as simply "given"; i.e., the technology and its context are understood as being open to determination by conscious human effort and as malleable over time. In each situation there may be a number of dimensions to technology-practice which might not be immediately apparent and which require critical analysis to be properly understood.

Secondly, the Appropriate Technology critique is radical in that it does not assume the *status quo* in a given set of circumstances to be either inviolable or optimum. Accordingly, it is aimed at reform of the *status quo* in cases where prevailing circumstances do not measure up well from a normative point of view. Appropriate technologies are seen as means towards reform of the *status quo* in accordance with the ideals of the movement. This reformist theme does not necessarily apply only to the whole of societies, but is directed at particular communities where reform is both needed and possible - irrespective of the need for more universal reforms. The local focus in Appropriate Technology points to the value of local reforms even if such reforms do not become widespread.

Thirdly, the Appropriate Technology critique incorporates a structural perspective. In other words, it is understood that historical events and particular examples of technology-practice are acted upon by structural forces in the society - forces which exhibit a dynamism and influence which extend far beyond the particular circumstances under consideration and which may not be significantly altered by the actions of a small number of individuals alone. The reforms which may be envisaged for a community, as part of the Appropriate Technology innovation strategy, are constrained by these structural forces, but may also be strengthened by them. The critical assessment of social structures is important to enable proper understanding of the forces of technology-practice.

Fourthly, Appropriate Technology may be seen as a response to the observation that the dynamics of the *whole* of society may influence the *parts*, and vice versa. The structural perspective just mentioned implies that actions directed at solving particular problems within a community also have implications for the community as a whole, because of the way they relate to the structures which exist in that community or within the broader society. Hence, advocates of Appropriate Technology normally have broader social and environmental objectives in mind when advocating particular appropriate technologies for particular purposes. Appropriate Technology is based upon a view of reality which portrays everything as somehow interrelated to everything else; viz., an ecological view.

The radical critique indicated by Appropriate Technology implies that technology choice at the level of local technology-practice has important implications for structural forces which influence prospects for either reform or consolidation of the *status quo* in a community.

Human Compatibility

At the center of Schumacher's notion of Intermediate Technology, and throughout the streams of the Appropriate Technology movement surveyed earlier, may be found a *conscious concern for people*. In Chapters Four and Five it was demonstrated that, for Schumacher, this is much more than a platitude: it is an explicit policy focus which is translated into operational guidelines for economic development and technological innovation. Thus, the content of technology-practice ought to embody in a tangible way the characteristics of the human context in which it is intended to operate.

Appropriate Technology implies that human ends and technological means may not be randomly combined. Technological means may not properly serve given human ends unless they incorporate, in their constitution, qualities or features which are adequate for those ends. Technological means are not infinitely malleable vis-a-vis human ends. This understanding of means and ends is reflected in the use of the phrase "technology with a human face" by Schumacher and others as a general rubric for Appropriate Technology. The phrase fulfils a useful rhetorical function but it also does more than that: it expresses the basic principle of Appropriate Technology that technology ought to *fit* its context. The notion of the psychosocial and biophysical context which forms part of our technological fit concept implies that technologies ought to be compatible with their *human* context.

It is beyond the scope of this study to fully examine the nature of the human context referred to here. Such a task would require a comprehensive survey of all humanities and social science disciplines. The operational criteria adopted as part of the Appropriate Technology innovation strategy would be dependent upon the assessment of the human context made by each group of decision makers. Different schools of thought would probably adopt different criteria. The difficulty involved in making an assessment of the nature of the human context, at both the universal and particular level, does not *ipso facto* invalidate the requirement that such an assessment be made. Rather, it places the onus on each group of decision makers to make such an assessment for themselves. The difficulty of the task, and the fact that different groups of people may arrive at a different assessment of the circumstances and principles at stake, does not mean that the task is impossible or that consensus amongst groups of decision makers may not be achieved. It is quite possible for people within a local community to achieve some consensus, through political or other media, on some basic human concerns - without going through the process of comprehensive and rigorous scholarship. Within scholarship, furthermore, the existence of schools of thought indicates that some kind of cogent assessment of the human context is possible - even if schools of thought vary between each other. The practicability of the Appropriate Technology innovation strategy within a given community will be limited by the capacity for the technological decision makers of that community to achieve some basic assessment of the human context of that community.[23]

The human context of technology includes not only those objective factors which may be described in a straightforward empirical manner, but also the subjective preferences of people. To speak of technology being compatible with its human context therefore means that the technology must be adequate from the perspective of human subjectivity. These comments are not meant to imply that subjectivity and objectivity are mutually exclusive, but rather that *both* subjective *and* objective considerations are raised by the technological fit concept.

---

[23] An attempt to systematically apply considerations of the human context of science and technology for policy purposes has been conducted by the Canadian Government under the auspices of its Social Sciences and Humanities Research Council. The programme, directed by C. A. Hooker, illustrates how serious interdisciplinary work may provide a practical framework for the application of the "human compatibility" principle discussed here (see: C. A. Hooker, et al., *The Human Context for Science and Technology* [Ottawa: Canadian Social Sciences and Humanities Research Council, 1980]; C. A. Hooker, "Science, Technology and Australian Society: Shall We Follow the Lead of Our Cousins?", *Search*, 16, 5/6 (1985), 126-127.

The difficulty of ensuring that technologies are humanly compatible does not appear to have prevented the Appropriate Technology movement from articulating certain basic principles for guiding technology assessment and technological innovation. The fact that an emerging consensus may be identified in the movement is evidence that the complexities of the human context are not necessarily beyond the comprehension of most people; and, that achieving human compatibility of technology is not necessarily beyond the resources of most communities. On this point Schumacher writes:[24]

> No doubt, a price has to be paid for anything worth while: to redirect technology so that it serves man instead of destroying him requires primarily an effort of the imagination and an abandonment of fear.

The impact to date of the Appropriate Technology movement appears to stem substantially from the belief that, despite the obstacles, technology may be directed towards being humanly compatible.

Human compatibility involves both psychosocial compatibility and biophysical compatibility.

The biophysical health of people depends greatly upon the quality of the environment in which they live and upon which they depend for food and other essential inputs and for the processing of their wastes. The impact of technology upon the environment is therefore a potentially significant influence on human health. The environmental compatibility and the human compatibility of technology appear to be closely related. The biophysical aspects are important not only because of *indirect* human impacts mediated by the environment. The use of technologies in the home, in recreation and in the workplace, can have *direct* biophysical impacts on people (e.g., the absorption of household chemicals such as pesticides into the body, the development of unhealthy posture through use of industrial technology, or the development of repetitive strain injury through use of office or workshop technology). In humanly compatible technology the direct biophysical impacts of technology upon people are "acceptable" or benign.

The dominant aspect of human compatibility addressed by the Appropriate Technology movement concerns the capacity of technology to meet the psychosocial needs of human beings. The issue receives its fullest treatment in the movement's discussion of *work*. It is held almost universally throughout the movement that appropriate technolo-

---

[24] E. F. Schumacher, *Small is Beautiful: A Study of Economics as if People Mattered* (London: Blond and Briggs, 1973), p. 149.

gies ought to enable the *provision* of worthwhile work. This involves the provision of work opportunities for all people and the requirement that the *quality* of the work be such as to make it humanly worthwhile. This theme within the movement may not be derived solely from the general-principles concept of Appropriate Technology. The movement's discussion of work and technology involves the *advocacy* of certain normative purposes for human life and the evaluation of the status of work in terms of those purposes. The prime normative principle here is that work is considered to be a desirable human activity rather than a "necessary evil". Thus, for the Appropriate Technology movement, work (of a suitable kind) is considered to be both an end and a means: *technology ought to be a means of enhancing work and not a substitute for work.* This is not to say that all types of work carried out by people in a given community are intrinsically good and that work ought not to be altered by the introduction of new technology, but that, in the final instance, adequate work opportunities ought to be made available to all people and that the quality of the work experience ought to be heightened.

The approach to work adopted by the Appropriate Technology movement, and as contained in our synthesis of Appropriate Technology, draws upon the broader social vision of the movement. The movement places supreme importance on the attainment of general social wellbeing, spiritual-cum-cultural vitality and political freedom for individuals and communities - through structures which do not preclude the attainment of these goals by some people as a necessary precondition or consequence of their attainment by others. Such an ideal is not new and may be found repeatedly throughout Western intellectual history and political movements. A distinctive feature of Appropriate Technology is the claim that the pursuit of this ideal has tangible implications for the form of technology-practice adopted. This is recognized by Winner who argues that, at its most ambitious level, Appropriate Technology means setting the whole question of technology choice in the context of a theoretical understanding of what an emancipated society would look like.[25] In this respect, however, the movement is not without precedents. Winner writes:[26]

---

[25] L. Winner, "The Political Philosophy of Alternative Technology: Historical Roots and Present Prospects", *Technology in Society*, 1, 1, 80-82.

[26] *Ibid.*, p. 81. Note: Winner's use of "alternative technologists" is equivalent to the use of "proponents of Appropriate Technology" in this book.

... alternative technologists have revived a project which had been abandoned with the eclipse of 19th-century utopianism: the work of proposing a clear and systematic notion of the good life that can be translated into principles and criteria of institutional design.

Appropriate Technology requires the conscious articulation of design principles to ensure that technology-practice does not in fact frustrate the human ends which it ostensibly serves as means.

Environmental Compatibility

Parallel arguments to those just presented in relation to human compatibility may be applied to the environmental compatibility of technologies, recognizing that the latter may often be a precondition of the former.

Pointing to the need for technologies to be compatible with the environment (i.e., the biosphere and the biophysical context of human settlements) is at one level no more than a recognition of the functional requirements of sustainable biophysical systems; viz., the maintenance of the biophysical environment in which a technology is to operate requires that the technology's operation does not violate that environmental system beyond a point where it will no longer be viable. For example, if the use of agricultural technologies and concomitant management practices leads to excessive deforestation, soil erosion, destruction of soil micro-organisms and rising salt levels, that land in question may no longer be suitable for agricultural purposes and, therefore, the continued use of that technology.

At another level, however, the objective of environmental compatibility embodies normative goals. The collapse of a biophysical system might be accorded very little significance apart from the implications it holds for human society. If one does not place intrinsic value on the maintenance of biological diversity (or recognize its value - depending upon one's ethical philosophy) the *collapse* of a biophysical system might be considered, with equal justification, simply as a transformation towards a different system. Given that natural biophysical systems may not be viewed as static, the existence of technology-induced change is by itself of no great significance. The notion that technologies ought to be environmentally compatible implies that some form of *evaluation* of the biophysical environment is required - in terms of its possible instrumental, intrinsic or transcendental significance.

If it is assumed that the biophysical system which forms the context for technology-practice ought to be sustained (as a dynamic system), and if it is assumed that technology-practice does have environmental implications, then it follows that technologies ought to be made environmentally compatible. These suppositions are rather obvious and are not likely to be disputed. The Appropriate Technology movement's critique of the status quo, however, reveals its judgement that mainstream technology-practice is not grounded sufficiently in a recognition of these suppositions. The principles of Appropriate Technology require special and concerted attention - despite their seemingly common sense nature - precisely because modern technology-practice does not appear to effectively incorporate such common sense.

Schumacher offers a partial explanation for how such a gap could occur between common sense and common practice:[27]

> ... the changes of the last twenty-five years, both in the quantity and in the quality of man's industrial processes, have produced an entirely new situation - a situation resulting not from our failures but from what we thought were our greatest successes. And this has come so suddenly that we hardly noticed the fact that we were very rapidly using up a certain kind of irreplaceable capital asset, namely the tolerance margins which benign nature always provides.

According to Schumacher the scale and impact of human technology-practice, until very recently, was generally limited in relation to the carrying capacity of the environment. The seemingly exponential growth during this century of industrial activity, and associated factors such as energy usage and population growth, has brought the impact of human civilization close to the point where many of the homeostatic processes of the natural environment cease to operate properly. The harmful impact of environmentally "incompatible" technologies may have been tolerable when their magnitude was relatively small in relation to local ecosystems and the biosphere as a whole.[28] This may have made it easier for modern technological society to evolve in a mood of corporate indifference to what now appears common sense.

---

27 Schumacher, *Small is Beautiful*, p. 15.

28 This theme does not warrant extensive discussion here. Basic evidence for the argument has been assembled elsewhere (e.g.: D. L. Meadows, et al., *Dynamics of Growth in a Finite World* [Cambridge, Mass.: Wright-Allen, Inc., 1974]; A. H. Ehrlich, P. R. Ehrlich and J. R. Holdren, *Ecoscience: Population, Resources and Environment* [San Francisco: Freeman, 1977]; P. R. Ehrlich and A. H. Ehrlich, *Extinction: The Causes and Consequences of the Disappearance of Species* [New York: Random House, 1981].

Appropriate Technology represents a critical response to the dangers of a society being dependent upon a mode of technology-practice which tends to undermine the environment upon which that society ultimately depends.

The following practical principles accord with the environmental compatibility requirement of Appropriate Technology: that available natural resources be utilized as efficiently as possible, minimizing waste; that waste products be re-used and recycled as much as possible; that maximum use be made of locally available resources, with technology being tailored to match those resources; that local and distant environmental impact be minimized where possible, with technology-practice taking full account of ecological principles and local ecosystems; that renewable resource supplies be used wherever possible; and that a transition to a low-pollution, renewable-resource economy be pursued diligently.

A distinctive feature of the Appropriate Technology innovation strategy is that it emphasizes the role of local technology-practice as a key to the solution of global environmental problems. This is based upon the view that a reduction in local environmental damage is necessary for a reduction in global environmental problems, which are often exacerbated by the compound effects of locally produced pollutants. Evidence is mounting that the adoption of environmentally benign technology-practice at the local level may be pursued without weakening economic progress.[29]

Integrated Problem Solving

A final feature of the Appropriate Technology innovation strategy is the integrated approach to problem solving which it embodies. This incorporates two dimensions.

The first dimension to integrated problem solving is the adoption of a *systems approach* to analyzing phenomena, understanding problems and designing practical solutions. This is implied by the foregoing discussion. The systems approach is based upon the general principle that phenomena do not exist in isolation but as part of larger systems which exhibit internal dynamics of their own and which interact in a dynamic

---

[29] See: W. U. Chandler, *Energy Productivity: Key to Environmental Protection and Economic Progress*, Worldwatch Paper #63 (Washington, D.C.: Worldwatch Institute, 1985); H. E. Daly, "Introduction to the Steady-State Economy", in *Economics, Ecology, Ethics: Essays Toward a Steady-State Economy*, ed. by H. E. Daly (San Francisco: Freeman, 1980), pp. 1-31.

way with other systems. According to this principle a reductionistic and atomistic form of analysis is incapable of providing a comprehensive and fully reliable explanation of phenomena. The internal dynamics of individual systems (which, in the "real world", may normally be taken to be open systems) are affected by their relationship to other systems. The interaction of systems or subsystems may lead to results which could not have occurred had those systems existed in isolation.

Systems theory can become very formal and abstract, and seemingly distant from the complexities of the real world. Appropriate Technology, however, is an example of how the theory may be applied in a practical way. At its most basic level it means that technology choice must be based upon an assessment of the main factors in the operating environment of technology which will affect how that technology operates, and the main effects upon that environment which are likely to occur. The systems approach means that indirect as well as direct impacts ought to be considered. For example, by applying a systems analysis to four different methods of rice cultivation in the Third World setting, with a special focus on thermodynamic and cultural factors, Freedman has provided forceful evidence for the productive superiority of a new alternative to either traditional labour-intensive methods and the modern "green revolution" methods.[30] Another example of the application of systems oriented research may be found in the work of the New Alchemy Institute, a Massachusetts based Appropriate Technology group. The Institute has applied advanced biological science to the development of an integrated horticulture/agriculture/space-heating system for cold climates, incorporating the use of solar absorption and wind technologies for energy supply.[31]

The second dimension to integrated problem solving in Appropriate Technology is what may be called the *strategy of simultaneous problem solving*. This strategy is based upon the recognition that problems from ostensibly unrelated fields (e.g., energy policy and employment policy) may in fact be closely related, and that the appropriate technological response to one problem might be related to the appropriate response to another. A single technology or, strictly speaking, a single package (or system) of technology-practice, may *simultaneously* solve more than

---

[30] S. Freedman, "Agricultural Development in the Less Developed World: Energy Limitations and Planning Strategies", in *Technology Choice and Change in Developing Countries: Internal and External Constraints*, ed. by B. G. Lucas and S. Freedman (Dublin: Tycooly, 1983), pp. 143-155.

[31] T. Cashman, "The New Alchemy Institute: Small-Scale Ecosystem Farming", *Appropriate Technology*, 2, 2 (1975), 20-22..

one problem. The Appropriate Technology innovation strategy aims to implement systems of technological means which deliberately serve a range of human ends *simultaneously*.

This theme is most apparent in the literature and programmes which are promulgated under the rubrics of "community development" and "local self-reliance". The local community level provides a nexus for most of the issues and factors outlined so far in this study. The local community or region provides a high enough level of aggregation for the major dynamics of culture, political-economy, people-environment interaction and technology-society interaction, to become apparent. It also provides a low enough level of aggregation for the "real life" content of these dynamics (or structures) to exhibit a comprehensible meaning against which local people may respond. When only high levels of regional aggregation are considered, it is difficult to identify effective strategies for dealing with local dynamics intertwined with national and international structures. Within the Appropriate Technology movement, action at the local community level is rarely viewed as a substitute for action addressed to solving structural-cum-political problems at a higher level of aggregation; it is viewed as a complement to the latter and as a realistic medium through which actual individuals may make a meaningful contribution to political and economic life.

A good example from the Appropriate Technology movement of simultaneous problem solving is the promotion of "humanly scaled energy systems" and "energy efficient community planning". Through a judicious combination of technology choice, community planning, responsive government action, life-style management, and entrepreneurship, it is possible to simultaneously reduce pollution, recycle waste, conserve energy, create new employment opportunities, vitalize local business activity, save money and develop greater community cohesiveness.[32] A second example is the Intermediate Technology approach to economic development advocated by Schumacher. The synthesis of Schumacher's work in Chapters Four and Five shows that Intermediate Technology is an integrated strategy for the simultaneous solution of the six major components of the "development problematique". Schumacher's analysis of the vicious circle of problems which constitute underdevelopment is based upon the systems approach.

---

[32] See, e.g.: D. Morris, *Self-Reliant Cities: Energy and the Transformation of Urban America* (San Francisco: Sierra Club Books, 1982); J. Ridgeway and C. S. Projansky, *Energy-Efficient Community Planning* (Emmaus, PA: The JG Press, n.d. [1980?]) Note: George McRobie's survey of the Appropriate Technology movement in the North places great emphasis on local, community based initiatives which serve several purposes at once (cf., McRobie, *Small is Possible*, esp. pp. 86-191, 247-280).

The combination of systems thinking and the simultaneous problem solving strategy leads to an integrated assessment of human and environmental problems. For the Appropriate Technology movement, human problems and environmental problems may not readily be separated, if at all. A distinguishing feature of the emerging consensus in Appropriate Technology, and a principle which is contained in the concept of the technological fit, is that technologies ought to be simultaneously humanly compatible and environmentally compatible.

The integrated problem solving which is part of Appropriate Technology opens up a wide scope of resources for problem solving. Mainstream approaches to economic development and technological development which ignore the importance of systems and multifactorial approaches to technology choice, in favour of simplistic and unidimensional approaches, may fail to acknowledge and utilize all available resources for the development and maintenance of local communities.

For example, when faced with a situation where effective demand within a community for electricity outstrips available supply, engineers in a public utility not sympathetic to Appropriate Technology may look at the problem as one of how to generate higher absolute amounts of electricity. If substantial new installations were required for this purpose and if the availability of local capital for investment was low, combined with a relatively high cost of credit, that community's energy problems might either remain unsolved or be solved at the price of redirecting financial resources away from some other area of concern to the community (such as education). Adopting a wider view, in contrast, could reveal the potential for conservation as a cost-effective means for resolving the community's energy supply problems.[33] This would involve the adoption of a range of technologies more suited to energy efficiency for domestic, industrial and other purposes; it would also involve greater participation by people, in their capacity as citizens or as professional officers, in producing, installing and using new technologies, and in making decisions and plans regarding lifestyle and management practices. It might also involve an increase in the use of alternative sources of energy requiring alternative technologies to those already in use. The important point here is that the conservation approach would draw upon a range of human resources irrelevant to the narrower approach. From the point of view of an engineer experienced only with the mainstream energy-supply technologies, reliance upon increased use of *human* resources may appear unattractive. From the

---

[33] Cf., section on "Energy Pathways" in Chapter Seven.

point of view of integrated community development, however, the mobilization of possibly underutilized human resources with the concomitant stimulus this would provide throughout the community, may appear very attractive. The arguments for the latter option could be even more compelling for communities with high unemployment levels.

The conservation oriented approach also has the advantage that, because it aims to serve more than energy-supply objectives alone (e.g., provision of community welfare services, employment training and employment generation, environmental protection, improvement of building stock, and environmental education), it may be integrated with other community programs and draw upon resources (people, waste materials, infrastructure, money, etc.) which would otherwise not be available for addressing energy policy problems.

The integrated problem solving approach which characterizes Appropriate Technology may reveal new possibilities for community development which are affordable because of the emphasis upon mobilizing underutilized community resources which would not be relevant to the more simplistic mainstream approaches.

The term "practical holism" has been coined to denote the approach which combines radical criticism and integrated problem solving with concern for the human and environmental compatibility of technologies. It denotes not only holistic ways of understanding reality, but a mode of praxis which incorporates holistic analysis and an integrated package of complementary activities.

In contrast to the critiques of the technological society which portray the imperatives of technology, the requirements of environmental conservation and the needs of humanity as being in some kind of intrinsic conflict, Appropriate Technology points to the possibility of these three factors being harmonized, but on the condition that technology is chosen judiciously in accordance with the principles of the technological fit, endogenous technological development and practical holism.

Conceptual Clarification

Before proceeding further, certain concepts contained herein deserve further clarification.

Misconceptions

Repeated reference has been made to the notions of *mainstream technology,* *alternative technology* and *appropriate technology.* It is not uncommon for these three notions to be confused and misinterpreted.

The misdirected view, which appears to predispose some people against serious consideration of Appropriate Technology, begins with a simplistic distinction between "mainstream technology" and "alternative technology" and assumes that these categories are mutually exclusive. Thus, a technology is depicted as either mainstream or alternative, but not as a combination of both.

It is also assumed that "mainstream technologies" are by nature superior from the point of view of efficiency and are the quintessence of modernity. "Alternative technologies", in contrast, are assumed to be inherently inefficient, and based upon outmoded knowledge and practice. It is common for "mainstream technologies" to be lauded as "high" and for "alternative technologies" to be dubbed "low".

When stated explicitly in this manner this "misdirected" view appears untenable; nevertheless, the attitudes which it embodies are quite widespread. The extremely simplistic nature of this view is generally not apparent to its adherents because it is not normally articulated explicitly. It normally takes the form of a tacit perspective which is only revealed indirectly through policy decisions, action programmes and incidental discussion about technological matters. Occasionally such views, which normally remain tacit, are expressed openly in published literature; this is illustrated by the following quote from a paper by an urban physical planner:[34]

> There is some talk, which seems to have some standing as an intellectual 'fashion', about abstaining from the use of new 'hard' technology in favour of a retreat to earlier technologies or of inventing alternative ones. This fashion attracts followers who try personally to realize such 'escapist' attitudes, but it is scarcely applicable as a solution for society as a whole. It is almost impossible to induce people to produce *less* for the same input of labour and capital, or to spend *more time* to reach the same destination, or even to abstain voluntarily from some com-

---

[34] E. Brutzkus, "Technological Advance Beyond the Optimum", *Ekistics,* 284 [Sept/Oct 1980], 385). An ironic feature of the paper by Brutzkus is that, despite his disparaging remarks, the substantive position he adopts is remarkably similar to that advocated by the Appropriate Technology movement as a whole.

fort which makes life 'easier' when it is offered by modern 'hard' technology.

It is important to raise these misconceptions explicitly because, by remaining as tacit views, they appear to be a major obstacle to bridging the gap discussed earlier between common sense and common practice.

By tacitly assuming that a dichotomy exists between the two modes of technology practice ("alternative" and "mainstream") adherents of the misdirected view tend to automatically equate "appropriate technology" with "alternative technology". The term "appropriate technology" consequently connotes a type of technology-practice incompatible with the mainstream, and appropriate technologies (when labelled as such) are therefore assumed to be "low", "outmoded", "inefficient" and therefore not worthy of serious consideration. The misdirected view leads to the ironic situation where the term "appropriate technology" is used to describe technologies which by definition would have little chance of qualifying as appropriate.

Preferred View

The view of Appropriate Technology adopted herein rejects the assumption that technology-practice may be divided into two discrete modes, "alternative technology" and "mainstream technology". The preferred view is portrayed in Figure 10.2. Alternative Technology, Mainstream Technology and Appropriate Technology are taken to be three overlapping categories of technology-practice. Appropriate technologies may include technologies from the mainstream, alternative technologies and a range of new technologies required to fit contexts for which accepted alternative and mainstream technologies are inadequate.

Thus Appropriate Technology is not predisposed towards "low", "inefficient", "old fashioned" or "unpopular" technologies, on one hand, or "high", "efficient", "modern" or "popular", on the other hand - although, where efficiency is defined in broad, systems terms, Appropriate Technology is by definition concerned with the attainment of maximum efficiency.

Figure 10.2 *Preferred View of Appropriate Technology*

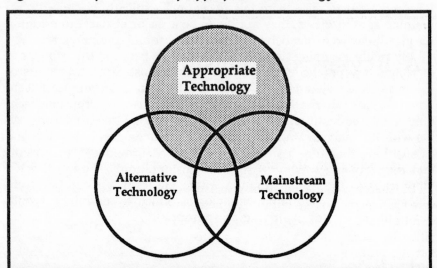

In summary, our critique of the Appropriate Technology movement and its literature leads to a view of Appropriate Technology as a mode of technology-practice rather than a particular collection of artefacts. Although proponents of Appropriate Technology have shown more of an interest in some fields rather than others, it is a mode of technology-practice applicable to most, if not all, fields.[35] It is primarily an innovation strategy aimed at achieving a good fit between technologies and the contexts in which they are intended to operate. This strategy also aims at endogenous technological development within local communities and regions, and depends upon the adoption by technological decision makers and the communities in question of holistic modes of praxis.

---

[35] In addition to the fields addressed in earlier chapters, there is an increasing tendency for the rationale of Appropriate Technology to be applied to new or emerging fields such as information technology (see, e.g., M. Elmandjra, et al., *Informatics: Is There a Choice?*, special edition of *Development: Journal of the Society for International Development*, 1 [1985], 1-85).

11

A Review of Plausible Criticisms

Socio-political Obstacles to Appropriate Technology

In Chapter Nine a range of criticisms of Appropriate Technology were examined. A large number of these were shown to be misdirected or unfounded. Some criticisms, however, those labelled "political" criticisms, were found to possess a degree of plausibility. It was found that the profusion of implausible criticisms has probably been facilitated by the lack of an integrated framework for construing Appropriate Technology. Rejoinders to the implausible criticisms were raised in Chapter Nine, and most of these were conditional upon Appropriate Technology being understood in terms of the integrated framework subsequently outlined in Chapter Ten. A full response to the plausible political criticisms was not possible prior to the articulation of an systematic model of Appropriate Technology.

The integrated framework has confirmed the earlier analysis of the implausible criticisms by indicating how the normative concerns of the Appropriate Technology movement might be adequately addressed without having to neglect considerations of technical efficiency, economic viability, physical practicability or intellectual-cum-cultural sophistication. Rejoinders to the implausible criticisms depend in a number of cases upon empirical evidence and not just logical analysis; Chapter Ten provides a framework for incorporating surveys of technical-empirical evidence from earlier chapters into a forceful defence of the Appropriate Technology concept. Many of the implausible criticisms are unconvincing precisely because of a failure to acknowledge the tripartite make-up of Appropriate Technology and a failure to take full cognizance of the difference between technological artefacts and Appropriate Technology as a mode of technology-practice.

A number of the political criticisms considered in Chapter Nine were shown to be unfounded. Most of these looked upon Appropriate Technology as if its protagonists were largely concerned only with technical-empirical considerations. Part Two indicated that the Appropriate Technology movement, with or without a universally attested conceptual model, has always addressed factors other than technical-empirical ones (notwithstanding some exceptions to the general rule). By formally and explicitly incorporating socio-political and ethical-personal factors the integrated framework strengthens our refutation of the "narrow technicism" and "technological determinism" criticisms. As it stands, however, Chapter Ten does not provide a full response to the relatively plausible criticisms of Appropriate Technology.

The plausible criticisms of Appropriate Technology, as indicated earlier, do not invalidate the concept and movement but rather raise questions about the prospects for the Appropriate Technology innovation strategy being taken up on more than a marginal basis. This chapter will examine whether the integrated framework provides a way of assessing the grounds for hope that the applicability of Appropriate Technology may be enhanced sufficiently to overcome the constraints to its successful diffusion.

Before proceeding further, the scope of the obstacles to Appropriate Technology will be described more precisely. The term "political" has been used rather loosely in earlier chapters and in much of the literature on the politics of technology. Politics may be understood in a narrow sense where it concerns the professional activities of politicians in their official capacity as representatives of their constituencies, the functions of formal political parties and associated institutions, and the interaction between these entities. For most proponents of Appropriate Technology the political dimension of Appropriate Technology extends far beyond this narrow conception of politics. Schumacher's comments are illustrative:[1]

> I have no hope in politicians. Politicians are the executive committee
> of the majority. ... On the whole the change that is necessary will
> never come from the majority.

---

[1] E. F. Schumacher: "Alternative Technology: Gordon Laing, Dorothy Emmet and Anthony Appiah talk to Fritz Schumacher, Founder of the Intermediate Technology Development Group", *Theoria to theory*, 9, 1 (1975), 15; *Technology with a Human Face* (Perth, Aust.: Campaign to Save Native Forests, 1977), p. 5.

... I would invite you to think about politics. We have developed a strange type of politics, where we occasionally change the crew, and the new crew does exactly the same as the old crew, except in favour of slightly different people. As long as the technology doesn't change, they can't do anything ...

Politics may also be understood in a broader sense which takes account of the structural forces of society, the conflict of interest between different social power groups, and the manner in which these forces and conflicts bear upon the formal entities mentioned above. A further, even broader conception of politics may be envisaged, which may or may not incorporate the formal institutions which characterize the narrow conception, viz.: all human activities which involve the exertion of power in society for the purpose of either maintaining or reforming the environment (psychosocial or biophysical) and its structure.

This broadest conception of politics is the most relevant from the point of view of assessing the prospects for Appropriate Technology, and it cannot be isolated from factors normally denoted by terms such as "social", "cultural" or "institutional". A more useful term for describing the obstacles to Appropriate Technology outlined in the criticisms chapter would be "socio-political", rather than just "political". *The integrated framework for Appropriate Technology developed here stresses that, in the main, socio-political change cannot be effectively achieved without concomitant and suitable technological change. The framework incorporates the main insights of the "political" critics but adds that technology deserves attention as a factor in its own right and not just as a minor adjunct to "politics".* Portraying Appropriate Technology as a mode of technology-practice makes it possible to avoid the problem of artificially separating the technological and the political (a problem encountered with specific-characteristics definitions of Appropriate Technology).

Socio-technical Obstacles to Appropriate Technology

Many of the socio-political criticisms canvassed earlier were based upon the imputation that Appropriate Technology embodies a crude doctrine of technological determinism. This imputation was shown to be superficial and misdirected. In Chapter Nine it was argued that the term "technological determinism" is frequently used in an ambiguous and misleading way, but it was agreed that the adoption of an absolute technological determinist stance would be inconsistent with the main tenets of Appropriate Technology. It was also argued that a common weakness of the "political" criticisms is their failure to address properly the characteristics and dynamics of technology, and their tendency to artificially separate the political from the technological. It was suggested that the Appropriate Technology innovation strategy ought to be seen as a form of politics rather than as an attempt to bypass politics. Elaboration of this theme was left, however, to follow the outlining of an integrated framework for Appropriate Technology.

In Part Two it was indicated that the bulk of the impetus for Appropriate Technology in the North appears to have come from responses to the perceived growth of the technological society. According to a number of commentators on the technological society, technology is subject to laws of progress which are independent, or at least semi-independent, of human judgements as to what is desirable.

Chapter Ten listed as a corollary of Appropriate Technology that technology-practice may act as a determining factor in society. It was pointed out that this does not amount to a disguised doctrine of technological determinism. Nevertheless, if considered in isolation, this corollary could be construed as consistent with technological determinism and, in view of the ambiguity with which "technological determinism" is employed in the literature, the subject deserves further discussion.

While most of the published criticisms of Appropriate Technology stem from a socio-political perspective (where "socio-political" is taken to be discrete vis-a-vis "technological"), earlier chapters have hinted at another possible perspective from which the practicability of Appropriate Technology might be questioned: the socio-technical perspective. While Appropriate Technology has been criticized from both the socio-political and the socio-technical perspectives, the two schools of thought associated with each perspective tend to be at theoretical loggerheads.

The school of thought associated with the socio-technical perspective is exemplified by the writings of people such as: Ellul, Marcuse, Mumford, Heidegger, Galbraith, Habermas, Horkheimer, and, in a different sense, Weiner or Steinbuch. These and other writers in the school are united by a view that technology seems to have developed in an apparently indomitable manner in urban-industrialized societies.[2] It is

---

[2] See, e.g.: J. Ellul,*The Technological Society*, trans. by J. Wilkinson (New York: Knopf, 1964); J. Ellul, *The Technological System* (New York: Continuum, 1980); H. Marcuse, *One Dimensional Man* (Boston: Beacon Press, 1964); L. Mumford, *Technics and*

claimed that while technology has always been a part of human culture, it has gained such prominence, centrality and complexity that it may no longer be thought of as subservient to conscious human control or traditional human institutions. In other words, technology has become "autonomous technology". Winner, in his widely cited review of this school of thought, summarizes the notion of autonomous technology as follows:[3]

> At the outset, the development of all technologies reflects the highest attributes of human intelligence, inventiveness and concern. But, beyond a certain point, the point at which the efficacy of the technology becomes evident, these qualities begin to have less and less influence upon the final outcome; intelligence, inventiveness and concern effectively cease to have any real impact on the ways in which technology shapes the world.

Ellul, perhaps the most trenchant exponent of this perspective, writes:[4]

> Everything takes place as if the technological phenomenon contained some force of progression that makes it move independently of any outside interference, of any human interference, of any human decision ... The technological phenomenon chooses itself by its own route ... [If] man produces the self-augmentation of technology (which could not generate itself, of course), he does so by assuming only an occasional and not a creative role. He cannot help but produce this augmentation; he is conditioned, determined, destined, adjusted, and preformed for it.

Civilization (New York: Harcourt Brace Jovanovich, 1963; originally published in 1934); M. Heidegger, *The Question Concerning Technology - and Other Essays* (New York: Harper Colophon, 1977); J. Habermas, *Toward a Rational Society*, trans. by J. Shapiro (London: Heinemann, 1971); M. Horkheimer, *Critique of Instrumental Reason* (New York: Seabury, 1984); J. K. Galbraith, *The New Industrial State* (2nd. ed.; Harmondsworth: Penguin, 1972); N. Weiner, *The Human Use of Human Beings: Cybernetics and Society* (Cambridge, Mass.: The Riverside Press, 1950); K. Steinbuch, *Automat und Mensch: Über Menschliche und Maschinelle Intelligenz* (Berlin: Springer Verlag, 1965). Note: a comparative review of Habermas, Weiner and Steinbuch has been conducted by E. Schuurman in *Technology and the Future: A Philosophical Challenge* (Toronto: Wedge, 1980), pp. 17-260. An interesting anthology of recently published articles which may be loosely grouped under the rubric of "socio-technical", but not all of which adopt the "autonomous technology" notion, has been edited by P. L. Bereano (*Technology as a Social and Political Phenomenon* [New York: John Wiley and Sons, 1976].

[3] L. Winner, *Autonomous Technology:Technics-Out-of-Control as a Theme in Political Thought* (Cambridge, Mass.: MIT Press, 1977), pp. 313-314.

[4] Ellul, *Technological System*, p. 233.

Elsewhere he succinctly concludes:[5]

> Technology develops not in terms of goals to be pursued but in terms
> of already existing possibilities of growth.

To the extent that the historical development of technology exhibits
the kind of autonomy expressed here the concept of Appropriate Technology appears impracticable.

Appropriate Technology relies upon the capacity of human beings
to tailor technology to fit its psychosocial and biophysical context -
that is, using Ellul's words, to ensure that technology develops in terms
of goals to be pursued. Ellul argues that in the technological society
such a function may only be fulfilled "occasionally" and not
"creatively". He also argues that technology is intrinsically systemic
in nature and that, as society becomes increasingly technological, it
also takes on the characteristics of technical systems. Thus, in the
technological society the context of technology becomes increasingly
technological. The frame of reference for assessing technology consequently exhibits similar features to the technology to be assessed. According to Ellul's portrayal of modern urban-industrialized society,
human beings lack operable reference points other than technical ones
for judging technical phenomena.[6] It is this perceived difficulty of
transcending the sphere of technical rationality that lies behind Marcuse's portrayal of modern industrial society as "one dimensional".[7]

The Appropriate Technology innovation strategy requires the
multi-dimensional assessment of technology. The multi-dimensional
assessment of technology, however, would appear immensely difficult
to achieve in a society with a preponderance of a one-dimensional mode
of rationality. Amongst the corollaries of the Appropriate Technology
concept are the additional requirements that human beings have the
power to *control* technology and *choose* between alternative technologies. If technology is indeed autonomous, as the socio-technical perspective appears to indicate, then Appropriate Technology is under-
mined.

There are, nevertheless, some thinkers in opposition to the notion of
autonomous technology. The existence of autonomous technology is
strongly disputed by some political theorists, particularly those in the
Marxist tradition, because it conflicts with conventional concepts of po-

---

[5] *Ibid.*, p. 256.
[6] See, esp., *ibid.*, pp. 310-325.
[7] H. Marcuse, *One Dimensional Man* (Boston: Beacon Press, 1964).

litical hegemony.[8] For most Marxists the idea that technological change might exhibit some form of internal objective logic is a negation of a basic tenet of Marxism; viz., that the "social relations of production", together with the "productive forces" (a combination of labour-power and the means of production, broadly defined), provide the decisive explanation for the dynamics of a society.[9] It is frequently unclear in Marxist publications whether technology is equivalent to the means of production, whether it is part of the means of production or whether it is something different again. Nevertheless, the dominant Marxist view, represented by Braverman or Thompson, for example, holds that technological change is determined by social relations; technology is placed in a subservient position to non-technological factors in society.[10]

It is not possible to debunk the apparent threats to Appropriate Technology posed by autonomous technology, however, simply by citing the above Marxist view. Firstly, Marxism itself is inconsistent on these matters.[11] Furthermore, as demonstrated by Elster, this inconsistency may be traced back to Marx himself, and it is possible to argue plausi-

---

[8] E.g., D. MacKenzie and J. Wajcman, "Introductory Essay", *The Social Shaping of Technology*, ed. by D. MacKenzie and J. Wajcman (Milton Keynes: Open University Press, 1985), pp. 2-25.

[9] H. Braverman, *Labour and Monopoly Capital: The Degadation of Work in the Twentieth Century* (New York: Monthly Review Press, 1974); H. Thompson, "The Social Significance of Technical Change", *The Journal of Australian Political Economy*, 8 (July 1980), 57-68. Cf., A. Zvorikine, "Technology and the Laws of its Development", paper presented to the Encyclopaedia Britannica Conference on the Technological Order, March 1962, Santa Barbara, California, in *The Technological Order*, ed. by C. F. Stover (Detroit: Wayne State University Press, 1963), pp. 59-74. A recent publication, edited by M. Dubofsky, brings together papers which deal with this topic (*Technological Change and Workers' Movements* [London: Sage, 1985]).

[10] Thompson admits at one stage ("Social Significance of Technical Change", p. 57) that technology might also affect the forces and relations of production as well as be affected by them. The whole force of his argument, however, is to deny the mutuality of determining influences. A similar tendency is found in the work of R. Johnston (whose views on Marx are less apparent) who, in his critique of Collingridge's work, appears to acknowledge the two-way determination between technological change and broader social change; yet, the force of his argument is that technology is quite malleable in the face of social pressures whereas the reverse does not hold (R. Johnston: "Controlling Technology: An Issue for the Social Studies of Science", *Social Studies of Science*, 14 [1984], 97-113; "The Social Character of Technology [Reply to Collingridge]", *Social Studies of Science*, 15 [1985], 381-383).

[11] See the two works by Marxist scholar, A. Gouldner (*The Two Marxisms: Contradictions and Anomalies in the Development of Theories* [New York: Seabury, 1980]; *The Dialectics of Ideology and Technology* [New York: Seabury Press, 1976]).

bly that Marx too actually held technological change to be the prime mover of history.[12]

Secondly, the weight of analytical and practical evidence for the phenomenon of autonomous technology, provided by writers from the socio-technical perspective, is not easily dismissed. When understood in its broadest sense, as the growing dominance of technical rationality in society, the theme of autonomous technology has been addressed in sociology for some time.[13] The theme of autonomous technology has also recurred extensively throughout Western literature.[14]

The *mutual* interaction of social relations and technology is increasingly being recognized by Marxists.[15] Some recent Marxist publications even openly acknowledge the phenomenon of autonomous technology. Mathews, for example, writes:[16]

> We now have to recognize that the socialist dream of liberation from nature has become for us the capitalist nightmare of technology grown out of control.

In short, there are insufficient grounds to conclude that Marxist polemic against the doctrine of technological determinism constitutes a substantive rebuttal of Appropriate Technology.

There is a prima facie case for believing that some kind of force operates in technological society which accords with the notion of autonomous technology. This is not equivalent to adopting the view that technology is *absolutely* autonomous or independent of social influences. It would appear that the obstacles to Appropriate Technology which were labelled earlier as "socio-political" bear some relationship to the

---

[12] J. Elster, *Explaining Technical Change* (Cambridge: Cambridge University Press, 1983), esp. pp. 158-184, 209-228.

[13] The work of M. Weber is an example (*The Theory of Social and Economic Organization* [New York: Oxford University Press, 1947]; cf., *Economy and Society* [New York: Bedminster Press, 1968]). A review of academic literature from Weber onwards on the topic of the spread of technical rationality has been conducted by S. Cotgrove ("Technology, Rationality and Domination", *Social Studies of Science*, 5 [1975], 55-78).

[14] See, Winner, *Autonomous Technology*, esp. pp. 13-43.

[15] See, R. Dunford, "Politics and Technology: Unravelling the Connections", in *Public Sector Administration: New Perspectives*, ed. by A. Kauzmin (Melbourne: Longman Cheshire, 1983), pp. 183-199. This is also apparent in a collection of essays sub-edited by A. Huyssen under the rubric of "Machines, Myths, and Marxism" published in *The Technological Imagination: Theories and Fictions*, ed. by T. de Lauretis, A. Huyssen and K. Woodward (Madison, Wisconsin: Coda Press, 1980), pp. 77-131; this collection also reveals the inconsistency within Marxism on this subject.

[16] J. Mathews, "Marxism, Energy and Technological Change", in *Politics and Power*, *One*, ed. by D. Adlam, et al. (London: Routledge and Kegan Paul, 1980), p. 30.

obstacles labelled here as "socio-technical". Therefore, a useful response to the criticisms of Appropriate Technology is not to attempt a complete confutation of the ideas in each school of thought, but to understand the *extent* to which the obstacles raised are indomitable and the *manner* in which they are interrelated. The weight of evidence from the literature on technology and society, when reviewed as a whole, points to the need to simultaneously consider socio-political and socio-technical factors in a review of the criticisms of Appropriate Technology. A review which focussed on only one of these types of factors would, by that fact, be unreliable.[17]

While the socio-political and socio-technical schools of thought tend to argue against each other's perspectives, there are certain common features in their perspectives on Appropriate Technology. Unfortunately, however, the writings within each of these schools are marred by semantic confusion similar to that which was identified within the Appropriate Technology movement. To some extent, therefore, the plausible criticisms of Appropriate Technology may be dealt with by a careful analysis of definitions. This matter will be addressed in the next section, but first it is necessary to introduce a new idea.

Addressing the problem of whether the socio-technical obstacles to Appropriate Technology might be surmounted amounts to addressing the problem of whether the purported phenomenon of autonomous technology might be surmounted. To address this problem it is proposed here that Appropriate Technology and Autonomous Technology be viewed as two antithetical technological phenomena. In other words, Appropriate Technology and Autonomous Technology should be viewed as diametrically opposed modes of technology-practice. The implications of this will now be explored.[18]

There is some disagreement amongst those who point to the existence of Autonomous Technology as to its normative significance. One pole of opinion views Autonomous Technology as inimical to human wellbeing and as intrinsically violent towards nature. Adherents to

---

[17] MacKenzie and Wajcman, in one of the more informed reviews of this subject ("Introductory Essay", passim.), reveal their preference for interpretations which emphasize socio-political factors as more fundamental than socio-technical factors. The substance of their arguments, nevertheless, points to the dynamics of society and technology as deriving from the complex mutual interactions of technological, social, economic and political factors. This accords with the position we have adopted here and which is embodied in the integrated framework from the previous chapter.

[18] For reasons which will be explained in the following sections the general notion of autonomous technology will hereafter be denoted with capitals as follows: "Autonomous Technology".

this view may be labelled "techno-pessimists". A romantic anti-technological stance is the typical response, characterized by either attempted transcendental withdrawal from intentional involvement in technological activity or by nihilistic acquiescence. We may include such thinkers as Roszak, Reich, Jünger or the members of the "beat generation" as representative.[19] Another pole of opinion views technology (understood in terms similar to Autonomous Technology) as the guarantee of a salubrious future for humankind. Adherents to this view may be labelled "techno-optimists". A romantic pro-technological stance is the typical response characterized by ardent advocacy of technology as the solution to human and environmental problems, or by faith in technology *per se*. Some hold this faith while being fully aware of the potential dangers and violence of technology.[20] Others exhibit a dogmatic faith in the goodness of technology and a commitment to allowing "technological development" to proceed autonomously, unimpeded by human attempts to control or restrain it.[21] Despite the conflict of opinion between the techno-pessimists and the techno-optimists, they may still be viewed as belonging to one school of thought - because of their common assumption about the autonomous na-

---

[19] See, e.g.: T. Roszak, *The Making of a Counter Culture: Reflections on the Technocratic Society and its Youthful Opposition* (London: Faber and Faber, 1968); C. Reich, *The Greening of America* (New York: Random House, 1970); F. G. Jünger, *The Failure of Technology* (Chicago: Regnery, 1956). For a critical review of the "beat generation" see the work of O. Guiness (*The Dust of Death: A Critique of the Counterculture* [Downers Grove, Ill.: Inter Varsity Press, 1973], esp. pp. 114-274).

[20] The writings of E. G. Mesthene illustrate this perspective (e.g.: "How Technology Will Shape the Future", *Science*, 161 [July 1968], 135-143; *Technological Change* [Cambridge, Mass.: Harvard University Press, 1970]). Mesthene holds that the undesirable aspects of technological change stem primarily from changes in human values lagging behind technological change; he claims that the solution lies in human values being "brought into better accord" (*Technological Change*, p. 62) with contemporary technology. Mesthene employs rhetoric against the notion of Autonomous Technology (e.g., *ibid.*, pp. 40-41) but most of his analysis is based upon the assumption that technological change is "given" and proceeds autonomously.

[21] Examples of the perspective include: H. Kahn, *The Next 200 Years: A Scenario for America and the World* (New York: Morrow, 1976); H. Kahn and T. Pepper, *Will She Be Right? The Future of Australia* (Brisbane: The University of Queensland Press and Prentice-Hall International, 1980). Cf., the remarkable series of booklets ("Dialogues on Technology") published by the American company, Gould Inc., Rolling Meadows, Illinois; booklet #1 (*Technology: Abandon, Endure or Advance?*), for example, concludes with the statement: "Can we be sure that science and technology will find the answers? Can we be sure that solutions to our problems exist? No, but we can be sure that nothing but science and technology can find them, if they do exist. To put it as bluntly as possible: science and technology must answer our problems. If they don't, nothing else will" [p. 12]). Faith in the benign nature of unimpeded technological change is apparent in recent publications aimed at demonstrating the social value of so-called "high technology" (see, e.g., R. W. Riche, "The Impact of Technological Change", *Economic Impact*, 41, 1 [1983], 13-18).

ture of technological change (an assumption which may be either tacit or overt).

The optimism of the techno-optimists ought not to be confused with optimism as to the future prospects for Appropriate Technology. The former type of optimism is based upon the confidence that technology itself will bring about a salubrious future and that human beings will adapt to the new technological milieu with minimal social cost. It embodies a normative stance that untrammeled technological change is good and that the imperatives of technology are automatically in accord with the needs of human beings and the purposes of human existence (insofar as such purposes may be held to exist or possess validity). Optimism vis-a-vis Appropriate Technology, in contrast, is based upon different normative premises. Appropriate Technology, on principle (according to our definition and integrated framework), points to the importance of technological change within a community not proceeding independently of the efforts by people in that community to control it and choose its direction. The question of whether there are strong grounds for optimism regarding the future of society vis-a-vis technology may not be reduced to technical-empirical considerations. The meaning of the question depends upon what normative goals are adopted or recognized for society. Thus, the availability of firm empirical evidence for the possibility of the sort of future society envisaged by the techno-optimists does not provide grounds of hope for the success of the Appropriate Technology innovation strategy.

In summary, the plausible criticisms of Appropriate Technology embrace both socio-political considerations and socio-technical considerations. The literature we have surveyed points to the growing prominence of technology in modern societies and in so-called traditional societies. While some commentators view technology simply as an aspect of society and others attribute it greater independence from other factors, most serious commentators point to the mutual interaction and overlapping nature of technology and society; in other words, technology incorporates social factors and society incorporates technological factors. Politics must therefore be vitally concerned with socio-technical factors if it is to adequately address prevailing "real life" circumstances.[22] Political choices need to be understood as requiring comcomitant technological choices - and technological choices need to be

---

[22] Our term "technology-practice" embodies this insight. It also appears that recognition of this point is the main reason for Winner's use of the term "technological politics" (*Autonomous Technology*, pp. 237-278). Cf., L. Winner, "*Techné* and *Politeia*: The Technical Constitution of Society", *Philosophy and Technology*, ed. by P. T. Durbin and F. Rapp [Dordrecht: D. Reidel, 1983], pp. 97-111.

understood as either explicit or implicit political-cum-social choices. The "political" criticisms identified in Chapter Nine must be treated as possessing only a limited degree of plausibility if they are understood as excluding the socio-technical perspective outlined above. By emphasizing *both* the socio-technical and the socio-political objections which may be raised against Appropriate Technology a firmer basis has been developed for addressing the problem which was left unresolved at the end of Chapter Nine; i.e., that of identifying whether there are any grounds for hope that the goals of the Appropriate Technology movement may be substantially achieved.

Appropriate Technology and Autonomous Technology

Having reviewed the criticisms of Appropriate Technology we are now in a position to further consider the capacity of the integrated framework to reveal how the obstacles to Appropriate Technology might be surmounted.

A major strength of Appropriate Technology, and a characteristic which features prominently in the synthesis of the concept in this book, is that it is grounded in a recognition of the fundamental role of technology and technology-practice in modern society. Many of the politically oriented criticisms of Appropriate Technology are flawed by a failure to recognize that technology plays a dominant role in the modern world - probably more dominant than it has in the past. While it would be difficult to find a serious attempt in the literature to refute such an observation about technology, many political theorists do not effectively incorporate this notion into the structure of their theories and strategies. One atypical political scientist has noted this in the following manner:[23]

> If it is clear that the social contract implicitly created by implementing a particular generic variety of technology is incompatible with the kind of society we would deliberately choose, then that kind of device or system ought to be excluded from society altogether. A crucial failure of modern political theory has been its inability or unwillingness even to begin this project: critical evaluation of society's technical constitution. The silence of modern liberalism on this issue is matched by an equally obvious neglect in Marxist theory. Both persuasions have en-

---

23 Winner, *"Techné and Politeia"*, pp. 109-110.

thusiastically sought freedom in sheer material plenitude, welcoming whatever technological means (or monstrosities) seemed to produce abundance fastest.

The Appropriate Technology innovation strategy may be understood *inter alia* as a political strategy which is superior to other techno-political strategies insofar as these treat technology as a largely passive, malleable and *automatically subservient* factor in the dynamics of society. The political criticisms reviewed in Chapter Nine may be understood as a reflection of the limitations of dominant political theory as much as a reflection of the actual obstacles encountered by Appropriate Technology.

Another major strength of Appropriate Technology (construed in the terms of our integrated framework) is that it is capable of clarifying much of the debate over Autonomous Technology. The integrated framework is based upon the semantic conventions and conceptual categories specified in Chapter Two, in which the importance of distinguishing between technology-practice, technology, technicity, technique and technological science was stressed. In most of the debate over Autonomous Technology - especially in the English language - these categories are frequently confused. Consequently, when commentators disagree over "autonomous technology" or "technological determinism" it is not immediately obvious whether there are substantive differences between their viewpoints or whether the disagreements stem from semantic ambiguity.

The reader is referred to Table 2.1 at the end of Chapter Two for a list of technology-related nomenclature used in this book. A few of the most important definitions will nevertheless be reiterated here. *Technology* is the ensemble of artefacts intended to function as relatively efficient means. *Technology-practice* is the ensemble of operations, activities, situations or phenomena which involve technology to a significant extent. *Technicity* is the distinguishing factor or quality which makes a phenomenon technical. *"Technical"* is an adjective or adverb used to qualify phenomena *dedicated* to efficient, rational, instrumental, specific, precise, and goal-oriented operations. It is important to recognize that there is a distinction between "technological" and "technical" in these semantic conventions.

If the word "technology" in "autonomous technology" is given the meaning stipulated in this study then the concept of Autonomous Technology is clearly difficult to defend. This is because artefacts, in the final analysis, cannot be absolutely autonomous or completely separated from the sphere of human activity - artefacts, after all, are defined as the products of human art and workmanship.[24] This observation appeals so readily to common sense that it is hardly surprising that the notion of Autonomous Technology has been so widely disputed. Even Ellul readily admits that the "technological phenomenon" is generated by human beings and that it could not generate itself.[25] He writes, for example, as follows:[26]

> To speak of a machine that lives and thinks, or even reproduces itself is infantile anthropocentrism ... But technology is inevitably part of a world that is not inert. It can develop only in relation to that world. No technology, however autonomous it may be, can develop outside a given economic, political, intellectual context.

Despite passages such as this one, Ellul is frequently castigated for purportedly ignoring the social and political context of technology.[27] Eberhard's description of Ellul is typical:[28]

> Confusion arises in Ellul's work from a failure to treat social consequences of technological progress within the context of a more general phenomenon - namely the attempt to shape behavior according to political and economic interests.

One could be tempted to think that Ellul is a rather inconsistent or incompetent scholar - to be known so widely as the most extreme proponent of "context-free" interpretations of technology and as a promulgator of the Autonomous Technology doctrine, on the one hand, while making statements, on the other hand, such as the one just cited.

A more satisfactory review of the evidence ought to take into account the explanation offered by Ellul himself. Firstly, as indicated by the foregoing quote, Ellul does not equate technological autonomy with the complete independence of technology from socio-political influences. Secondly, as stressed throughout his writings, Ellul makes a distinction between technologies (*techniques*, French) and what he labels

---

[24] J. B. Sykes, ed., *The Concise Oxford Dictionary of Current English* (6th ed.; Oxford: Oxford University Press, 1976), p. 52

[25] Ellul, *Technological System*, p. 233.

[26] *Ibid.*, pp. 30-31.

[27] E.g., H. Rose and S. Rose, "The Incorporation of Science", in *The Political Economy of Science*, ed. by H. Rose and S. Rose (London: Macmillan, 1976), p. 31.

[28] A. Eberhard, *Technological Change and Development: A Critical Review of the Literature*, Occasional Paper in Appropriate Technology, School of Engineering Science, University of Edinburgh, 1982, p. 68.

"*La Technique*" (translated into English as the ambiguous "technology").[29] Thus, the concept and phenomenon labelled here as "Autonomous Technology" does not mean that actual technologies are autonomous. Ellul's writings are highly idiomatic and it is not always easy to identify just what it is that he is arguing; he also frequently places discussion of *La Technique* in close proximity to discussion of particular fields of technology-practice. Nevertheless, even a relatively cursory reading of his works makes it apparent that by "*La Technique*" Ellul intends to denote a phenomenon similar to that which many other writers have recognized and which Cotgrove has dubbed as "technological rationality".[30] In his later writings Ellul clarifies the situation by explaining that *La Technique* is a system characterized by "technological rationality". While Ellul doesn't explain this concisely in one place, it is apparent that, for him, *La Technique* is the totality of all technical systems.[31] It would appear that the phenomenon translated into English from Ellul's writing as "technology" would best be denoted by the term "technicity" which was adopted in Chapter Two. Ellul appears to argue, to use our terminology, that systems characterized by a high degree of technicity *ipso facto* exhibit a propensity for autonomy. This is not meant to imply, however, that the autonomy of such systems is absolute in relation to other systems. The following quote from Ellul's work illustrates this point:

> In reality, we must not confuse the technological system and the technological society. The system exists in all its rigor, but it exists within the society, living in and off the society and grafted upon it. There is a duality here exactly as there is between nature and the machine. The machine works because of natural products, but it does not transform nature into a machine. ... At a certain level, culture and nature overlap, forming society, in a totality that becomes a nature for man. And into this complex comes a foreign body, intrusive and unreplaceable: the technological system. It does not turn society into a machine. It fashions society in terms of its necessities; it uses society as an underpinning; it transforms certain of society's structures. But there is always something unpredictable, incoherent, and irreducible in the social body. ... It is only at an extreme point that we can view the society

---

[29] *Technological Society; Technological System*, esp. pp. 23-33; "Technological Order", pp. 10-37.
[30] Cotgrove, "Technology, Rationality", pp. 55-78.
[31] Cf.: *Technological Society*, p. xxv; *Technological System*, passim.

and the system as one and the same. But nobody can seriously maintain that this extreme has been reached.

By making a distinction between technology-practice, technology and technicity, and by examining the writings of Ellul in a less cursory manner than appears normal amongst his critics, we are in a better position to understand the significance of Autonomous Technology.

Technicity is a type of process or mode of rationality which, by definition, is autonomous vis-à-vis human purposes. Thus, Autonomous Technology ought to be understood as a mode of technology-practice characterized by the dominance of technicity.[32] The term "appropriate technicity" is absurd because technicity, as understood here, is not open to transformation to anything other than itself. Technicity is not ontologically differentiable - to use philosophical terminology. Qualification of "technicity" with an adjective for the purposes of differentiating it is therefore not semantically acceptable. On this view the only options available to human beings vis-à-vis technicity are to place *limits* or constraints on the degree to which technicity has a role in fields of human endeavor or to increase reliance upon technicity through either conscious commitment to such an option or by a *de facto* failure to place limits on it. "Autonomous Technology" ought therefore to be viewed as a short-hand term for a mode of technology-practice in which no limits have been placed upon the scope and dominance of technicity.

Appropriate Technology, according to our integrated framework, requires that human beings be capable of controlling technology and of adopting one technological option rather than another. This, in turn, requires that there are options available in the kinds of rationale with which a particular case of technology-practice may be imbued. Technology-practice completely dominated by technical rationality (leading to technologies which are a perfect manifestation of technicity) would preclude the possibility of such control and choice. It is for this reason that Appropriate Technology and Autonomous Technology are portrayed here as antithetical modes of technology-practice.

If one fails to make the distinction between technology-practice, technology and technicity, one is forced to reject the idea of Appropriate Technology. This is reflected in the debate between Habermas and

---

[32] The term "Technology", as used here in "Autonomous Technology" is, strictly speaking, a misnomer; "technology-practice" would be more consistent with the rest of our analysis. The former term is employed here, however, in keeping with the established use of "Autonomous Technology" (after Winner, *Autonomous Technology*).

Marcuse on the possibility of a "new technology".[33] Both of these thinkers appear to confuse the above concepts, with the end result that their writings fail to point to clear grounds for hope for an "Appropriate Technology". Marcuse does point to the possibility of a qualitatively new mode of technology-practice [our terminology] which is imbued with independently constituted human values, viz.:[34]

> ... the historical achievement of science and technology has rendered possible the *translation of values into technical tasks* - the materialization of values. Consequently, what is at stake is the redefinition of values in *technical terms*, as elements in the technological process. The new ends, as technical ends, would then operate in the project and in the construction of the machinery, and not only in its utilization.

Marcuse's idea here appears to coincide closely with our version of Appropriate Technology - as an innovation strategy aimed at matching the intrinsic and extrinsic ends of technology.[35] Marcuse, however, proposes his vision of technology imbued with human values against a backdrop of having argued earlier that, in advanced industrial society, the only effective value is the one-dimensional value of technical rationality. Speaking of the "project" of technological progress, for example, he writes:[36]

> As the project unfolds, it shapes the entire universe of discourse and action, intellectual and material culture. In the medium of technology, culture, politics and the economy merge into an omnipresent system which swallows up or repulses all alternatives. The productivity and growth potential of this system stabilize the society and contain technical progress within the framework of domination.

Marcuse points to the possibility of alternative modes of technology-practice but his analysis of society seems to preclude such alternatives.

---

[33] J. Habermas, "Technology and Science as 'Ideology'", in his *Toward a Rational Society*, trans. by J. Shapiro (London: Heinemann, 1971), pp. 81-122; Marcuse, *One-Dimensional Man*.

[34] Marcuse, *One-Dimensional Man*, pp. 231-232.

[35] See Chapter Ten.

[36] Marcuse, *One-Dimensional Man*, p. xvi.

This contradiction in Marcuse's work makes his stated hope seem rather gratuitous.[37]

Habermas argues against Marcuse's notion of a new mode of technology-practice [our terminology] on the grounds that technological progress requires the suppression of ethics as a category of human experience.[38] Habermas holds that two mutually exclusive modes of activity are possible, one corresponding to technical rationality and the other corresponding to what is traditionally thought of as the authentic domain of the human spirit. The former he labels "purposive-rational action" (equivalent to our "technicity") and the latter he labels "communicative interaction" (which is the sphere of normative human interest). He argues that the human problems of the technological society stem from a failure to actively maintain the distinction between the "technical" and the "practical" (i.e., communicative interaction within a normative order, ethics and politics).[39] For Habermas, technology may not be "humanized", "transformed" or imbued with non-technical human values; there is an intrinsic contradiction between the two modes of activity and the full pursuit of "practical" human interests may only be achieved if the technical mode is constrained. From this point of view it would be absurd to attempt to make "technology" humanly appropriate - placing limits on the profusion of technology would be the only humanly appropriate option.[40]

This discussion of Habermas and Marcuse may be concluded by noting that, in the final analysis, while both writers appear to write from a normative perspective similar to that which undergirds the Appropriate Technology movement, neither provides satisfactory grounds for confidence in the achievement of the movement's objectives. Marcuse perceives the need for "Appropriate Technology" and he even articulates something of the principles involved, yet his theoretical criti-

---

37 This conclusion here is reinforced by Fromm's insistence that the apparent optimism in Marcuse's work is a thin veil over his complete "hopelessness" (E. Fromm, *The Revolution of Hope: Toward a Humanized Technology* (Perennial Library; New York: Harper and Row, 1964), pp. 8-9). Fromm writes, "Marcuse is essentially an example of an alienated intellectual, who presents his personal despair as a theory of radicalism" (ibid., p. 9).

38 Habermas, "Technology and Science as 'Ideology'", pp. 81-122, passim., esp. pp. 112-113.

39 Note: Habermas uses "practical" with a meaning which is quite different to that given to the term in current English (where it often means "expedient" or "efficacious"); for Habermas "technical" and "practical" are mutually exclusive opposites.

40 This dichotomy between the spheres of the "technical" and the"human" is also apparent in the work of M. Horkheimer (*Eclipse of Reason* [New York: Columbia University Press, 1947]).

cism of the technological society (upon which his prognosis depends) appears to also undermine his hopes. Habermas adopts a seemingly more rigorous intellectual framework than Marcuse, but his recommendations on how a humanly desirable future may be achieved do not include a positive role for technology. Given that both writers point to the domination of all fields of human endeavor by technological modes of operation and by the imperative of technological progress, it is hard to see how their analyses may provide a sanguine view of the future. Both Marcuse and Habermas point to the need for a complete reversal in the dynamics of society, but their abstract polemic does not provide an indication of how this might be achieved at the level of technology-practice.[41] Both writers observe the phenomenon of Autonomous Technology, but neither is prepared to accept it as a *fait accompli*.[42] This apparently unresolved tension in their work appears to derive from their failure to embrace and rigorously apply the distinction between technology-practice, technology and technicity.

This discussion of Marcuse and Habermas has not been included here merely for reasons of pedantry. Rather, their writings, the interpretation of which appears to have generated a notable academic industry, have significant ramifications for Appropriate Technology.[43] If we accept Habermas' theories and his critique of Marcuse, then we are forced to conclude that the achievement of the Appropriate Technology movement's normative goals has very little to do with technology as such - this conclusion amounts to a rejection of a chief tenet of the movement.

---

[41] This hope for a complete "reversal" in the direction of social reality (or more completely, in metaphysical terms, in the direction of being) is also apparent in the work of Heidegger. Cf.: Heidegger, *Question Concerning Technology*; Lovitt, "Techne and Technology - Heidegger's Perspective on What is Happening Today", *Philosophy Today* (Spring 1980), 62-72.

[42] E.g., Habermas speaks of "systems of purposive-rational action that have taken on a life of their own" ("Science and Technology as 'Ideology'", p. 113) yet elsewhere he refers to this phenomenon as the "quasi-autonomous progress of science and technology" (*ibid*, p. 105).

[43] For evidence of this "industry" see, e.g.: J. B. Thompson and D. Held, eds., *Habermas: Critical Debates* (London: Macmillan, 1982); J. J. Shapiro, "The Dialectic of Theory and Practice in the Age of Technological Rationality: Herbert Marcuse and Jürgen Habermas", in *The Unknown Dimension: European Marxism Since Lenin*, ed. by D. Howard and K. E. Klare (New York: Basic Books, 1972), pp. 276-303; N. Stockman, "Habermas, Marcuse and the *Aufhebung* of Science and Technology", *Philosophy of the Social Sciences*, 8 (1978), 13-55. In view of the extensive debate in the social science literature over Habermas and the "critical theorists" of the Frankfurt School, the space which has been devoted to the relevant theoretical issues in this chapter is quite appropriate. To pursue the debate any further, however, would be beyond the purview of this study.

The formulation of Appropriate Technology presented herein, however, provides a basis for resolving the apparent tension just identified in the writing of Habermas and Marcuse. It is possible to maintain the view that "technology" (i.e., technology and technology-practice) is ontologically differentiable while simultaneously accepting the power of Habermas' claim that "technology" (i.e. technicity) is not ontologically differentiable. This means, in less philosophical language, that it is possible to seriously maintain the possibility of choices between alternative technologies and between alternative modes of technology-practice, without disregarding the propensity for autonomy (and the consequential lack of practical choice)[44] within technical systems - that is, within technological systems which are completely dominated by technicity. In other words, *so long as technology-practice is not dominated by technicity the Appropriate Technology innovation strategy is a realistic possibility.*

In view of the foregoing analysis and, in particular, the claim that Autonomous Technology and Appropriate Technology ought to be understood as antithetical modes of technology-practice, we are now in a position to restate these two concepts in a form which will clarify the problems under consideration.

To point to the existence of Autonomous Technology is not to deny the socio-political determination of technology-practice nor to suggest that individual technologies may be generated entirely independently of human decision. Instead, it is proposed here that "Autonomous Technology" be employed to denote a mode of technology-practice dominated by technical systems and by the failure of human beings to place limits on the scope of technicity. Technology-practice is not intrinsically autonomous vis-à-vis human control but, because of the propensity of highly technical systems for autonomy, technology-practice may exhibit a degree of contingent autonomy in the context of the technological society and where limits have not been placed on the dominance of technicity. According to this view human beings can exert control over technological systems but only when deliberate constraints have been placed upon technicity. In other words, human autonomy vis-à-vis technology is not automatic in the technological society but requires deliberate and concerted effort.

To point to the possibility of Appropriate Technology is not to deny the existence of Autonomous Technology, even though the two have been portrayed herein as antithetical. *It is proposed here that*

---

[44] "Practical" is used here with Habermas' meaning rather than with the dominant meaning it possesses in current English.

"Appropriate Technology" be employed to denote a mode of technology-practice aimed at achieving a good technological fit and based upon the deliberate imposition of limits on technicity. The imposition of limits on technicity does not in itself guarantee that technology-practice is "appropriate" but creates an essential pre-condition for the implementation of Appropriate Technology.[45]

It may be concluded, at this point, that the analyses of the writers within the "socio-technical" perspective do not constitute an effective rebuttal of Appropriate Technology. The only defensible notion of Autonomous Technology is one which views it as a contingent phenomenon rather than as an absolute and immutable phenomenon. The critics of "technological determinism" and, often by association, of Appropriate Technology, do not generally address the crucial distinction between absolute and contingent technological autonomy (or technological determinism). Consequently, as reflected in Chapter Nine, much of the emerging technology studies literature is spent debating a pseudo-problem; i.e., protagonists from the socio-political perspective criticize protagonists from the socio-technical perspective for holding to a position (absolute technological determinism) which they do not in fact embrace. The "politics" oriented thinkers frequently argue as if political hegemony is an immutable, absolute and eternally "given" fact and consequently interpret *any* acknowledgment of technological autonomy as completely incompatible with a creative role for human beings and with the processes of politics. The Appropriate Technology movement may be interpreted as an attempt to get beyond this ideological impasse by acknowledging the tendency towards autonomy in technological systems yet also pointing to the possibility of such a tendency being negated.

By taking into account certain, at times philosophical considerations, this chapter has shown how no sustainable fundamental theoretical objections to Appropriate Technology have been seriously argued in the literature. This does not, however, provide any guarantee that Appropriate Technology will be successfully diffused as the dominant mode of technology-practice. Whether or not there are adequate

---

[45] It appears that, despite his criticisms of Appropriate Technology, Ellul is actually in agreement with the basic tenets of Appropriate Technology (as per our integrated framework). He makes this explicit in a footnote: "The sole act of authentic, verifiable, and concrete control of technology would be to set limits to its development. But this is the very contradiction of the system. Contrary to what many people may think, setting limits creates freedom. Illich's thinking here coincides with mine. And I feel that nothing is as fundamental as this problem of voluntary limits" (*Technological System*, p. 355). The point we are making here is dependent, however, upon understanding that it is to technicity rather than technology-practice or technology that Ellul refers.

grounds for hope that this may occur will be the main focus of the final chapter.

12

Facing the Future

The Problem

This book is grounded in the observation that technology is a key factor in the contemporary world. Technology is central to economic activity and is bound up in most current critical social and environmental problems. The public status of technology is ambiguous: it is seen as a source of problems and not just as a source of solutions. Dissension over the normative value of technological change abounds in public policy debate and academic discourse. While there has recently been some renewal of public confidence in technology as a source of social benefit, this has not brought about a fundamental renewal of public confidence in the prospects for a salubrious world for future generations.

Appropriate Technology emerged during the last few decades as a fresh approach to the problems of technology, society and the environment. By pointing to the constructive role which enlightened technology choice might play in the economic sphere, and in practical human affairs more generally, the Appropriate Technology movement started a new wave of technological experiments inspired by a convivial, humanistic vision of the future. Appropriate Technology represented new hope that the poor of the world (from both rich and poor countries) could take control of their economic fortunes and gradually work towards an abundant future which was both environmentally and socially sustainable. For many people Appropriate Technology meant facing the future with hope. It might even be said that it was the technological embodiment of hope.

Over a decade has now passed since the death of the Appropriate Technology movement's internationally acclaimed leader, Fritz Schumacher. It is now appropriate to reflect on how widespread the

331

movement's impact on society has been. This book has shown that Appropriate Technology was not just an intellectual fad of the 1960s and 1970s. Rightly understood, it is a serious concept accompanied by a sizable international social movement and a notable record of practical accomplishments, especially given the fact that it has stemmed largely from the independent activities of grass roots organizations and committed individuals. Despite this success, however, and despite the obvious appeal of the Appropriate Technology approach (rightly understood), Appropriate Technology has failed to be adopted into the mainstream as the dominant form of technology-practice. This leads us back to the central problem of this book: are there sufficient and reasonable grounds for hope that the goals of the Appropriate Technology movement may be realized in practice?

To many, the hope offered by Appropriate Technology looks increasingly vacuous. Does this mean that the promise of Appropriate Technology - a reasonable livelihood and a sustainable environment for most communities - must now be abandoned? Some sympathizers of Appropriate Technology appear to have come to such a conclusion. We should ask, however, whether the abandonment of hope is the only defensible response to this situation. Our analysis of the concept of technology choice in the preceding chapters suggests that it is not. To see the possibility of intellectually sustainable grounds for hope for Appropriate Technology, however, it is necessary to reflect a little on the idea of hope itself and to re-examine some of the ideas expounded in earlier chapters.

As part of this exercise it is instructive to reflect on the way in which Appropriate Technology is now commonly viewed by informed professionals in various relevant fields of policy. Personal observations of the author suggest that most policy makers have been exposed to the ideas of the Appropriate Technology movement at some stage, but that the intensity of debate over the subject which was apparent several years ago appears to have subsided.

To some extent this reflects the movement's success. While the Appropriate Technology idea initially generated intense opposition among orthodox professionals it is now almost impossible to find anybody who cares to mount a concerted attack on the idea. This is partly because, as suggested in an earlier chapter, nobody would seriously wish to advocate an "inappropriate technology". It is also because the idea, when properly articulated, appeals to common sense. There have also been some important shifts in the thinking of senior business managers and economists in recent times away from the old "Fordist" emphasis on centralized production and economies of scale, towards an emphasis on decentralized, flexible production systems and the "economies of scope". In other words, mainstream policy in business and technological affairs has shifted somewhat in ways which accommodate more readily certain themes previously championed mainly by Appropriate Technology advocates. Appropriate Technology, therefore, is now far less "alternative" than it used to be. Appropriate Technology has not so much been rejected by mainstream policy makers, as it has come to be thought of as a little passe.

The term "Appropriate Technology" is now used much less frequently than it used to be. This is not necessarily because the concept it represents has been superseded. Rather, many protagonists of Appropriate Technology who used to be leaders in special-interest Appropriate Technology organizations have now taken up significant positions in mainstream organizations. In their new roles they are frequently applying their Appropriate Technology principles and skills in a new context, under a different rubric and with the advantage of a number of years of practical experience to guide their present work.

Another typical response which the notion of Appropriate Technology receives from mainstream policy makers, especially in industrialized countries, is acknowledgement that it is legitimate and feasible, but only for special circumstances such as might prevail in Third World villages or certain remote locations in industrialized countries. Appropriate Technology is not disputed as such; rather it is assumed not to be universally applicable.

Appropriate Technology has been partially incorporated into the mainstream, thereby providing some affirmation of the hope which it represents. Nevertheless, it has not become the dominant mode of technology-practice in most organizations and communities, even under another name.

Addressing the Problem

The problem of whether or not there are grounds for hope that Appropriate Technology will be adopted on a major scale has already been addressed to some extent in the preceding chapters. First, the substantial proportions of the Appropriate Technology movement and its global scope provide some reasonable grounds for hope. Second, our analysis has shown that there appear to be no *intrinsic* obstacles to the uptake of the Appropriate Technology as we have defined it. The fact that we have been able to articulate an integrated framework for

Appropriate Technology and provide rejoinders to the theoretical objections which have been levelled against the idea suggests that there are no intrinsic *conceptual* obstacles to its uptake. There have been sufficient practical experiments with the Appropriate Technology approach by now, furthermore, to indicate that there are little in the way of intrinsic *technical* obstacles to its uptake.

The serious obstacles to Appropriate Technology are meta-technical: they are primarily social and political in nature. While these are certainly very widespread, our analysis has revealed that there is no reason to assume that such social and political factors are immutable or absolute. In other words, just because obstacles to Appropriate Technology may be real it does not follow that they constitute *fatal negations* of it as a mode of technology-practice.

The issue of whether Appropriate Technology may be successfully practiced as a normal part of the mainstream may be illustrated by using the competition between two sports teams as an allegory. Each team may be viewed as an obstacle to the other team winning the game. In this sense, the interests of the two teams are antithetical. Because the visiting team, for example, may be committed to preventing the home team from winning the competition, sports commentators do not automatically conclude that there are therefore no grounds for hope that the home team will win. Supporters of the home team, furthermore, do not expect absolute proof that their team is destined to win in order for them to maintain reasonable hope that it may in fact do so. If such proof was actually possible (although such a notion is actually absurd), then there would in fact be no real sport. The outcome of a game depends upon how it is played, which in turn depends upon a variety of factors such as the motivation of the players, the capricious nature of their interactions during the game, strategies, accidents, and the complex interactions of contrasting forces at work in the teams. The possible success or failure of the home team in a game is contingent upon its performance. The likelihood of the Appropriate Technology movement succeeding in its quest, similarly, depends upon how well the "team members" of the movement face up against their "opponents". There are various political and social forces which are antithetical to Appropriate Technology. There is no guarantee that Appropriate Technology will "win" against such forces, and no way of knowing the outcome in advance; and, should the "Appropriate Technology team" lose the game, it would not necessarily follow that it would not improve its performance in other games as the season progressed. What this book has done is to indicate some of the important factors likely to affect the outcome of the "game".

Hope does not require certainty about the future in order for it to be reasonable. Hope does not require proof that a certain desired course of events is the only feasible outcome of present circumstances. Rather, hope is a personal characteristic which exists when the legitimacy of a normative goal has been determined by a person and when a commitment has been maintained by that person to bring about a course of events in keeping with the attainment of that goal.[1] Something of this understanding of hope is contained in the way Schumacher viewed the prospects for Appropriate Technology:[2]

> What we need are what I call optimistic pessimists who can see clearly that we can't continue as before, but who have enough vigour and joyfulness to say all right, so we change course. I know it can be done and it will be done if people are not paralyzed by either the optimists or the pessimists.

Schumacher avoided simplistic pronouncements on the likely success or otherwise of the movement of which he was a leader. He was more concerned with determining what actions were desirable and with facilitating the development of organizations committed and equipped towards the pursuit of such actions.

The core argument of this book is that the general applicability of Appropriate Technology will depend upon it being construed, not as a particular collection of technologies (such as solar cookers or methane generators), but as a *mode of technology-practice* which emphasizes the concept of technology choice and which is characterized by the harmonious integration of technical-empirical, socio-political and ethical-personal factors. In other words, Appropriate Technology needs to be understood and implemented along the lines of the integrated framework outlined herein. Whether or not a sufficiently large number of people will do so, and whether or not a sufficiently large number of institutions and structures embodying and reinforcing "enlightened" Appropriate Technology practices, will emerge cannot be known in advance. All we can reasonably prognosticate about are the ingredients necessary for success, not the likelihood of their being assembled.

The future prospects for Appropriate Technology could be enhanced if an integrated framework along the lines of the one outlined in

---

[1] For an exposition of this interpretation of hope see, E. Fromm, *The Revolution of Hope: Toward a Humanized Technology* (Perennial Library; New York: Harper and Row, 1968).

[2] E. F. Schumacher, Interview, *The Futurist*, 8 (1974), 281-284.

Chapter Ten was embraced throughout the movement. This would have the two advantages of making the Appropriate Technology notion appear more *cogent*, and of making it *universally applicable*.

A more cogent understanding of Appropriate Technology would provide the basis for more effective planning and decision making within the Appropriate Technology movement. A consensus on the parameters of the movement's activities could help it to achieve more efficient use of the resources at its disposal, and facilitate more effective access by the Appropriate Technology movement to resources and institutions not currently at its disposal. The lack of a *readily identifiable* and consistently articulated framework for Appropriate Technology currently acts as an obstacle to the diffusion of Appropriate Technology, because it is an obstacle to the idea being taken seriously by those currently indifferent to it.

The integrated framework makes Appropriate Technology more cogent by allowing conjoint discussion of actual technologies (i.e., artefacts), broader technological principles, alternative categories of technology-practice, social goals, physical parameters and environmental impacts, among other things, without encountering the semantic problems which have plagued the Appropriate Technology literature. Making a distinction between technology and technology-practice leads to a parallel distinction between appropriate technology and Appropriate Technology. This, in turn, has implications for the distinction between the general-principles and specific-characteristics approaches to defining appropriate technology. Through the integrated framework we have been able to incorporate the insights of both approaches without abandoning normal intellectual and semantic standards. Defining Appropriate Technology as a mode of technology-practice leads directly to the adoption of the general-principles approach as the more universally applicable of the two, while still allowing for the legitimate application of specific-characteristics criteria for specific circumstances. It shows that the general-principles approach has the capacity to provide useful decision making guidelines and that it need not be reduced to the level of formally correct but unimportant platitudes.

Appropriate Technology proponents have a tendency to engage concurrently in more than one mode of discourse and to concurrently discuss fields of study which are normally separated in orthodox scholarship. For example, the literature includes discussion of engineering design principles in the manufacture of industrial technologies together with analysis of political-economy and the dynamics of culture; it includes ecological assessments alongside consideration of market forces, and philosophical reviews of historical forces together with outlines of practical artefacts for domestic food supply or household energy management; it also combines descriptions of scientific principles together with reflections on the subjective experience of people in the workplace. The integrated framework makes it feasible for such complex discourse to be conducted more coherently.

Viewing Appropriate Technology as a mode of technology-practice enables the foregoing diversity of factors to be linked, cross-examined and categorized, without loss of the particular integrity belonging to each sphere of endeavor. The idea of the harmonious integration of technical-empirical, socio-political and ethical-personal considerations, is a meta-concept which is compatible with the majority of other concepts embodied in Appropriate Technology. It may only be applied to each of the fields we have surveyed, however, if it is mediated through some form of praxis. In this respect the integrated framework makes explicit an assumption which is implicit throughout the Appropriate Technology movement: viz., that *a diverse range of human concerns and environmental issues may be viewed from the common perspective of technology-practice.* Because it recognizes the centrality, ubiquity and pervasive impact of technological processes and objects in the modern world, the framework can be a tractable tool for inter-disciplinary policy analysis.

The second advantage of our integrated framework is its potential to improve the prospects for Appropriate Technology to be seen as universally applicable. In contrast to the view that intermediate technologies are inferior to other technologies or that "Appropriate Technology" is more advanced and valid than "Intermediate Technology", we have argued that Intermediate Technology is entirely consistent with the Appropriate Technology innovation strategy. This is not to say that Appropriate Technology and Intermediate Technology are identical notions, but rather that the latter is a specific application of the former. The integrated framework requires that the Appropriate Technology rationale be applied to different situations with the intention that different technologies be developed or selected to suit those situations. Schumacher looked upon intermediate technologies as appropriate technologies for circumstances which he perceived as prevalent in the rural areas of the South.

When Appropriate Technology is understood as a collection of artefacts with a set of specific characteristics, Intermediate Technology tends to be looked upon as an embarrassing aberration of Appropriate Technology, which ought to be mentioned as little as possible for fear of harming the Appropriate Technology "cause". Understood in this way, the scope for applying Intermediate Technology would be quite limited. Adopting this narrow and static interpretation of Intermediate Technology would also force one to conclude that Schumacher was naive, lacking basic economic experience and a realistic insight into international affairs: how else could he have advocated such a regressive and simplistic option? Such an interpretation of Schumacher cannot be justified by his writings; we demonstrated in Chapter Four that Schumacher was a highly trained, experienced and top-level economist with a substantial international background. If one is to treat the seminal Appropriate Technology literature seriously and objectively one is forced to acknowledge the plausibility of Intermediate Technology (in contrast to the caricatures of it which frequently appear in the polemic of critics) and to recognize its continuity with Appropriate Technology. The plausibility of Intermediate Technology is, as we have demonstrated, dependent upon it being viewed as an application of Appropriate Technology. The integrated framework enables us to acknowledge the continuity and compatibility of Intermediate Technology with Appropriate Technology without damaging the integrity of each notion and without crudely distorting Schumacher's work and ideas.

The integrated framework not only allows for a more realistic assessment of Schumacher and Intermediate Technology, it also provides a basis for a more *comprehensive evaluation* of Appropriate Technology as a whole. Evaluation of the prospects for Appropriate Technology ought to include evaluation of its present and past manifestations. Programs and experiments in the South, or activities based in the North and directed towards the problems of the South, related to Intermediate Technology, have been amongst the most abundant within the Appropriate Technology movement. Material surveyed in Chapters Three and Four provides reasonable grounds for viewing much of this activity as both desirable and feasible. Thus, our study of Schumacher, Intermediate Technology and the Appropriate Technology movement in the South provides substantial evidence for the desirability and feasibility of Appropriate Technology in general. Intermediate Technology, rightly understood, reinforces rather than undermines the universal applicability of Appropriate Technology. This conclusion only holds, however, if Appropriate Technology is comprehended in terms similar to those of our integrated framework.

Appropriate Technology, rightly understood, is compatible with recognizing that the conditions of countries of the North are different to those which prevail in the South. Our approach indicates how Appropriate Technology might provide a framework for policy and action which is desirable and feasible for the North, without requiring that the economies of the North regress to some "lower" level of development or complexity.

The synthesis of Appropriate Technology in this book is flexible enough to embrace the fact that, in the final analysis, nations may not be divided discretely into two homogeneous categories - the North and the South. Each country is unique in terms of natural resource endowments, geopolitical status, cultural ambience, economic infrastructure and demography. Consequently, completely standardized patterns of technological development for either the South or the North are impracticable. Furthermore, there are a number of countries which occupy an interesting "middle" or dualistic position in the world economy, such as Australia, Ireland, New Zealand, and various outlying provinces of otherwise industrialized nations. They exhibit some characteristics of the South (e.g., heavy reliance upon export of unprocessed primary products as a means of obtaining foreign exchange, and a tendency for technologies to be either imported or produced locally by subsidiaries of foreign multinational companies) and some characteristics of the North (e.g., relatively high per capita wealth levels, low population growth rates and relatively high standards of health care, education and social welfare benefits). Approaches to technological development evolved within the dominant nations of the North or promulgated for the typical conditions within countries of the South, might not be very successful in these "middle" countries/regions. The Appropriate Technology innovation strategy we have articulated provides a framework which is capable of incorporating countries which differ from the dominant country-types in the world economy. Failure by policy makers in non-dominant countries to evolve technological development patterns which are tailored to fit the unique conditions which prevail in their economies may lead to a deteriorating position vis-à-vis international trade and competitiveness.

In addition to its general applicability and desirability in the contexts of the North, the South and of the "middle" countries, the integrated framework we have articulated for Appropriate Technology is applicable across a variety of fields and categories of technology-practice. For example, the analysis in Part Two of this book covered, among other things, both the rural/agricultural sector and the urban/industrial sector, both the formal and the informal economy, so-called "high technology" and "low technology", and human and environmental matters. The universal application of Appropriate Technology is untenable if it is comprehended simply in technical-empirical terms. The polymorphic nature of Appropriate Technology relies upon the incorporation of socio-political and ethical-personal factors into the framework for policy and action.

The Practice of Technology Choice

Besides having the potential for improving the prospects for Appropriate Technology by making it more cogent and universally applicable, the integrated framework evokes a key principle for the practice of technology choice. Because Appropriate Technology is a three dimensional mode of technology-practice it follows that the efficacy of *technology choice* as a policy concept will depend upon all three of the dimensions of technology-practice being taken seriously. That is, the choice of technology requires the simultaneous incorporation of technical-empirical, socio-political and ethical-personal factors in planning and decision making. The failure of Appropriate Technology to be disseminated more widely appears to be linked to the fact that its tripartite nature has not been properly appreciated.

Accounts of failed "appropriate technologies" seem to arise from situations where only one or two of the three dimensions of technology-practice have been given serious attention. For example, stories abound of foreign aid organizations introducing parabolic-dish reflecting solar cookers to rural Third World villages, as an "appropriate technology" substitute for wood-burning stoves or open fires. While there may indeed be technical-empirical evidence that such devices may be applied for the purpose of cooking food, it does not follow that such an application will be feasible *in practice*. Socio-political factors may make it impossible for such devices to be afforded by poor villagers, especially when highly specialized materials are required, and ethical-personal factors (i.e., lifestyle preferences or patterns) may mean that such devices would not be used even if they were affordable.

The Appropriate Technology approach requires that all of these dimensions be taken into account in design and implementation. It is a misnomer to call a solar energy device an "appropriate technology" simply because, from the technical-empirical point of view, it may be compatible with the principle of environmental conservation. Unless that device is economically feasible, unless the local social or market environment is conducive to the widespread application of the technology, and unless it addresses real human needs in the community, then, according to the principles of Appropriate Technology, it is inappropriate. Adopting the Appropriate Technology approach does not mean that the socio-political environment must be treated as simply "given", and our integrated framework for Appropriate Technology does not mean that its practitioners should abstain from political activity aimed at reform of the existing socio-political environment. Rather, it implies that political change will have implications for local technology-practice, and that the introduction into a community of new technology-practice (whether a household appliance, new farming equipment, or a new manufacturing system) might only be feasible if organized as part of a larger program of community development and planning - or, at the very minimum, if it is consonant with the prevailing socio-political and ethical-personal factors at work in the community. Technology choice may be used as a primary tool of a larger political strategy, but it cannot be isolated from the larger political context.

A similar argument applies to the situation where members of a community may advocate the introduction of a new type of technology-practice on ethical grounds (e.g., pesticide-free methods of horticulture), but where the technical-empirical support is not sufficiently developed to ensure success. The right socio-political environment may be in place (consumer demand for pesticide-free produce, and legislation to discourage food with pesticide residues), and the right ethical-personal conditions might be in existence ("organic growing" enthusiasts prepared to work hard and make considerable personal sacrifices in efforts to grow the produce), but the "technology" may fail because of insufficient technical-empirical support from the relevant research and development professionals. It may be that certain biological methods of pest control would be feasible, if more funds were put into improving them and into disseminating the expertise in their use through local agricultural extension systems; but that biases built into the existing systems effectively prevent resources being allocated to enable such improvements from being properly realized. The integrated framework for Appropriate Technology suggests that the successful choice of appropriate technology, whether for energy supply in the South or food production in the North, requires that the three dimensions of technology-practice be carefully managed simultaneously.

The simultaneous management of all three dimensions of technology-practice probably requires unusual skills, and this may be one reason why Appropriate Technology has not yet become a dominant part of the mainstream. The practice of technology choice requires: technical specialists capable of incorporating socio-political data and ethical-personal considerations into the technical design process; individuals in the community capable of expressing community needs and values in terms which may be made operational by technologists and planners;

and, planners, policy makers, and business managers capable of empathizing with both technical specialists (e.g., mechanical engineers) and lay community leaders/members, and facilitating the harmonizing of the interests of both groups. Appropriate Technology requires teams of people able to manage the challenging task of assimilating complex information about the psychosocial and biophysical context of a community, and expressing it in terms which are relevant to the mode of operation of technologists; such teams also need to include technologists capable of effectively communicating technical information to people beyond their field of technical expertise.

The prospects for Appropriate Technology in a given society or community will be dependent upon the extent to which its people become skilled in the above sense, and upon the extent to which it cultivates the kind of institutions which facilitate activities based around teams of people such as the ones just described. The continuing work of organizations such as Appropriate Technology International, discussed in Chapter Eight, provides some grounds for hope that such endeavors may be within the reach of communities throughout the world.

The integrated framework for Appropriate Technology outlined in Chapter Ten provides one final insight on the problem of the grounds of hope for enlightened technology choice. The inclusion of an ethical-personal dimension as part of technology-practice emphasizes that there is always a human factor associated with technological processes. While people are constrained by the structures and forces which constitute their environment, they provide the dynamic element in technology practice. Appropriate Technology is a mode of technology-practice which requires the participation of the people whose interests it is intended to serve. This is both its strength and its weakness. Schumacher considered the human factor in technology-practice to be the critical factor affecting its future development, as illustrated by the following quote.[3]

> The future ... is largely predictable, if we have a solid and extensive knowledge of the past. Largely, but by no-means wholly; for into the making of the future there enters that mysterious and irrepressible factor called human freedom.

It is not possible to deduce with certainty the likelihood of Appropriate Technology eventually becoming normal practice in most

---

[3] E. F. Schumacher, *Small is Beautiful: A Study of Economics as if People Mattered* (London: Blond and Briggs, 1973), p. 213.

countries. It will depend upon the extent to which that "mysterious and irrepressible factor" comes into play. As to whether there are reasonable grounds for hope that it will, Schumacher's words provide a fitting close to our study:[4]

> It is widely accepted that politics is too important a matter to be left to experts. Today, the main content of politics is economics, and the main content of economics is technology. If politics cannot be left to the experts, neither can economics and technology. The case for hope rests on the fact that ordinary people are often able to take a wider view, and a more "humanistic" view, than is normally being taken by experts. The power of ordinary people, who today tend to feel utterly powerless, does not lie in starting new lines of action, but in placing their sympathy and support with minority groups which have already started.

[4] *Ibid.*, p. 147.

Index

technological dependency, 254
technological determinism, 246, 247, 248, 249, 251, 252, 275, 310, 311, 312, 316, 329
technological development, 173, 276, 285, 292, 308, 318
technological ends, 278
technological fit, 269, 270, 271, 272, 285, 292, 296, 305
technological means, 8, 246, 269, 270
technological milieu, 319
technological mix, 285, 292
technological niche, 269-271
technological order, 27
technological phenomena, 26
technological progress, 172, 325
technological rationality, 323
technological science, 36, 41, 274, 321
technological society, 3, 28, 171, 173, 183, 197, 305, 314
technological system, 199, 272
technological systems
technologie, 26, 27, 41
technology, 25, 38, 39, 42, 95, 215, 321
technology assessment, 194, 282, 283
technology choice, 5, 8, 11, 14, 120, 123, 227, 232, 279, 281, 340, 341
technology policy, 14
technology-practice, 13, 14, 36, 40, 44, 172, 226, 238, 272, 280, 292, 321, 324, 335
technostructure, 27
textiles, 119, 127, 206
Third World, 86, 120, 340

three dimensional mode of technology-practice, 266
tractors, 128
traditional technologies, 123
TRANET, 51
transport, 119, 134, 157, 289
trickle down, 291
Undercurrents, 179
underdevelopment, 87, 96, 100, 102, 142, 303
unemployment, 72, 76, 191, 229, 287
United States, 22, 86, 159, 187, 190, 193, 216
universities, 212
unsuitable technology, 107
urban-industrialized societies, 312
utopian technology, 17
value system, 218
vernacular technology, 17
vested interests, 251
VITA, 51, 150
vicious circle, 100, 142, 303
village development, 160
village technology, 17, 149
Ward, 210
water storage/supply, 125, 156
wealth creation, 105
weaving looms, 125
windpower, 129
Winner, 298
wood-based fuels, 129
work, 101, 297, 298
workplace democracy, 197
Zambia, 125